# Functional Polymers for Controlled Drug Release

# Functional Polymers for Controlled Drug Release

Special Issue Editor
**Umile Gianfranco Spizzirri**

MDPI • Basel • Beijing • Wuhan • Barcelona • Belgrade

*Special Issue Editor*
Umile Gianfranco Spizzirri
University of Calabria
Italy

*Editorial Office*
MDPI
St. Alban-Anlage 66
4052 Basel, Switzerland

This is a reprint of articles from the Special Issue published online in the open access journal *Pharmaceutics* (ISSN 1999-4923) from 2019 to 2020 (available at: https://www.mdpi.com/journal/pharmaceutics/special_issues/Functional_Polymers_Drug_Release).

For citation purposes, cite each article independently as indicated on the article page online and as indicated below:

LastName, A.A.; LastName, B.B.; LastName, C.C. Article Title. *Journal Name* **Year**, *Article Number*, Page Range.

**ISBN 978-3-03928-490-0 (Pbk)**
**ISBN 978-3-03928-491-7 (PDF)**

© 2020 by the authors. Articles in this book are Open Access and distributed under the Creative Commons Attribution (CC BY) license, which allows users to download, copy and build upon published articles, as long as the author and publisher are properly credited, which ensures maximum dissemination and a wider impact of our publications.

The book as a whole is distributed by MDPI under the terms and conditions of the Creative Commons license CC BY-NC-ND.

# Contents

About the Special Issue Editor . . . . . . . . . . . . . . . . . . . . . . . . . . . . . . . . . . . . . . . . . vii

**Umile Gianfranco Spizzirri**
Functional Polymers for Controlled Drug Release
Reprinted from: *Pharmaceutics* 2020, 12, 135, doi:10.3390/pharmaceutics12020135 . . . . . . . . . 1

**Xinxu Han, Peipei Huo, Zhongfeng Ding, Parveen Kumar and Bo Liu**
Preparation of Lutein-Loaded PVA/Sodium Alginate Nanofibers and Investigation of Its Release Behavior
Reprinted from: *Pharmaceutics* 2019, 11, 449, doi:10.3390/pharmaceutics11090449 . . . . . . . . . 4

**Fabian Ávila-Salas, Adolfo Marican, Soledad Pinochet, Gustavo Carreño, Oscar Valdés, Bernardo Venegas, Wendy Donoso, Gustavo Cabrera-Barjas, Sekar Vijayakumar and Esteban F. Durán-Lara**
Film Dressings Based on Hydrogels: Simultaneous and Sustained-Release of Bioactive Compounds with Wound Healing Properties
Reprinted from: *Pharmaceutics* 2019, 11, 447, doi:10.3390/pharmaceutics11090447 . . . . . . . . . 15

**Dina Guzmán-Oyarzo, Tanya Plaza, Gonzalo Recio-Sánchez, Dulcineia S. P. Abdalla, Luis A. Salazar and Jacobo Hernández-Montelongo**
Use of nPSi-βCD Composite Microparticles for the Controlled Release of Caffeic Acid and Pinocembrin, Two Main Polyphenolic Compounds Found in a Chilean Propolis
Reprinted from: *Pharmaceutics* 2019, 11, 289, doi:10.3390/pharmaceutics11060289 . . . . . . . . . 34

**Justyna Odrobińska, Katarzyna Niesyto, Karol Erfurt, Agnieszka Siewniak, Anna Mielańczyk and Dorota Neugebauer**
Retinol-Containing Graft Copolymers for Delivery of Skin-Curing Agents
Reprinted from: *Pharmaceutics* 2019, 11, 378, doi:10.3390/pharmaceutics11080378 . . . . . . . . . 51

**Ioannis Partheniadis, Paraskevi Gkogkou, Nikolaos Kantiranis and Ioannis Nikolakakis**
Modulation of the Release of a Non-Interacting Low Solubility Drug from Chitosan Pellets Using Different Pellet Size, Composition and Numerical Optimization
Reprinted from: *Pharmaceutics* 2019, 11, 175, doi:10.3390/pharmaceutics11040175 . . . . . . . . . 68

**Wan Wang, Changrim Lee, Martha Pastuszka, Gordon W. Laurie and J. Andrew MacKay**
Thermally-Responsive Loading and Release of Elastin-Like Polypeptides from Contact Lenses
Reprinted from: *Pharmaceutics* 2019, 11, 221, doi:10.3390/pharmaceutics11050221 . . . . . . . . . 92

**Dorota Neugebauer, Anna Mielańczyk, Rafał Bielas, Justyna Odrobińska, Maria Kupczak and Katarzyna Niesyto**
Ionic Polymethacrylate Based Delivery Systems: Effect of Carrier Topology and Drug Loading
Reprinted from: *Pharmaceutics* 2019, 11, 337, doi:10.3390/pharmaceutics11070337 . . . . . . . . . 105

**Mhd Anas Tomeh, Roja Hadianamrei and Xiubo Zhao**
Silk Fibroin as a Functional Biomaterial for Drug and Gene Delivery
Reprinted from: *Pharmaceutics* 2019, 11, 494, doi:10.3390/pharmaceutics11100494 . . . . . . . . . 122

**Giuseppe Cirillo, Umile Gianfranco Spizzirri, Manuela Curcio, Fiore Pasquale Nicoletta and Francesca Iemma**
Injectable Hydrogels for Cancer Therapy over the Last Decade
Reprinted from: *Pharmaceutics* 2019, 11, 486, doi:10.3390/pharmaceutics11090486 . . . . . . . . . 144

# About the Special Issue Editor

**Umile Gianfranco Spizzirri** received his degree in Chemistry cum laude in 2001, and a Ph.D. degree in 2005 from University of Calabria. In 2004, he was visiting researcher at the University of Texas at Austin (USA). He is currently a member of the Department of Pharmacy, Health, and Nutrition Science at the University of Calabria. His research activities are mainly related to polymer chemistry for the preparation of stimuli-responsive drug delivery systems, polymers with tailored biological activity for biomedical applications, and functional polymers for the food industry. He is author and co-author of 130 publications, including research and review articles as well as invited book chapters (total citations more than 2500 and H-index of 32).

*Editorial*
# Functional Polymers for Controlled Drug Release

Umile Gianfranco Spizzirri

Department of Pharmacy, Health and Nutritional Sciences, University of Calabria, 87036 Rende (CS), Italy; g.spizzirri@unical.it; Tel.: +39-0984-493-298

Received: 31 January 2020; Accepted: 31 January 2020; Published: 5 February 2020

---

In the last decade, the pharmaceutical application of hydrophilic materials has emerged as one of the most significant trends in the biomedical and pharmaceutical areas [1]. This Special Issue serves to highlight and capture the contemporary progress recorded in this field.

In this Special Issue, two articles were devoted to exploiting the employment of polyvinyl alcohol in the developing of useful drug delivery tools. Polyvinyl alcohol is one of the most popular water-soluble, non-carcinogenic, biocompatible, biodegradable synthetic polymers, and is largely employed to prepare hydrogels useful as artificial organs, drug delivery devices, and wound dressings [2]. Specifically, Han et al. explored the release properties of lutein-loaded polyvinyl alcohol combined with sodium alginate nanofibers, prepared by electrospinning [3]. The release profiles were analyzed by mathematical models, highlighting that the employment of the electrospinning in the encapsulation of the carotenoid molecule is an effective method to achieve the sustained lutein release. Additionally, Avila-Salas et al. crosslinked polyvinyl alcohol with different dicarboxylic acids to synthesize dressing hydrogels [4]. These formulations were suggested as multi-target therapies in wound healing, as a consequence of the sustained release of simultaneous bioactive compounds, such as dexpanthenol, allantoin, caffeic acid, and resveratrol.

The transport and sustained release of bioactive polyphenols, usually extracted from plants or food, was also dealt with by Guzman-Oyarzo et al. [5]. In order to avoid degradation reactions of these bioactive molecules, the authors proposed a synthetic strategy involving a flexible and soft β-ciclodextrin polymer within the highly porous inorganic matrix of nanoporous silicon [6] as a substrate. This device was tested as carrier for the controlled release of caffeic acid and pinocembrin, two of the main components of a *Chilean propolis* with anti-atherogenic and anti-angiogenic activities.

The fast release of selected antioxidants and skin-lightening agents by suitable micellar systems, was exploited by Odrobinska et al. for applications in cosmetology as components of masks, creams, and wraps [7]. The authors proposed the synthesis of an innovative material obtained by "click" chemistry reaction of azide-functionalized polyethylene glycol onto multifunctional polymethacrylates containing alkyne units, and using bromoester-modified retinol as the initiator. The tendency of the designed amphiphilic graft copolymers to form micelles allowed them to record a high effective encapsulation of arbutin or vitamin C and in vitro experiments highlighted the maximum release in few minutes.

Intelligent polymeric devices able to undergo morphological modifications in response to an internal or external stimulus, such as pH, redox balance, temperature, magnetic field, and light have been actively pursued [8–10]. In particular, in this Special Issue, Partheniadis et al. [11] synthesized pharmaceutical pellets [12] of different sizes, using an extrusion/spheronization technique, and medium viscosity chitosan for the pH-dependent delivery of piroxicam. The authors suggested that a remarkable reduction in pellet size influenced the release rate, avoiding the need to employ hydrophilic excipients such as lactose [13].

In another paper of this Special Issue, Wang et al. explored a novel strategy to drive the reversible adsorption of peptide-based therapeutics using commercially available contact lenses [14]. To accomplish this, thermo-sensitive elastin-like polypeptides, alone or tagged with a candidate ocular therapeutic, were characterized. This research suggests that elastin-like polypeptides may be useful to control loading or release from suitable formulations, with the aim to deliver appropriate biologically active peptides to the ocular surface via contact lenses.

Finally, this Special Issue was completed by three reviews exploiting the employment of particular materials and/or analyzing specific route of drug administration. In particular, Neugebauer et al. investigated the synthesis of ionic polymethacrylate-based delivery systems, including conjugates and self-assemblies [15]. The influence of the hydrophilic/hydrophobic content on physicochemical and delivery properties of the polymer carriers were exploited, by analysing how the topology and architecture of the macromolecular devices regulate the physical entrapment or chemical attachment of the specific drugs.

Furthermore, Tomeh et al. analyzed the use of silk fibroin to prepare versatile drug delivery devices [16]. Mild aqueous possessing conditions, high biocompatibility and biodegradability, and the ability to enhance the stability of the loaded active pharmaceutical ingredients, justify the increased use of these natural polymers in the pharmaceutical and biomedical fields [17].

Finally, Cirillo et al. proposed a review focused on the recent advances in the development of highly engineered injectable delivery vehicle systems, suitable for combined chemo- and radio-therapy, as well as thermal and photo-thermal ablation, with the aim of finding effective solutions to overcome the current obstacles of conventional therapeutic protocols [18].

**Conflicts of Interest:** The author declares no conflicts of interest.

## References

1. Daly, A.C.; Riley, L.; Segura, T.; Burdick, J.A. Hydrogel microparticles for biomedical applications. *Nat. Rev. Mater.* **2020**, *5*, 20–43. [CrossRef]
2. Alves, M.-H.; Jensen, B.E.B.; Smith, A.A.A.; Zelikin, A.N. Poly(vinyl alcohol) physical hydrogels: New vista on a long serving biomaterial. *Macromol. Biosci.* **2011**, *11*, 1293–1313. [CrossRef] [PubMed]
3. Han, X.; Huo, P.; Ding, Z.; Kumar, P.; Liu, B. Preparation of Lutein-Loaded PVA/Sodium Alginate Nanofibers and Investigation of Its Release Behavior. *Pharmaceutics* **2019**, *11*, 449. [CrossRef] [PubMed]
4. Ávila-Salas, F.; Marican, A.; Pinochet, S.; Carreño, G.; Valdés, O.; Venegas, B.; Donoso, W.; Cabrera-Barjas, G.; Vijayakumar, S.; Durán-Lara, E.F. Film Dressings Based on Hydrogels: Simultaneous and Sustained-Release of Bioactive Compounds with Wound Healing Properties. *Pharmaceutics* **2019**, *11*, 447. [CrossRef] [PubMed]
5. Guzmán-Oyarzo, D.; Plaza, T.; Recio-Sánchez, G.; Abdalla, D.S.P.; Salazar, L.A.; Hernández-Montelongo, J. Use of nPSi-CD Composite Microparticles for the Controlled Release of Caffeic Acid and Pinocembrin, Two Main Polyphenolic Compounds Found in a Chilean Propolis. *Pharmaceutics* **2019**, *11*, 289. [CrossRef] [PubMed]
6. Gidwani, B.; Vyas, A. A Comprehensive Review on Cyclodextrin-Based Carriers for Delivery of Chemotherapeutic Cytotoxic Anticancer Drugs. *BioMed Res. Int.* **2015**, *2015*, 198268. [CrossRef] [PubMed]
7. Odrobinska, J.; Niesyto, K.; Erfurt, K.; Siewniak, A.; Mielanczyk, A.; Neugebauer, D. Retinol-Containing Graft Copolymers for Delivery of Skin-Curing Agents. *Pharmaceutics* **2019**, *11*, 378. [CrossRef] [PubMed]
8. Spizzirri, U.G.; Altimari, I.; Puoci, F.; Parisi, O.I.; Iemma, F.; Picci, N. Innovative antioxidant thermo-responsive hydrogels by radical grafting of catechin on inulin chain. *Carbohyd. Polym.* **2011**, *84*, 517–523. [CrossRef]
9. Cirillo, G.; Curcio, M.; Spizzirri, U.G.; Vittorio, O.; Valli, E.; Farfalla, A.; Leggio, A.; Nicoletta, F.P.; Iemma, F. Chitosan-quercetin bioconjugate as multi-functional component of antioxidants and dual-responsive hydrogel networks. *Macromol. Mater. Eng.* **2019**, *2019*, 1800728. [CrossRef]
10. Cao, Z.-Q.; Wang, G. Multi-Stimuli-Responsive Polymer Materials: Particles, Films, and Bulk Gels. *Chem. Rec.* **2016**, *16*, 1398–1435. [CrossRef] [PubMed]
11. Partheniadis, I.; Gkogkou, P.; Kantiranis, N.; Nikolakakis, I. Modulation of the Release of a Non-Interacting Low Solubility Drug from Chitosan Pellets Using Different Pellet Size, Composition and Numerical Optimization. *Pharmaceutics* **2019**, *11*, 175. [CrossRef] [PubMed]
12. Nikolakakis, I.; Partheniadis, I. Self-Emulsifying Granules and Pellets: Composition and Formation Mechanisms for Instant or Controlled Release. *Pharmaceutics* **2017**, *9*, 50. [CrossRef] [PubMed]

13. Tapia, C.; Buckton, G.; Newton, J.M. Factors influencing the mechanism of release from sustained release matrix pellets, produced by extrusion/spheronisation. *Int. J. Pharm.* **1993**, *92*, 211–218. [CrossRef]
14. Wang, W.; Lee, C.; Pastuszka, M.; Laurie, G.W.; MacKay, J.A. Thermally-Responsive Loading and Release of Elastin-Like Polypeptides from Contact Lenses. *Pharmaceutics* **2019**, *11*, 221. [CrossRef] [PubMed]
15. Neugebauer, D.; Mielanczyk, A.; Bielas, R.; Odrobinska, J.; Kupczak, M.; Niesyto, K. Ionic Polymethacrylate Based Delivery Systems: Effect of Carrier Topology and Drug Loading. *Pharmaceutics* **2019**, *11*, 337. [CrossRef] [PubMed]
16. Tomeh, M.A.; Hadianamrei, R.; Zhao, X. Silk Fibroin as a Functional Biomaterial for Drug and Gene Delivery. *Pharmaceutics* **2019**, *11*, 494. [CrossRef] [PubMed]
17. Crivelli, B.; Perteghella, S.; Bari, E.; Sorrenti, M.; Tripodo, G.; Chlapanidas, T.; Torre, M.L. Silk nanoparticles: From inert supports to bioactive natural carriers for drug delivery. *Soft Matter* **2018**, *14*, 546–557. [CrossRef] [PubMed]
18. Cirillo, G.; Spizzirri, U.G.; Curcio, M.; Nicoletta, F.P.; Iemma, F. Injectable Hydrogels for Cancer Therapy over the Last Decade. *Pharmaceutics* **2019**, *11*, 486. [CrossRef] [PubMed]

© 2020 by the author. Licensee MDPI, Basel, Switzerland. This article is an open access article distributed under the terms and conditions of the Creative Commons Attribution (CC BY) license (http://creativecommons.org/licenses/by/4.0/).

*Article*

# Preparation of Lutein-Loaded PVA/Sodium Alginate Nanofibers and Investigation of Its Release Behavior

Xinxu Han [1], Peipei Huo [2,*], Zhongfeng Ding [3], Parveen Kumar [2] and Bo Liu [2,*]

1. School of Materials Science and Engineering, Shandong University of Technology, Zibo 255000, China
2. Laboratory of Functional Molecules and Materials, School of Physics and Optoelectronic Engineering, Shandong University of Technology, Xincun West Road 266, Zibo 255000, China
3. College of Life Sciences, Shandong University of Technology, Zibo 255000, China
* Correspondence: peipeihuo@sdut.edu.cn (P.H.); liub@sdut.edu.cn (B.L.); Tel.: +86-053-3278-3909 (B.L.)

Received: 6 July 2019; Accepted: 28 August 2019; Published: 2 September 2019

**Abstract:** This investigation aims to study the characteristics and release properties of lutein-loaded polyvinyl alcohol/sodium alginate (PVA/SA) nanofibers prepared by electrospinning. In order to increase PVA/SA nanofibers' water-resistant ability for potential biomedical applications, the electrospun PVA/SA nanofibers were cross-linked with a mixture of glutaraldehyde and saturated boric acid solution at room temperature. The nanofibers were characterized using scanning electron microscopy (SEM) and X-ray diffractometer (XRD). Disintegration time and contact angle measurements testified the hydrophilicity change of the nanofibers before and after cross-linking. The lutein release from the nanofibers after cross-linking was measured by an ultraviolet absorption spectrophotometer, which showed sustained release up to 48 h and followed anomalous (non-Fickian) release mechanism as indicated by diffusion exponent value obtained from the Korsmeyer–Peppas equation. The results indicated that the prepared lutein-loaded PVA/SA nanofibers have great potential as a controlled release system.

**Keywords:** lutein; nanofibers; polyvinyl alcohol; sodium alginate

## 1. Introduction

Lutein, also known as "plant lutein", is a natural pigment and an excellent antioxidant [1] widely found in vegetables, flowers, fruits, and certain algae organisms. The intake of a certain amount of lutein as a food supplement can prevent a series of organ aging-related diseases [2]. Lutein is susceptible to light, heat, and pH [3], the property of which compromises its bioavailability and limits its storage and human administration [4]. The delivery of drugs via nano-carriers is a highly effective and proven method to improve the bioavailability and until now, numerous nano-carriers including nanofibers, nanocapsules, liposomes, polymer micelles, and nanogels have been widely investigated for the delivery of various drugs.

Electrospinning has attracted more and more attention from the past decade due to its potential use in biomedical materials, filtration, catalysis, optoelectronics, food engineering, cosmetics, and drug delivery devices [5–7]. Among which drug delivery is one of the most promising applications. Nanofibers produced by electrospinning exhibit several interesting properties, including high surface area to volume ratio, and void fraction [8], which make electrospinning nanofibers an appropriate candidate as a drug delivery system. Polymeric matrices such as polyvinyl alcohol (PVA), fibrinogen, chitosan, polycaprolactone, and polyvinylpyrrolidone provide an excellent source for electrospinning based on their biocompatibility [9,10]. PVA nanofibers have been widely utilized as potential biomaterials owing to its extraordinary hydrophilicity, biocompatibility and mechanical properties [11–14]. This type of material readily composes into the film due to the fact that it contains a large amount of –OH groups, which provide a platform for hydrogen bond formation with water molecules. Based on the excellent properties of PVA, much interest in research has been devoted to its electrospinning for utilization in different areas such as a biosensor [15], antimicrobial fibers [16–18], composite films [19,20], nanoporous films, and filtration membranes [21–23]. Sodium alginate (SA) is a non-toxic, biodegradable, compatible, and sustained-release material. SA shows a good hemostatic effect in combination with calcium ions, therefore, is widely used in hemostatic dressings and wound permeate absorption dressings [24]. However, the fact that SA alone in aqueous solution is not readily electrospun into a nanofiber mat and its brittleness largely restricts its application. Blending PVA and SA is an effective polymer solution that can be electrospun into nanofiber mats and in addition, their mechanical property and thermal stability can be improved probably owing to hydrogen bond formation [25]. Various studies have been reported on the development of wound dressing with PVA and SA [26–30]. The nanofibers made from blending PVA and SA are highly hydrophilic.

Electrospun PVA/SA nanofibers as drug carriers are limited due to the burst release of drugs. For example, Li et al. [31] reported the preparation of a fast-dissolving drug delivery system using PVA as a polymeric carrier, in which the drug was released from the nanofiber matrix in an explosive manner. In order to realize sustained release as required for some drug delivery system, it is necessary to adjust the hydrophilicity of the nanofiber. Cross-linking is one of the methods that allow drugs to be released in a controlled manner by adjusting the hydrophilicity of nanomaterials [32–34]. For example, Zhang et al. [35] cross-linked electrospun gelatin nanofibers with glutaraldehyde saturated steam at room temperature to improve nanofibers thermal and mechanical properties. Kenawy et al. [36] studied the controlled release of ketoprofen from electrospun PVA nanofibers with methanol cross-linking. Zhang et al. [37] used salicylic acid-loaded collagen (COL)/PVA electrospun nanofibers cross-linked with UV-radiation or glutaraldehyde to control the release of salicylic acid.

In this study, a controlled drug delivery system was developed from electrospun PVA/SA nanofibers. Lutein was utilized as a model drug. Lutein-loaded nanofibers were cross-linked by using glutaraldehyde (GA) and saturated boric acid solution as a cross-linking agent. The properties of nanofibers before and after cross-linking and the lutein release behavior were investigated.

## 2. Experimental Methods

### 2.1. Materials

Polyvinyl alcohol (PVA, 9002-89-5(CAS No.), 87%–89%(purity), Mw = 72600–81400), sodium alginate (SA, 9005-38-3), glutaraldehyde (GA, 111-30-8, 50% aqueous solution), boric acid (10043-35-3, 99.8%), sodium phosphate dibasic ($Na_2HPO_4$, 10039-32-4, 99%), potassium phosphatemonobasic ($KH_2PO_4$, 7778-77-0, 99%), potassium chloride (KCl, 7447-40-7, 99.98%), sodium chloride (NaCl, 7647-14-5, 99.9%), and N,N-dimethylformamide (DMF, 4472-41-7, 99.5%) were purchased from Aladdin, Shanghai China. Lutein (127-40-2, 82.35%) was provided by Shandong Tian Yin Biotechnology co.ltd, Zibo, China.

### 2.2. Preparation of Nanofibers

#### 2.2.1. Preparation of Polymer Solutions

In a typical preparation, 1.6 g of PVA and 0.1 g of SA were dissolved in 15 mL of deionized water at 60 °C under constant stirring for 6 h, then cooled to room temperature. Lutein (51 mg, 3% weight ratio PVA and SA) was dissolved in 5 mL DMF at room temperature. The blended above PVA/SA solution and lutein solution were mixed and stirred for 5 h at room temperature to ensure homogeneous distribution.

#### 2.2.2. Electrospinning Lutein-Loaded PVA/SA Nanofibers

Electrospinning was performed to fabricate lutein-loaded PVA/SA nanofibers at room temperature. The mixed solution was poured into a 10 mL plastic syringe with a needle having an inner diameter of

0.41 mm. Output voltage applied to the solution was 15 kV. Besides, the flow rate of the injection pump was set to 0.3 mL/h. The nanofibers collector was cylindrical and covered by aluminum foil. At the same time, the distance from the syringe needle to the receiver collector was 150 mm. After completing the electrospinning process, the nanofibers were placed in a vacuum oven for 12 h to remove residual traces of solvents.

2.2.3. Cross-Linking of Lutein-Loaded PVA/SA Nanofibers

Cross-linking of the electrospun lutein-loaded nanofibers were carried out using a mixture of GA and saturated boric acid solution. Electrospun lutein-loaded nanofibers were carefully peeled from the aluminum foil and weighed exactly using a digital balance. A weight of 0.020 g of each sample was immersed in the cross-linking fluid at room temperature for various times (1 h, 3 h, and 5 h) to carry out cross-linking. After completing the required cross-linking time for each sample, all the samples were dried with filter paper and then exposed to a vacuum oven for 12 h at room temperature to remove residual GA and water.

2.3. Characterization

2.3.1. Morphology

The morphology of the lutein-loaded PVA/SA nanofibers before and after cross-linking was observed by quanta 250 field emission environment scanning electron microscope (SEM). The average diameter of the lutein-loaded PVA/SA nanofibers before cross-linking was calculated based on the SEM image. The distribution of lutein was observed by fluorescence microscope (CKX41, Olympus, Tokyo, Japan).

2.3.2. Water Contact Angle Analysis

The surface static contact angles of the lutein-loaded PVA/SA nanofibers before and after cross-linking were investigated using a contact angle meter analysis system (JY-82, Dingsheng, Chengde, China).

2.3.3. X-Ray Diffractometer Analysis

X-ray diffractometer (XRD, Bruker AXS, Karlsruhe, Germany) was used to observe the physical state of lutein in PVA/SA nanofibers in the range from 5° to 50°.

2.3.4. FTIR Spectroscopy

The cross-linking effectiveness of lutein-loaded PVA/SA nanofiber mat was analyzed using Fourier transform infrared (FTIR) spectroscopy (Nicolet5700, Waltham, MA, USA).

2.4. Pharmacotechnical Properties

2.4.1. Determination of Drug Encapsulation Efficiency

To determine the encapsulation efficiency (EE) of lutein in PVA/SA nanofibers, the lutein-loaded PVA/SA nanofibers were completely dissolved in water and lutein content was measured using UV-Vis spectrophotometer. The encapsulation efficiency of lutein was determined using the following equation:

$$EE\ (\%) = \text{real lutein content in nanofibers} / \text{theoretical lutein content in nanofiber} \times 100. \quad (1)$$

2.4.2. In Vitro Drug Release

The release profile of lutein from the cross-linked lutein-loaded PVA/SA nanofibers was studied in phosphate buffer saline (PBS, pH = 7.4) solution. Lutein-loaded PVA/SA nanofibers (20 mg) were placed in 50 mL of PBS solution at 37 °C with constant stirring. At defined time intervals, 1 mL of sample was taken from the release medium and replaced with fresh PBS to maintain the original volume. The amount of lutein released at different time intervals in PBS solution was measured by a UV-Vis spectrophotometer (UV-3600plus, Shimadzu, Kyoto, Japan). The lutein release percentage was calculated and release profile was drawn. All the measurements were performed in triplicates.

## 3. Results and Discussion

### 3.1. Morphology Characterization and XRD Spectroscopy of PVA/SA Nanofibers

SEM images of un-crosslinked lutein-loaded PVA/SA nanofibers are shown in Figure 1a, displaying uniform one dimensional nanofibers with no beads and diameters in the range of 240 nm to 340 nm. Figure 1b shows the fluorescence micrograph of un-crosslinked lutein-loaded PVA/SA nanofibers. It can be observed that lutein was uniformly distributed along the axis of the nanofibers. Figure 1c shows the morphology of lutein-loaded PVA/SA nanofibers with 1 h cross-linking. Nanofibers collapse during the cross-linking, after which nanofibers were not smooth and appeared in an independent form. The morphology change indicates that the cross-linking agent induced adhesion between the electrospun nanofibers, which could be attributed to the nanofibers that tend to swell in the presence of a crosslinker and adhere with each other. In addition, visual observation showed that after the cross-linking treatment, the nanofiber mats became yellowish and shrank slightly in size, which could be due to the interaction of hydroxyl groups on the PVA with GA of the cross-linking agent [31,38]. It should be noted that the residual trace amount of GA after cross-linking treatment can induce toxicity due to its reaction with proteins. The toxicity can be eliminated via reaction with glycine [34,39].

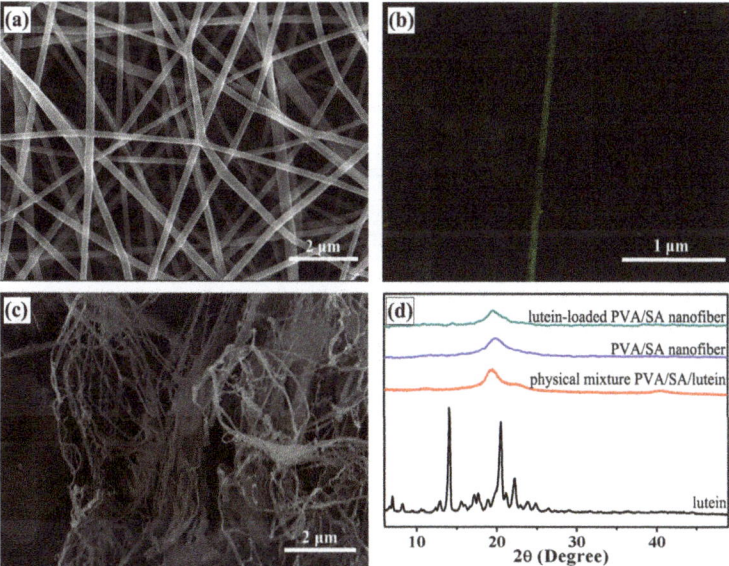

**Figure 1.** (**a**) SEM and (**b**) fluorescence microscope image of lutein-loaded polyvinyl alcohol/sodium alginate (PVA/SA) nanofibers before cross-linking, (**c**) SEM of lutein-loaded PVA/SA nanofibers after cross-linking for 1 h, (**d**) XRD diffraction pattern of free lutein, PVA/SA nanofibers, physical mixture of lutein and PVA/SA, and lutein-loaded PVA/SA nanofibers.

X-ray diffraction (XRD) measurements were used to study the physical state and distribution of the lutein-loaded PVA/SA nanofibers. The XRD patterns of free lutein, PVA/SA nanofibers, and lutein-loaded PVA/SA nanofibers are shown in Figure 1d. Free lutein exhibited two strong crystal diffraction peaks between 10° and 25° (2θ = 20.54° and 14.06°), which were attributed to the crystallinity of lutein and consistent with the literature [40,41]. As shown in Figure 1d, the crystallization peaks were not observed in the XRD pattern of PVA/SA nanofibers, which might be due to the sensitivity of the measurement being too low to detect the crystalline drug. Since in a separate XRD measurement for physical mixture of lutein and PVA/SA in the same mass ratio as in the electrospun lutein-loaded PVA/SA nanofiber, as shown in Figure 1d, the characteristic peak of lutein crystalline structure did not appear as well.

## 3.2. Contact Angle Measurement

Hydrophilic assessment of biological materials is a very important parameter in the field of drug delivery. Therefore, in order to determine the hydrophilicity of the nanofibers, the water contact angle of the lutein-loaded PVA/SA nanofibers before and after cross-linking was measured using a contact angle meter analysis system (Figure 2). The un-crosslinked PVA/SA nanofibers exhibited the contact angle of 18.6 ± 0.29°, which indicates that they had good hydrophilicity. Whereas the contact angle values of 33.5 ± 0.22°, 56.2 ± 0.65°, and 78.2 ± 0.37° were observed for the lutein-loaded PVA/SA nanofibers cross-linked for 1 h, 3 h, and 5 h, respectively. The cross-linked PVA/SA nanofibers showed an increase in contact angle compared to un-crosslinked PVA/SA nanofibers indicating that cross-linking could improve the hydrophilicity of nanofibers and thereby improve the stability of nanofibers in aqueous media.

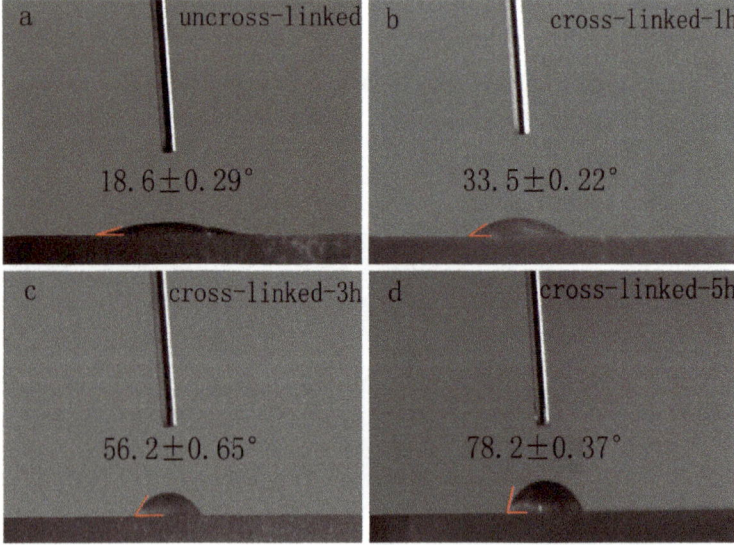

**Figure 2.** The shape of water drops and contact angle measurement for lutein-loaded PVA/SA nanofibers: (**a**) Un-crosslinked, (**b**) cross-linked for 1 h, (**c**) cross-linked for 3 h, and (**d**) cross-linked for 5 h.

## 3.3. FTIR Analysis for Cross-Linking Degree

The cross-linking effectiveness of lutein-loaded PVA/SA was analyzed using FTIR. Infrared scanning was performed in the range of 4000 to 600 cm$^{-1}$. The FTIR spectra of the lutein-loaded PVA/SA nanofibers with different cross-linking time are presented in Figure 3. It can be observed that spectra of lutein-loaded PVA/SA nanofibers consists prominent peaks at 3373 cm$^{-1}$ is ascribed to a hydroxyl group, at 2939 cm$^{-1}$ to C–H stretching (CH$_2$), at 1728 cm$^{-1}$ and 1268 cm$^{-1}$ to acetate groups (C=O and C–O, respectively), and at 1427 cm$^{-1}$ to C–H stretching (CH$_3$) [42]. The broad peak of the hydroxyl group at 3373 cm$^{-1}$ is due to the hydrogen bonding between hydroxyl groups of PVA and SA [43]. During the cross-linking of lutein-loaded PVA/SA with GA, the amount of hydroxyl functions decreases to create acetal functions, while the peak at 1728 cm$^{-1}$ (C=O) remains constant [42]. Therefore, the ratio between signal intensity at 3373 and 1728 cm$^{-1}$ could be an indicator of cross-linking degree [42,44]. The ratio between the maximum intensity of hydroxyl ($h_{OH}$) and carbonyl functions ($h_{CO}$) decreased from 2.02 to 0.60 as the nanofibers cross-linking time increased from 1 to 5 h. This reflects a higher cross-linking effectiveness in the lutein-loaded PVA/SA nanofibers.

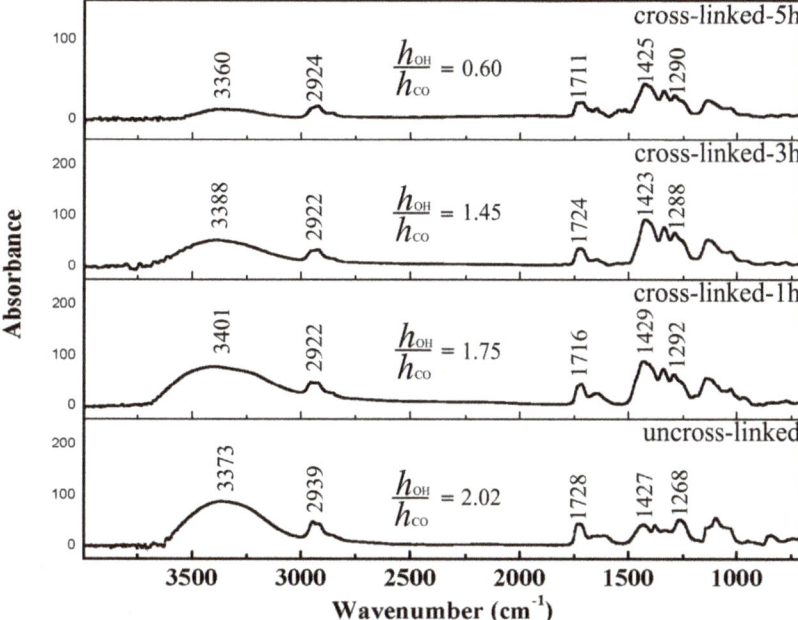

**Figure 3.** FTIR spectra for lutein-loaded PVA/SA nanofibers as un-crosslinked, cross-linked for 1 h, cross-linked for 3 h, and cross-linked for 5 h.

### 3.4. Disintegration Characterization and In Vitro Drug Dissolution

The lutein-loaded PVA/SA nanofibers were cut into a size of 1 cm × 1 cm and dissolved in deionized water to verify its disintegration time. As shown in Figure 4a–d, when the lutein-loaded PVA/SA nanofibers were placed in deionized water, they first floated on the surface. Then the color of the lutein-loaded nanofibers became darker and the size contracted within 1 s, as the water molecules rapidly penetrated into the PVA/SA nanofibers scaffold, indicating the lutein release in a rush manner. As the nanofibers scaffold continued to immerse in deionized water, it rapidly disintegrated and dispersed into hundreds of small pieces (disintegration of about 3 s), which gradually dissolved in deionized water. The whole dissolving process was completed within about 7 min. The visual observation of immediate dissolving of lutein-loaded PVA/SA nanofibers is consistent with the strong hydrophilicity of PVA/SA composite.

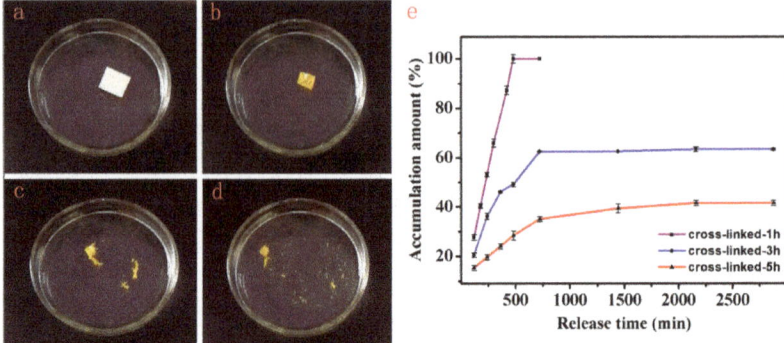

**Figure 4.** (**a**–**d**) The visual observation of the disintegration of lutein-loaded PVA/SA nanofibers, and (**e**) percentage released of lutein from electrospun PVA/SA nanofibers with different cross-linking time (1 h, 3 h, and 5 h).

The EE and in vitro drug release assessment is necessary in order to determine the bioavailability and extent of drug assimilation, which subsequently determines the drug therapeutic efficiency [45]. The EE of lutein-loaded PVA/SA nanofibers was found to be 91.9% ± 2.58%, which is comparable with ciprofloxacin-loaded PVA/SA formulation (EE = 98%) [43].

Since both PVA and SA polymers are hydrophilic, the composite of these two elements (i.e., PVA/SA nanofibers) by electrospinning is readily soluble in water, as shown by the disintegration test. The slow release of lutein can be achieved by cross-linking the PVA/SA nanofibers with a cross-linking agent. In order to study the effect of different levels of cross-linking on the release behavior of lutein from PVA/SA nanofibers, an in vitro release study was performed for 48 h in PBS (pH = 7.4; Figure 4e). Drug release from nanofibers can be attributed to three channels including drug desorption from the surface, proliferation of pores, and/or matrix degradation. All these steps are likely to get affected by the choice of polymer, porosity, morphology, and geometry of nanofibers [46]. For the lutein-loaded PVA/SA nanofibers cross-linked for 1 h, lutein was released in a controlled manner with an average release rate of 12.5%/h and complete lutein was released in 10 h. The controlled release performance achieved with lutein-loaded PVA/SA nanofibers cross-linked for 1 h was better in terms of release time span compared to complete release within 7 h in literature [43], which could be attributed to the usage of cross-linking agent. For the lutein-loaded PVA/SA nanofibers cross-linked for 3 h, the release rate was apparently decreased to around 9.4%/h, which was further decreased to 0.85%/h for the nanofibers cross-linked for 5 h. It can be seen from the above experimental results that drug release was dependent on cross-linking time i.e., the longer the cross-linking time, the better the sustained release of drug, which was consistent with the previous results by Zhang et al. [37].

### 3.5. The Release Kinetics Studies

The release curves were fitted to kinetic models to analyze the kinetics of in vitro drug release. It was proved that lutein was uniformly distributed in the PVA/SA nanofibers. Lutein was loaded into the polymer matrix by a simple packaging of the polymer, the kinetics of drug release in PVA/SA nanofibers was analyzed using the Korsmeyer–Peppas kinetic model and the Higuchi model (matrix system) respectively.

The Korsmeyer–Peppas equation is as follows:

$$M_t/M_\infty \times 100\% = kt^n, \qquad (2)$$

where, $M_t$ is the mass of the released drug at $t$ time, $M_\infty$ is the mass of the released drug when the time approaches infinity, $k$ is a constant, and $n$ is the diffusion exponent. This expression depicts a proportional mass release out of the polymer matrix with time. The value of $n$ is dependent on the type of drug delivery mechanism, geometry, and polydispersity. When $n < 0.5$, $0.5 < n < 1.0$, and $0.5 < n < 1.0$, the type of release follows Fickian diffusion, non-Fickian diffusion, and Case-II transport, respectively.

Higuchi model is another widely used pattern for analyzing the mechanism of drug release:

$$M_t/M_\infty \times 100\% = k_1 t^{1/2}, \qquad (3)$$

where $k_1$ is the diffusion rate constant.

Table 1 shows values of $n$, $k$, $k_1$, and the correlation coefficient ($R^2$) from fitting curves with two models. The $R^2$ values of the Korsmeyer–Peppas model (matrix system) were closer to 1, compared to that of the Higuchi model for all the three nanofibers with different cross-linking time. From the Korsmeyer–Peppas model, $n$ values were found to be 0.7398, 0.5840, and 0.6278 for lutein release from PVA/SA nanofibers cross-linked for 1 h, 3 h, and 5 h, respectively, indicating that lutein release from PVA/SA nanofibers occurred through non-Fickian (anomalous) diffusion suggesting that more than one mechanism process was involved in lutein release. These mathematical models could be purely empirical through the insignificant changes of $R^2$. Although the release mechanism requires further clarification, the phenomenon of the study proves that drugs can be released from the electrospun PVA/SA nanofibers matrices in a continuous manner.

Table 1. The release of lutein from PVA/SA nanofibers cross-linked for 1 h, 3 h, and 5 h.

| PVA/SA Nanofibers Cross-Linking Time | Korsmeyer–Peppas Model | | Higuchi Model | |
|---|---|---|---|---|
| 1 h | $k$ | 24.97 ± 2.713 | $k_1$ | 36.19 ± 1.736 |
|  | $n$ | 0.7398 ± 0.0653 | $R^2$ | 0.8993 |
|  | $R^2$ | 0.9707 | | |
| 3 h | $k$ | 23.08 ± 2.056 | $k_1$ | 25.94 ± 0.7468 |
|  | $n$ | 0.5840 ± 0.0608 | $R^2$ | 0.9560 |
|  | $R^2$ | 0.9641 | | |
| 5 h | $k$ | 12.81 ± 1.358 | $k_1$ | 15.32 ± 0.5497 |
|  | $n$ | 0.6278 ± 0.0709 | $R^2$ | 0.9334 |
|  | $R^2$ | 0.9540 | | |

The release of the drug in the drug-loaded nanofibers is controlled by the diffusion of drug within the polymer matrix and/or matrix degradation, which involves bulk and surface-polymer erosion, depending on the polymer composition [47]. When the penetration of water into the nanofiber matrix is slower than the matrix degradation, surface erosion predominates and when matrix degradation is faster than water penetration, bulk erosion predominates. In addition, water penetration into the individual nanofibers can affect the drug release as well. In our case, the initial drug release from the hydrophilic PVA/SA nanofibers cross-linked for 1 h, was primarily determined by the diffusion of drug and dissolution of the polymer due to the penetration of water. However, nanofibers become more hydrophobic when cross-linked for 5 h, which was verified by water contact angle measurements. Due to the enhanced hydrophobicity of the PVA/SA nanofibers, the degradation of the polymer matrix occurred relatively slow, at this point the release of the drug in the nanofibers was primarily determined by the diffusion of drug.

## 4. Conclusions

In this research, the lutein-loaded PVA/SA nanofibers with uniform and smooth morphology were obtained by electrospinning. The PVA/SA nanofibers were cross-linked for different time points and their hydrophilicity was measured with a contact angle measurement experiment. XRD analysis showed that lutein was present in the stable amorphous state in the PVA/SA nanofibers. The sustained release was achieved after the tuning the hydrophilicity of the lutein-loaded PVA/SA nanofibers as it released the lutein in a controlled manner extending up to 48 h. The drug release kinetics revealed that the release of the lutein was through non-Fickian diffusion mechanism. The results indicate that encapsulation of lutein utilizing polymer matrices by electrospinning is an effective method in drug delivery and cross-linking could further help to achieve the sustained lutein release by tuning the hydrophilicity. Therefore, the drug-loaded PVA/SA nanofibers developed in this study has great potential to be used as the delivery system in the near future.

**Author Contributions:** Conceptualization, P.H.; data curation, X.H. and Z.D.; writing—original draft preparation, X.H.; writing—review and editing, X.H., P.H., and B.L.; supervision, B.L.; project administration, P.H.; funding acquisition, B.L.

**Funding:** This research was funded by A Project of Shandong Province Higher Educational Science and Technology Program, grant number: J18KB098.

**Acknowledgments:** High quality lutein used in the experiment was a generous gift from Shandong Tian Yin Biotechnology co.ltd, Zibo, China.

**Conflicts of Interest:** The authors declare no conflict of interest.

## References

1. Silva, J.T.P.; Geiss, J.M.T.; Oliveira, S.M.; Brum, E.S.; Sagae, S.C.; Becker, D.; Leimann, F.V.; Ineu, R.P.; Guerra, G.P.; Gonçalves, O.H. Nanoencapsulation of lutein and its effect on mice's declarative memory. *Mater. Sci. Eng. C* **2017**, *76*, 1005–1011. [CrossRef] [PubMed]
2. Sato, Y.; Kobayashi, M.; Itagaki, S.; Hirano, T.; Noda, T.; Mizuno, S.; Sugawara, M.; Iseki, K. Protective effect of lutein after ischemia-reperfusion in the small intestine. *Food Chem.* **2011**, *127*, 893–898. [CrossRef] [PubMed]
3. Wang, Y.F.; Ye, H.; Zhou, C.H.; Lv, F.X.; Bie, X.M.; Lu, Z.X. Study on the spray-drying encapsulation of lutein in the porous starch and gelatin mixture. *Eur. Food Res. Technol.* **2012**, *234*, 157–163. [CrossRef]
4. Steiner, B.M.; Mcclements, D.J.; Davidov-Pardo, G. Encapsulation systems for lutein: A review. *Trends Food Sci. Technol.* **2018**, *82*, 71–81. [CrossRef]
5. Bhardwaj, N.; Kundu, S.C. Silk fibroin protein and chitosan polyelectrolyte complex porous scaffolds for tissue engineering applications. *Carbohydr. Polym.* **2011**, *85*, 325–333. [CrossRef]
6. Agarwal, S.; Greiner, A.; Wendorff, J.H. Functional materials by electrospinning of polymers. *Prog. Polym. Sci.* **2013**, *38*, 963–991. [CrossRef]
7. Nagarajan, S.; Pochat-Bohatier, C.; Balme, S.; Miele, P.; Kalkura, S.N.; Bechelany, M. Electrospun fibers in regenerative tissue engineering and drug delievery. *Pure Appl. Chem.* **2017**, *89*, 1799–1808. [CrossRef]
8. Opanasopit, P.; Sila-On, W.; Rojanarata, T.; Ngawhirunpat, T. Fabrication and properties of capsicum extract-loaded PVA and CA nanofiber patches. *Pharm. Dev. Technol.* **2013**, *5*, 1140–1147. [CrossRef]
9. Yang, J.M.; Yang, J.H.; Tsou, S.C.; Ding, C.H.; Hsu, C.C.; Yang, K.C.; Yang, C.C.; Chen, K.S.; Chen, S.W.; Wang, J.S. Cell proliferation on PVA/sodium alginate and PVA/poly (γ-glutamic acid) electrospun fiber. *Mater. Sci. Eng. C* **2016**, *66*, 170–177. [CrossRef]
10. Jia, Y.T.; Wu, C.; Dong, F.C.; Huang, G.; Zeng, X.H. Preparation of PCL/PVP/Ag Nanofiber Membranes by Electrospinning Method. *Appl. Mech. Mater.* **2012**, *268–270*, 580–583. [CrossRef]
11. Khatri, Z.; Wei, K.; Kim, B.-S.; Kim, I.-S. Effect of deacetylation on wicking behavior of co-electrospun cellulose acetate/polyvinyl alcohol nanofibers blend. *Carbohydr. Polym.* **2012**, *87*, 2183–2188. [CrossRef]
12. Vashisth, P.; Nikhil, K.; Roy, P.; Pruthi, P.A.; Singh, R.P.; Pruthi, V. A novel gellan-PVA nanofibrous scaffold for skin tissue regeneration: Fabrication and characterization. *Carbohydr. Polym.* **2016**, *136*, 851–859. [CrossRef] [PubMed]
13. Fathi-Azarbayjani, A.; Qun, L.; Chan, Y.W.; Chan, S.Y. Novel vitamin and gold-loaded nanofiber facial mask for topical delivery. *AAPS PharmSciTech* **2010**, *11*, 1164–1170. [CrossRef] [PubMed]
14. Deng, Y.; Zhang, X.; Zhao, Y.; Liang, S.; Xu, A.; Gao, X.; Deng, F.; Fang, J.; Wei, S. Peptide-decorated polyvinyl alcohol/hyaluronan nanofibers for human induced pluripotent stem cell culture. *Carbohydr. Polym.* **2014**, *101*, 36–39. [CrossRef] [PubMed]
15. Ren, G.; Xu, X.; Liu, Q.; Cheng, J.; Yuan, X.; Wu, L.; Wan, Y. Electrospun poly (vinyl alcohol)/glucose oxidase biocomposite membranes for biosensor applications. *React. Funct. Polym.* **2006**, *66*, 1559–1564. [CrossRef]

16. Ignatova, M.; Starbova, K.; Markova, N.; Manolova, N.; Rashkov, I. Electrospun nano-fibre mats with antibacterial properties from quaternised chitosan and poly (vinyl alcohol). *Carbohydr. Res.* **2006**, *341*, 2098–2107. [CrossRef]
17. Wang, Y.; Yang, Q.; Shan, G.; Wang, C.; Du, J.; Wang, S.; Li, Y.; Chen, X.; Jing, X.; Wei, Y. Preparation of silver nanoparticles dispersed in polyacrylonitrile nanofiber film spun by electrospinning. *Mater. Lett.* **2005**, *59*, 3046–3049. [CrossRef]
18. Lee, H.K.; Jeong, E.H.; Baek, C.K.; Youk, J.H. One-step preparation of ultrafine poly (acrylonitrile) fibers containing silver nanoparticles. *Mater. Lett.* **2005**, *59*, 2977–2980. [CrossRef]
19. Duan, B.; Yuan, X.; Zhu, Y.; Zhang, Y.; Li, X.; Zhang, Y.; Yao, K. A nanofibrous composite membrane of PLGA-chitosan/PVA prepared by electrospinning. *Eur. Polym. J.* **2006**, *42*, 2013–2022. [CrossRef]
20. Shao, C.; Kim, H.-Y.; Gong, J.; Ding, B.; Lee, D.-R.; Park, S.-P. Fiber mats of poly (vinyl alcohol)/silica composite via electrospinning. *Mater. Lett.* **2003**, *57*, 1579–1584. [CrossRef]
21. Hong, Y.; Shang, T.; Li, Y.; Wang, L.; Wang, C.; Chen, X.; Jing, X. Synthesis using electrospinning and stabilization of single layer macroporous films and fibrous networks of poly (vinyl alcohol). *J. Membr. Sci.* **2006**, *276*, 1–7. [CrossRef]
22. Li, L.; Hsieh, Y.-L. Chitosan bicomponent nanofibers and nanoporous fibers. *Carbohydr. Res.* **2006**, *341*, 374–381. [CrossRef] [PubMed]
23. Qin, X.-H.; Wang, S.-Y. Filtration properties of electrospinning nanofibers. *J. Donghua Univ.* **2006**, *102*, 1285–1290. [CrossRef]
24. Pasparakis, G.; Bouropoulos, N. Swelling studies and in vitro release of verapamil from calcium alginate and calcium alginate–chitosan beads. *Int. J. Pharm.* **2006**, *323*, 34–42. [CrossRef] [PubMed]
25. Ni, P.L.; Bi, H.Y.; Zhao, G.; Han, Y.C.; Wickramaratne, M.N.; Dai, H.L.; Wang, X.Y. Electrospun preparation and biological properties in vitro of polyvinyl alcohol/sodium alginate/nano-hydroxyapatite composite fiber membrane. *Colloids Surf. B* **2019**, *173*, 171–177. [CrossRef]
26. Kamoun, E.A.; Kenawy, E.R.S.; Tamer, T.M.; El-Meligy, M.A.; Eldin, M.S.M. Poly (vinyl alcohol)-alginate physically crosslinked hydrogel membranes for wound dressing applications: Characterization and bio-evaluation. *Arab. J. Chem.* **2015**, *8*, 38–47. [CrossRef]
27. Ishikawa, K.; Ueyama, Y.; Mano, T.; Koyama, T.; Suzuki, K.; Matsumura, T. Self-setting barrier membrane for guided tissue regeneration method: Initial evaluation of alginate membrane made with sodium alginate and calcium chloride aqueous solutions. *J. Biomed. Mater. Res.* **1999**, *47*, 111–115. [CrossRef]
28. Niranjan, R.; Kaushik, M.; Selvi, R.T.; Prakash, J.; Venkataprasanna, K.S.; Prema, D.; Pannerselvam, B.; Venkatasubbu, G.D. PVA/SA/TiO$_2$-CUR patch for enhanced wound healing application: In vitro and in vivo analysis. *Int. J. Biol. Macromol.* **2019**, *138*, 704–717. [CrossRef]
29. Kim, J.O.; Choi, J.Y.; Park, J.K.; Kim, J.H.; Jin, S.G.; Chang, S.W.; Li, D.X.; Hwang, M.-R.; Woo, J.S.; Kim, J.-A.; et al. Development of clindamycin-loaded wound dressing with polyvinyl alcohol and sodium alginate. *Biol. Pharm. Bull.* **2008**, *31*, 2277–2282. [CrossRef]
30. Kim, J.O.; Park, J.K.; Kim, J.H.; Jin, S.G.; Yong, C.S.; Li, D.X.; Choi, J.Y.; Woo, J.S.; Yoo, B.K.; Lyoo, W.S.; et al. Development of polyvinyl alcohol–sodium alginate gel-matrix-based wound dressing system containing nitrofurazone. *Int. J. Pharm.* **2008**, *359*, 79–86. [CrossRef]
31. Li, X.; Kanjwal, M.A.; Lin, L.; Chronakis, I.S. Electrospun polyvinyl-alcohol nanofibers as oral fast-dissolving delivery system of caffeine and riboflavin. *Colloid Surf. B* **2013**, *103*, 182–188. [CrossRef] [PubMed]
32. Hu, X.; Liu, S.; Zhou, G.; Huang, Y.; Xie, Z.; Jing, X. Electrospinning of polymeric nanofibers for drug delivery applications. *J. Control. Release* **2014**, *185*, 12–21. [CrossRef] [PubMed]
33. Nagarajan, S.; Soussan, L.; Bechelany, M.; Teyssier, C.; Cavaillès, V.; Pochat-Bohatier, C.; Miele, P.; Kalkura, N.; Janot, J.; Balme, S. Novel biocompatible electrospun gelatin fiber mats with antibiotic drug delievery properties. *J. Mater. Chem. B* **2016**, *4*, 1134–1141. [CrossRef]
34. Nagarajan, S.; Belaid, H.; Pochat-Bohatier, C.; Teyssier, C.; Iatsunskyi, I.; Coy, E.; Balme, S.; Cornu, D.; Miele, P.; Kalkura, N.S.; et al. Design of boron nitride/gelatin electrospun nanofibers for bone tissue engineering. *ACS Appl. Mater. Interfaces* **2017**, *9*, 33695–33706. [CrossRef] [PubMed]
35. Zhang, Y.Z.; Venugopal, J.; Huang, Z.M.; Lim, C.T.; Ramakrishna, S. Crosslinking of the electrospun gelatin nanofibers. *Polymer* **2006**, *47*, 2911–2917. [CrossRef]
36. Kenawy, E.R.; Abdel-Hay, F.I.; El-Newehy, M.H.; Wnek, G.E. Controlled release of ketoprofen from electrospun poly (vinyl alcohol) nanofibers. *Mater. Sci. Eng. A* **2007**, *459*, 390–396. [CrossRef]

37. Zhang, X.; Tang, K.; Zheng, X. Electrospinning and crosslinking of COL/PVA nanofiber-microsphere containing salicylic acid for drug delivery. *J. Biogic. Eng.* **2016**, *13*, 143–149. [CrossRef]
38. Vashisth, P.; Pruthi, V. Synthesis and characterization of crosslinked gellan/PVA nanofibers for tissue engineering application. *Mater. Sci. Eng. C* **2016**, *67*, 304–312. [CrossRef] [PubMed]
39. Li, X.C.; Yan, S.S.; Dai, J.; Lu, Y.; Wang, Y.Q.; Sun, M.; Gong, J.K.; Yao, Y. Human lung epithelial cells A549 epithelial-mesenchymal transition induced by PVA/collagen nanofiber. *Colloids Surf. B* **2018**, *162*, 390–397. [CrossRef]
40. Hadipour-Goudarzi, E.; Montazer, M.; Latifi, M.; Aghaji, A.A.G. Electrospinning of chitosan/sericin/PVA nanofibers incorporated with in situ synthesis of nano silver. *Carbohydr. Res.* **2014**, *113*, 231–239. [CrossRef]
41. Muhoza, B.; Zhang, Y.; Xia, S.; Cai, J.; Zhang, X.; Su, J. Improved stability and controlled release of lutein-loaded micelles based on glycosylated casein via Maillard reaction. *J. Funct. Foods* **2018**, *45*, 1–9. [CrossRef]
42. Gonzalez-Ortiz, D.; Pochat-Bphatier, C.; Gassara, S.; Camdedouzou, J.; Bechelany, M.; Miele, P. Development of novel h-BNNS/PVA porous membranes via Pickering emulsion templating. *Green Chem.* **2018**, *20*, 4319–4329. [CrossRef]
43. Kataria, K.; Gupta, A.; Rath, G.; Mathur, R.B.; Dhakate, S.R. In vivo wound healing performance of drug loaded electrospun composite nanofibers transdermal patch. *Int. J. Pharm.* **2014**, *469*, 102–110. [CrossRef] [PubMed]
44. M'barki, O.; Hanafia, A.; Bouyer, D.; Faur, C.; Sescousse, R.; Delabre, U.; Blot, C.; Guenoun, P.; Deratani, A.; Quemener, D.; et al. Greener method to prepare porous polymer membranes by combining thermally induced phase separation and crosslinking of poly (vinyl alcohol) in water. *J. Membr. Sci.* **2014**, *458*, 225–235. [CrossRef]
45. Vashisth, P.; Raghuwanshi, N.; Srivastava, A.K.; Singh, H.; Nagar, H.; Pruthi, V. Ofloxacin loaded gellan/PVA nanofibers-Synthesis, characterization and evaluation of their gastroretentive/mucoadhesive drug delivery potential. *Mater. Sci. Eng. C* **2017**, *71*, 611–619. [CrossRef] [PubMed]
46. Thakkar, S.; Misra, M. Electrospun polymeric nanofibers: New horizons in drug delivery. *J. Pharm. Sci.* **2017**, *107*, 148–167. [CrossRef] [PubMed]
47. Kajdič, S.; Planinšek, O.; Gašperlin, M.; Kocbek, P. Electrospun nanofibers for customized drug-delivery systems. *J. Drug Deliv. Sci. Technol.* **2019**, *51*, 672–681. [CrossRef]

© 2019 by the authors. Licensee MDPI, Basel, Switzerland. This article is an open access article distributed under the terms and conditions of the Creative Commons Attribution (CC BY) license (http://creativecommons.org/licenses/by/4.0/).

Article

# Film Dressings Based on Hydrogels: Simultaneous and Sustained-Release of Bioactive Compounds with Wound Healing Properties

Fabian Ávila-Salas [1,†], Adolfo Marican [2,7,†], Soledad Pinochet [3], Gustavo Carreño [2,7], Oscar Valdés [3], Bernardo Venegas [4], Wendy Donoso [4], Gustavo Cabrera-Barjas [5], Sekar Vijayakumar [6] and Esteban F. Durán-Lara [7,8,*]

1. Centro de Nanotecnología Aplicada, Facultad de Ciencias, Universidad Mayor, Huechuraba 8580000, Región Metropolitana, Chile
2. Instituto de Química de Recursos Naturales, Universidad de Talca, Talca 3460000, Maule, Chile
3. Vicerrectoría de Investigación y Postgrado, Universidad Católica del Maule, Talca 3460000, Maule, Chile
4. Department of Stomatology, Faculty of Health Sciences, University of Talca, Talca 3460000, Chile
5. Technological Development Unit (UDT), Universidad de Concepción, Av. Cordillera 2634, Parque Industrial Coronel, Coronel 4191996, Biobío, Chile
6. Nanobiosciences and Nanopharmacology Division, Biomaterials and Biotechnology in Animal Health Lab, Department of Animal Health and Management, Alagappa University, Science Campus 6th Floor, Karaikudi-630004, Tamil Nadu, India
7. Bio and NanoMaterials Lab, Drug Delivery and Controlled Release, Universidad de Talca, Talca 3460000, Maule, Chile
8. Departamento de Microbiología, Facultad de Ciencias de la Salud, Universidad de Talca, Talca 3460000, Maule, Chile
* Correspondence: eduran@utalca.cl; Tel.: +56-71-2200363
† These authors contributed equally to this work.

Received: 30 July 2019; Accepted: 26 August 2019; Published: 2 September 2019

**Abstract:** This research proposes the rational modeling, synthesis and evaluation of film dressing hydrogels based on polyvinyl alcohol crosslinked with 20 different kinds of dicarboxylic acids. These formulations would allow the sustained release of simultaneous bioactive compounds including allantoin, resveratrol, dexpanthenol and caffeic acid as a multi-target therapy in wound healing. Interaction energy calculations and molecular dynamics simulation studies allowed evaluating the intermolecular affinity of the above bioactive compounds by hydrogels crosslinked with the different dicarboxylic acids. According to the computational results, the hydrogels crosslinked with succinic, aspartic, maleic and malic acids were selected as the best candidates to be synthesized and evaluated experimentally. These four crosslinked hydrogels were prepared and characterized by FTIR, mechanical properties, SEM and equilibrium swelling ratio. The sustained release of the bioactive compounds from the film dressing was investigated in vitro and in vivo. The in vitro results indicate a good release profile for all four analyzed bioactive compounds. More importantly, in vivo experiments suggest that prepared formulations could considerably accelerate the healing rate of artificial wounds in rats. The histological studies show that these formulations help to successfully reconstruct and thicken epidermis during 14 days of wound healing. Moreover, the four film dressings developed and exhibited excellent biocompatibility. In conclusion, the novel film dressings based on hydrogels rationally designed with combinatorial and sustained release therapy could have significant promise as dressing materials for skin wound healing.

**Keywords:** sustained release; wound healing; crosslinking; allantoin; equilibrium swelling ratio; accumulative release; thermogravimetric analysis

## 1. Introduction

The primary function of the skin is to serve as a protective barrier against external hazards. The loss of integrity of large portions of the skin as a result of injury or disease may lead to a major disability or even death [1,2]. Therefore, wound healing is a fundamental physiological process that restores skin integrity, aiming to repair the damaged tissues [3]. Given its importance, the sequence of events of wound healing has been extensively studied for several decades [4]. The mechanism of wound healing is very complex, involving several physiological events such as coagulation, inflammation, cell proliferation, matrix repair, epithelization and remodeling of the scar tissue [5]. Interruption or deregulation of one or more phases of the wound healing process leads to non-healing (chronic) wounds [6].

Dressing materials, which are used for wounds or burns, are known as "artificial skin". They should possess properties of normal skin to accelerate the recovery of wounded or destroyed skin areas. One of the most studied is the hydrogel dressing [7]. Hydrogels are 3D, hydrophilic and polymeric networks capable of absorbing large amounts of water or biological fluids [8], but do not dissolve when brought into contact with water [9]. Due to their high water content, porosity and soft consistency, they closely simulate natural living tissue, more so than any other class of synthetic biomaterials [8]. The network crosslinked by covalent bonds is classified as a chemical gel, while the formation of a physical gel takes place via a physical association between polymeric chains [10]. Compared with other biomaterials, hydrogels have the advantages of increased biocompatibility, tunable biodegradability, and porous structure, among others. However, owing to the low mechanical strength and fragile nature of the hydrogels, the feasibility of applying hydrogels is still limited. Thus, novel hydrogels with stronger and more stable properties are still needed and remain an important direction for research [11].

The exclusive physical properties of hydrogels have aroused particular interest in their use in drug release applications. Their highly porous structure can be easily tuned by controlling the crosslink density (cross-linking degrees) in the gel matrix and the affinity of the hydrogels for the aqueous medium in which they are swollen [12]. Their porosity also allows the loading of drugs into the gel matrix and subsequent drug release at a rate dependent on the diffusion coefficient of the small molecule or macromolecule through the gel network [13]. The properties that the drug delivery (usually governed by passive diffusion mechanisms) has also depend on factors such as hydrogel mesh sizes, stimuli-sensitivity and hydrogel capacity, among others [14].

Therefore, the structural properties of hydrogels and their affinity for certain bioactive molecules will depend directly on the selection of constituent polymers and the type of crosslinker that will form the polymeric crosslinked mesh. Among the polymers most used for the preparation of hydrogels for the treatment of wound healing are chitosan [15] and polyvinyl alcohol (PVA) [16].

For this study, PVA was selected because it is a biocompatible polymer and non-toxic to humans [17,18]. The formation of hydrogels from PVA can be performed by chemical methods, which involve the formation of interactions and bonds between the PVA chains and the functional groups of the crosslinking agents [19]. The concentration of crosslinker affects the porous structure, swelling features and mechanical strength. By setting the suitable degree of crosslinking, it is possible to prepare super-porous hydrogels with the desired characteristics. This will provide a platform to design novel drug delivery systems [20]. There are scientific studies showing that hydrogels based on PVA formulations cross-linked with specific crosslinkers are excellent bioactive compound releasing agents, especially at the dermal level: PVA-Glutaraldehyde [21], PVA-Ethylenglycol [22], PVA-Chitosan [23], PVA-Collagen [24], PVA-Cellulose [25], PVA-starch [7], PVA-Gelatin [26], PVA-TEOS [27], PVA-Heparin [28], PVA-poly(AAm) [29], among others.

PVA crosslinked with organic acids generates flexible and transparent hydrogels [18] capable of interacting with water-soluble compounds [30]. Its flexibility allows it to be easily handled during the treatment of wounds, providing a stronger mechanical protection. On the other hand, its transparency allows evaluating the process of healing step by step [18].

The organic acids or bifunctional molecules of interest in this study have two carboxylic acid functional groups at both ends of its structure (dicarboxylic acid molecules, for example, succinic acid), which through an esterification process can generate covalent bonds with the hydroxyl groups present in the polymer chains of PVA, generating the crosslinking and porosity of the structure [18].

The hydrogels from PVA must meet a number of requirements including biocompatibility, suitable porosity, swelling, mechanical strength and degradation properties. As mentioned earlier, all these properties are affected by the kind and concentration of polymers employed in hydrogels as well as by the cross-linking type and density [13,31]. Therefore, the porosity is directly related to the structure of the bifunctional molecule with low molecular weight, the length of its skeleton and the amount of ester bonds it can generate with the PVA chains. For example, the chemical hydrogel has been synthesized with specific dicarboxylic acids (DCA) [18,32].

As is well known, wound healing is a complex and sequenced process formed by several phases. In this context, the properties of hydrogels allow utilizing a delivery system of drug combinations simultaneously (multi-target therapy) [33]. This property of hydrogels could play a key role as a simultaneous delivery system of therapeutic agents for the wound healing process (constituted by several coordinated stages). Congruent with the above, drug combination therapy (directed at multiple therapeutic targets) improves treatment response and minimizes adverse events [34,35]. Due to their inherent properties, hydrogels are able to efficiently encapsulate and deliver in a controlled release manner [14]. Additionally, these materials must possess properties similar to normal skin: not possess toxins, provide an environment that prevents drying of the wound, reduce the penetration of bacteria, avoid losses of heat, water, proteins and red blood cells, in addition to promoting a rapid healing [36,37]. Thus, the use of biomaterials for the treatment of wounds is an area of interest for the scientific and medical community.

Therefore, the goal of the present article was to rationally develop hydrogel polymer formulations based on PVA crosslinked with a series of crosslinkers to improve the wound healing process of complex injuries through a simultaneous and sustained-release of allantoin, resveratrol, dexpanthenol and caffeic acid in the skin mouse model.

## 2. Materials and Methods

*2.1. Theoretical Section*

2.1.1. Building Molecular Structures

The three-dimensional (3D) structures of allantoin (AL), resveratrol (RES), caffeic acid (CA), dexpanthenol (DEX), 20 different PVA hydrogel nanopores (PVAnp) (Table S1) and PVA chain (of five monomers long) were designed and built through MarvinSketch software version 19.1.0, ChemAxon Ltd., Budapest, Hungary [38]. For all 3D structures, their protonation states at pH 7.0 were considered. Their geometries were optimized using Gaussian software version 16, revision A.03, Inc., Wallingford, CT, USA [39] at Density Functional Theory level using the B3LYP method and 6-311+G(d,p) as the selected basis set. In this study, the DCA selected by Marican et al. 2018 [18] were evaluated.

2.1.2. In-Silico Calculation of Interaction Energies

Interaction energies ($\Delta E$) between the 20 nanopores and the compounds studied were calculated using a computational strategy implemented by Avila-Salas et al. 2012 [28] which couples a Monte Carlo conformational sampling [40] and $\Delta E$ calculations at the semi-empirical quantum mechanical (SQM) level [41]. With this methodology, it is possible to quickly evaluate the energy contribution of each component in the PVAnp-compound binding affinity. $\Delta E$ was calculated for molecule1-molecule2 complexes. In this case, molecule1 represents each one of the 20 PVAnp (Table S1) and molecule2 represents the four compounds studied (AL), RES, CA and DEX.

## 2.2. Experimental Section

### 2.2.1. Materials

Polyvinyl alcohol (PVA) 30-60 KDa, succinic acid (SA), aspartic acid (AA) malic acid (MALI), maleic acid (MALE), NaHCO$_3$, acetonitrile (HPLC grade), allantoin, dexphantenol, caffeic acid and resveratrol analytical standards were purchased from Sigma-Aldrich (St. Louis, MO, USA). HCl and methanol (HPLC grade) K$_2$HPO$_4$ and H$_3$PO$_4$ were purchased from Merck (Darmstadt, Germany). All solutions were prepared using MilliQ water.

### 2.2.2. Synthesis of Selected Hydrogels Based on PVA, Dicarboxylic Acids and Bioactive Compound Loading

For this study, twenty hydrogels based on PVA and dicarboxylic acids (PDCAH) were proposed. However, according to the theoretical analysis, four candidates present the best interactions with the bioactive compounds (BC) of interest. Therefore, four PDCAH with different crosslinkers were synthesized. The preparations of these platforms were performed through the esterification of PVA with DCA according to the method from Rodríguez Nuñez et al. 2019 with minor modifications [12]. Briefly, the reactions were performed by mixing an aqueous solution of PVA with an aqueous solution of a specific DCA (20 wt %) in presence of $1 \times 10^{-1}$ mol·L$^{-1}$ HCl (pH 1). After that, each reaction was carried out under reflux at 90 °C in a necked flask with magnetic agitation. After 3 h, each pre-hydrogel solution was poured into a new flask and a specific amount of BC (allantoin, dexpanthenol, caffeic acid, and resveratrol) was added for it encapsulation, as depicted in Table 1. Then, each solution was homogenized by stirring for 1 h and sonicated for 60 min until a homogenized solution was obtained. After that, each mixture solution was put in an oven at 45 °C overnight until the crosslinking was complete. Then, the PSAH, PAAH, PMALIH and PMALEH with encapsulated BC were washed several times with NaHCO$_3$ for removing the excess acid. Finally, the hydrogels were lyophilized in order to obtain the xerogel. Lastly, each formulation obtained was termed as PSAH-BC, PAAH-BC, PMALIH-BC and PMALEH-BC, respectively.

**Table 1.** Specifications of supramolecular PDCAH and amount of BC loading.

| PDCAH | Crosslinker | Crosslinker Ratio * | Bioactive Compounds | | | |
|---|---|---|---|---|---|---|
| | | | Allantoin * | Caffeic Acid * | Resveratrol * | Dexpanthenol * |
| PMALEH | Male | 20 | 5 | 2 | 2 | 2 |
| PAAH | AA | 20 | 5 | 2 | 2 | 2 |
| PMALIH | Mali | 20 | 5 | 2 | 2 | 2 |
| PSAH | SA | 20 | 5 | 2 | 2 | 2 |

(*) % w/w respect to PDCAH; Maleic acid (Male); Aspartic acid (AA); Malic acid (Mali) and succinic acid (SA).

### 2.2.3. Equilibrium Swelling Ratio of PDCAH

The water uptake process was estimated by equilibrium swelling ratio (% ESR) at desired time intervals. Each xerogel film was immersed in phosphate buffer saline (PBS) (pH 7.4) and acetate buffer (pH 3.0) at 25 °C for 21 h until swelling equilibrium was attained. The weight of the wet sample [$W_w$ (g)] was measured after carefully removing moisture on the surface with an absorbent paper. The weight of the dried sample [$W_d$ (g)] was determined after the freeze-drying process of the hydrogel. The ESR of the hydrogel samples was calculated as follows (Equation (1)):

$$\text{ESR (\%)} = \frac{W_w - W_d}{W_d} \times 100\% \qquad (1)$$

## 2.2.4. Infrared Spectroscopy

Fourier-Transform Infrared (FT-IR) spectra of PSAH, PAAH, PMALIH and PMALEH were recorded on a Nicolet Nexus 470 spectrometer (Thermo Scientific, Waltham, MA, USA) within the 4000–400 $cm^{-1}$ spectral intervals. All spectra were obtained in KBr pellets from an average of 32 scans with 4 $cm^{-1}$ resolution.

## 2.2.5. Mechanical analysis

Tensile tests were performed by means of a dynamometer model 4301, Instron (Canton, OH, USA) equipped with a 5 kN load cell. The measurements were performed on dumbbell-shaped films. The width and the length of the investigated films were 5 mm and 30 mm, respectively, while the thickness of each film was measured at five random points using a micrometer and the result was expressed as the average value. All the measurements were carried out at 25 ± 2 °C and 50 ± 5% relative humidity at a crosshead rate of 5 mm·$min^{-1}$. The reported data are the average values of five measurements. The obtained stress-strain curves were used to calculate tensile strength ($\sigma m$, MPa), elongation at break ($\varepsilon$, %) and Young's Modulus (E, MPa).

## 2.2.6. Scanning Electron Microscopy Analysis

The Scanning Electron Microscopy (SEM) analyses were performed for all four formulations. The films morphology was analyzed using a scanning electron microscope (JEOL-JSM 6380, Tokyo, Japan) operated at 15kV. Surface and side views of cryogenically fracture films were examined. All samples were sputtered with a gold layer, around 40 nm in thickness, previous to the analysis.

## 2.2.7. Sustained Release Kinetics of BC from PDCAH-BC

The BC content of each supramolecular PDCAH (PDCAH-BC) is depicted in Table 1. Each PDCAH-BC with a weight of 400 mg was disposed into a 10 mL tube and 5 mL of PBS (pH 7.4) was poured over the formulation as a release medium. The tubes were transferred to an orbital shaker incubator water bath (Farazteb, Iran) at 33.5 °C ± 0.1 °C (Skin Temperature) and shaken at 35 ± 2 rpm. At specific time intervals, the PBS was removed and replaced with an equal volume of PBS in order to maintain sink conditions throughout the study. The samples of each supramolecular formulation were analyzed by a Perkin Elmer series 200 HPLC system (Norwalk, CT, USA) with a UV-Vis detector. An YWG C-8 (250 mm × 4.6 mm i.d. × 10 μm) column was used for the analysis of samples. 20 μL of eluent was injected into the HPLC. The mobile phase used consisted of 20 mM $K_2HPO_4$ (pH 6.0, $H_3PO_4$)/Methanol (90:10, v/v), in isocratic mode, at a flow rate of 1.0 mL·$min^{-1}$. The samples were monitored at 210 nm (allantoin and dexpanthenol) and 300 nm (caffeic acid and resveratrol) by absorbance detection at 30 °C.

The release of each BC from each supramolecular formulation was determined by applying the amounts of released and loading BC to the following relationship (Equation (2)):

$$\text{Cumulative BC release (\%)} = \text{Cumulative amount of BC released} \times \frac{100}{\text{Inicial amount of BC}} \quad (2)$$

## 2.2.8. Wound Healing Testing on Dermal Models of Rats

Animals and Maintenance Conditions

The experiments were carried out in adult Sprague Dawley rats of 150–200 grs obtained from the animal facility from the Universidad de Talca. All the animal care and experimental protocol were reviewed and approved by Comité Institucional de Ética, Cuidado y Uso de Animales de Laboratorio (CIECUAL) of the Universidad de Talca (Project identification code: 11170155; approval date of the committee: 17 December 2017). The animals were maintained in standard environmental conditions (22 ± 2 °C, relative humidity 70–80%, 12-h light cycle). The animals were weighed at the beginning and at the end of the experimental period. In addition, the intake of water and food was recorded. The rats were fed a standard diet manufactured by Champion (6.4% moisture, 3.6% lipids, 6.7% protein, 7.3 ashes, 3.6 fibers, 72.4% carbohydrates). The animals had free access to water and food; the bed was changed three times a week. Each cage had a record of changes in behavior or intake that was filled daily by the personnel in charge.

Experimental Procedure

The animals were divided randomly into groups (5 animals per group). At the start of the surgical procedure, all animals were sedated with isofluorane and anesthetized with a mixture of ketamine (ketostop, DrangPharmainvetec S.A)/xylazine (Xylaret, Agroland) in a ratio of 3:1 (2.2 µL/g by weight). Once the anesthetized animals were in the surgical plane, the trichotomy of the inter scapular area was performed with a hair clipper (Oster gold) and the area washed with 0.25% chlorhexidine soap. Then, one skin segment in the area of the back between the scapulae was removed; surgery was performed with a special scalpel or punch. The diameter of the biopsy was around 1 cm. The excised wounds were covered with to-be-tested hydrogels (PDCAH-BC and controls, 1.2 cm × 1.2 cm) and affixed with an elastic adhesive bandage. Two groups control were used in this experiment, Madecassol™ a commercial product and PSAH (film dressing without BC). The commercial product was daily applied until day 14. The total duration of each test was 14 days. On day 7, the PDCAH-BC and control (film dressing) were removed to analyze their adhesion and the film dressing was not reapplied. From days 7–14, the natural wound healing process was analyzed, protocol modified by Murakami et al., 2010 [42]. The wounds were examined and photographed for measurement of wound size reduction. These results were expressed in area and were represented by the closure of the wound. Differences in wound closure between controls and treatments were compared macroscopically. Upon completion of wound-healing experiments, the animals were sacrificed by excess diethyl ether on day 14 after the surgery. The rate of wound closure, which represents the percentage of wound reduction from the original wound size, was estimated utilizing the following formula (Equation (3)):

$$\text{Wound healing reduction (\%)} = \frac{\text{wound area day 0} - \text{wound area day 14}}{\text{wound area day 0}} \times 100 \quad (3)$$

Values are expressed as a percentage of the healed wounds ± SD.

2.2.9. Histological Analysis

The histological analysis was oriented to the microscopic observation of the wound closure and was intended to compare the wound healing process. 5 µm thick sections from rat skin biopsies were used on silanized slides with 2% 3-aminopropyltriethoxysilane in acetone. The sample corresponded to rat skin affixed in 4% formaldehyde in 0.075 M sodium phosphate buffer pH 7.3, decalcified and embedded in paraffin. The sections were dewaxed and rehydrated following the routine protocol of the oral histopathology laboratory of the Universidad de Talca. The skin biopsies were stained with hematoxylin, eosin, Masson Trichrome and Giemsa.

2.2.10. Cytotoxicity and Cell Viability

The cytotoxicity of PDCAH was evaluated on fibroblast cells. For this purpose, the viability of fibroblasts was assessed using MTT assay according to the protocol of Mossman et al. [43]. Briefly, the cells were seeded in 24-well plates (5 µL, $1.6 \times 10^4$ cells per well) and 150 µL of Dulbecco's Modified Eagle Medium (DMEM)-High medium was added and incubated for 24 h at 37 °C in 5% $CO_2$. Then, the medium was substituted by 100 µL of fresh DMEM-High per well containing three different concentrations of PMALEH, PAAH, PSAH and PMALIH (500 µg·mL$^{-1}$, 1500 µg·mL$^{-1}$, and 2500 µg·mL$^{-1}$ per formulation). Fresh medium without any PDCAH was used as a control. Cell viability was evaluated after 24 h by the MTT assays. Specifically, 5 µL of MTT solution (3 mg·mL$^{-1}$

in PBS) and 50 µL of fresh medium were added to each sample and incubated for 4 h in the dark at 37 °C; formazan crystals were then dissolved in 100 µL dimethyl sulfoxide and incubated for 18 h. Supernatant optical density (o.d.) was evaluated at 570 nm (Spectrophotometer, Packard Bell, Meriden, CT, USA). Untreated cells were taken as control with 100% viability. The cell cytotoxicity of PDCAH was expressed as the relative viability (%), which correlates with the amount of liable cells compared with the negative cell control (100%).

2.2.11. Statistical Analysis

All experiments were performed in triplicate. Mean, standard deviation and Student's t-test was performed to test the statistical significance in MTT assay studies and graphs were prepared by using Graphpad Prism 6. Statistical significance was set at $p < 0.05$.

## 3. Results and Discussion

### 3.1. In-Silico Interaction Energy Study

20 nanopores of the PVA hydrogels crosslinked with different dicarboxylic acids were designed (Table S1). The interaction energy studies between the 20 nanopores and the compounds studied with healing activity were carried out: allantoin (AL), resveratrol (RES), caffeic acid (CA) and dexpanthenol (DEX). The results of these studies can be observed in Table 2.

**Table 2.** Interaction energy values calculated using SQM methods between the hydrogel nanopores (Hnp) and the different compounds studied.

| Id. | Hydrogel Nanopores (Hnp) | HNP-AL ΔE Kcal/mol | HNP-RES ΔE Kcal/mol | HNP-CA ΔE Kcal/mol | NPH-DEX ΔE Kcal/mol | Average ΔE Kcal/mol |
|---|---|---|---|---|---|---|
| 1 | PVAnp-Oxalic acid | −3.049 | −2.090 | −0.589 | −1.780 | −1.877 |
| 2 | PVAnp-Malonic acid | −3.083 | −2.093 | −0.876 | −1.824 | −1.969 |
| 3 | PVAnp-Succinic acid | −3.187 * | −2.172 * | −0.962 * | −1.963 * | −2.071 * |
| 4 | PVAnp-Malic acid | −3.127 * | −2.149 * | −0.986 * | −1.933 * | −2.049 * |
| 5 | PVAnp-Fumaric acid | −3.012 | −2.049 | −0.873 | −1.833 | −1.942 |
| 6 | PVAnp-Maleic acid | −3.153 * | −2.156 * | −0.950 * | −1.964 * | −2.056 * |
| 7 | PVAnp-Citraconic acid | −3.052 | −2.093 | −0.857 | −1.783 | −1.946 |
| 8 | PVAnp-Itaconic acid | −3.069 | −2.079 | −0.862 | −1.810 | −1.955 |
| 9 | PVAnp-Tartaric acid | −3.093 | −2.078 | −0.868 | −1.769 | −1.952 |
| 10 | PVAnp-Glutaric acid | −3.095 | −2.117 | −0.905 | −1.839 | −1.989 |
| 11 | PVAnp-Adipic acid | −3.089 | −2.105 | −0.868 | −1.791 | −1.963 |
| 12 | PVAnp-Pimelic acid | −3.009 | −2.012 | −0.813 | −1.720 | −1.889 |
| 13 | PVAnp-Suberic acid | −3.032 | −2.073 | −0.572 | −1.763 | −1.860 |
| 14 | PVAnp-Azelaic acid | −3.079 | −2.089 | −0.872 | −1.820 | −1.965 |
| 15 | PVAnp-Phtalic acid | −3.108 | −2.093 | −0.883 | −1.784 | −1.967 |
| 16 | PVAnp-Isophtalic acid | −3.114 | −2.136 | −0.924 | −1.858 | −2.008 |
| 17 | PVAnp-Terephtalic acid | −3.125 | −2.141 | −0.933 | −1.906 | −2.026 |
| 18 | PVAnp-2,5-pyridin acid | −3.117 | −2.079 | −0.842 | −1.858 | −1.982 |
| 19 | PVAnp-Aspartic acid | −3.130 * | −2.150 * | −0.951 * | −1.910 * | −2.035 * |
| 20 | PVAnp-Glutamic acid | −3.063 | −2.133 | −0.905 | −1.874 | −1.994 |

* Better interaction energy values (more negative).

According to the results obtained from ΔE, the pores generated between PVA and the succinic, malic, maleic and aspartic acids (marked in red in Table 2), have simultaneously better ΔE for the 4 compounds of interest. Therefore, hydrogels PVA cross-linked with these acids are good candidates to be evaluated experimentally.

### 3.2. Preparation of PDCAH

The preparation of formulations was performed as is depicted in Scheme 1 and Figure 1. Concisely, each hydrogel was prepared using polymerization by esterification in the presence of HCl as a catalyst. Once the pre-hydrogel was produced, the specific amount of each BC was added. With this methodology of loading, it is possible to obtain over 99% retention of the drug. The characterization analysis from FT-IR established the conjugation between PVA (–OH) and DCA (–COOH) into the PDCAH (The PDCAH characterization was performed without the loading drug (empty formulation). According to previous works, a crosslinking degree of 10:2 of PVA:DCA was prepared, which was kept constant due to its excellent features such as porosity, among others [12,18].

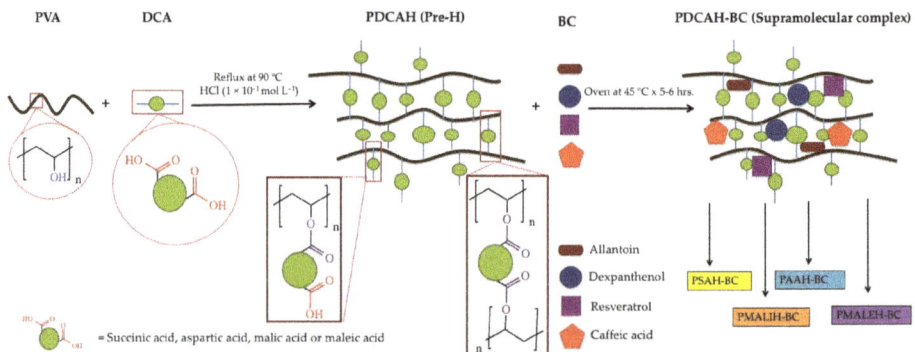

**Scheme 1.** Schematic representation of the synthesis of PDCAH and loading of BC.

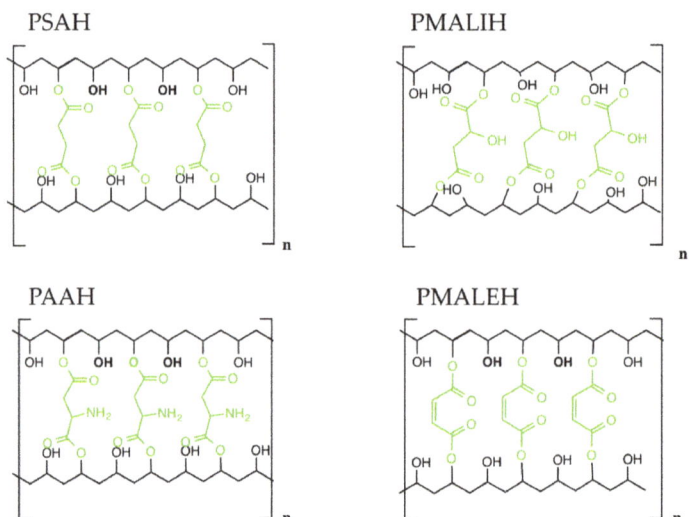

**Figure 1.** Proposed structures of PDCAH.

### 3.3. ESR Results

This characterization is very simple but very important at the same time since it confirms the hydrogel formation. In other words, if there is an increase in the swelling index, it means that the hydrogel matrix is absorbing the solvent and is not dissolved in the solvent; this being one of the most important features of a hydrogel [44]. In consequence, this characterization was made to confirm the preparation of the four hydrogel formulations with different crosslinker agents. Figure 2 shows the ESR for all four hydrogels. This figure displays an increase in the swelling index across time for all PDCAH. For all PDCAH, the swelling index in the first segment increased rapidly and afterwards slowly. This behavior may be due to the hydrogels reaching maximum constant swelling. The PSAH,

PMALEH, PAAH and PMALIH reached the swelling equilibrium (zero order) at about 4–5 h. After 5 h, the PSAH, PMALEH, PMALIH and PAAH reached about 300, 400, 500 and 600% or swelling index, respectively, at pH 7.4. This may be due to the crosslinker agent polarity, which can be explained by the available hydrophilic groups in their structure that form hydrogen bonds with water molecules. In this context, we may conclude that the polarity led to an increase in ESR. Also, a significant difference for the set of formulations was observed between the two pH models. In all cases, the swelling index is higher a 7.4 than 3.0, observing a difference of ~80% of ESR between pH models. The data previously mentioned confirming that swelling behavior of the prepared hydrogels are pH-dependent owing to their ionic networks. In this sense, the four PDCAH absorbed a higher amount of water at 7.4 than 3.0. The ionic networks from the prepared hydrogels are provided by containing ionic pendant groups from the crosslinkers (SA, MALE, MALE, and AA), which have different types of pKa (depends on each crosslinker agent) [45]. This feature at a certain pH provides higher ionization degree in the hydrogel matrix, producing an intensification of electrostatic repulsion between chains from the networks. This electrostatic repulsion causes a higher uptake of solvent into the matrix, which increases the size of the hydrogel [12,18]. The swelling index observed could be related to the diffusion process where the encapsulated bioactive compounds could diffuse through swollen hydrogel networks toward the outside [46].

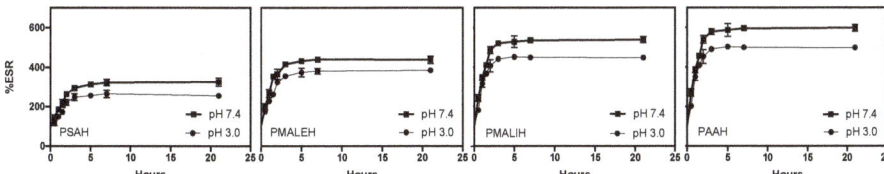

**Figure 2.** The swelling ratio of the PDCAH at 24 °C as a function of time, pH and crosslinker nature. Data are shown as mean ± SD ($n = 3$).

### 3.4. BC in Vitro Release Behavior of Supramolecular PDCAH

The encapsulation methodology carried out in this work was easier than conventional encapsulation by absorption. Specifically, the encapsulation was done through mixing BC with the pre-hydrogel solutions to encapsulate the BC into the PDCAH (PDCAH-BC), which decreased the encapsulation process time. On the other hand, this methodology allows for the loading of an exact amount of drugs (compared with the conventional method). The supramolecular PDCAH film was loaded with allantoin (5%), dexpanthenol (2%), caffeic acid (2%) and resveratrol (2%) according to the standard concentrations of bioactive compounds utilized in the dermatology area.

In order to analyze the in vitro release behavior of BC from PDCAH-BC, release profiles were obtained in physiological conditions (33.5 °C, PBS at pH 7.4). The samples were evaluated through HPLC method and the percent of the cumulative amount released was plotted over time. The BC cumulative release profiles are shown in Figure 3; the four PDCAH-BC provided a rapid release into the medium until 6 h, as shown for each BC and hydrogel. For example, 35% of allantoin, 38% of dexpanthenol, 52% of caffeic acid and 56% of resveratrol have been released from PMALEH-BC. After this initial fast release profile, PMALEH-BC showed a slower and steadier BC release into the medium for all cases.

The average release rate (%) of BC during the rapid phase (0 to 6 h) is depicted in Table 3 for each formulation. After 6 h, the rapid release rate changed toward a slower and sustained release. For PMALEH-BC, the average of the rapid-release phase was 0.58 mg/h, 0.63 mg/h, 0.87 mg/h and 0.93 mg/h for allantoin, dexpanthenol, caffeic acid and resveratrol, respectively. In contrast, the average of the slow-release phase was 0.11 mg/h, 0.11 mg/h, 0.08 mg/h and 0.08 mg/h for allantoin, dexpanthenol, caffeic acid, and resveratrol, respectively. In Tables 4 and 5, all the average release values of each formulation and BC are provided.

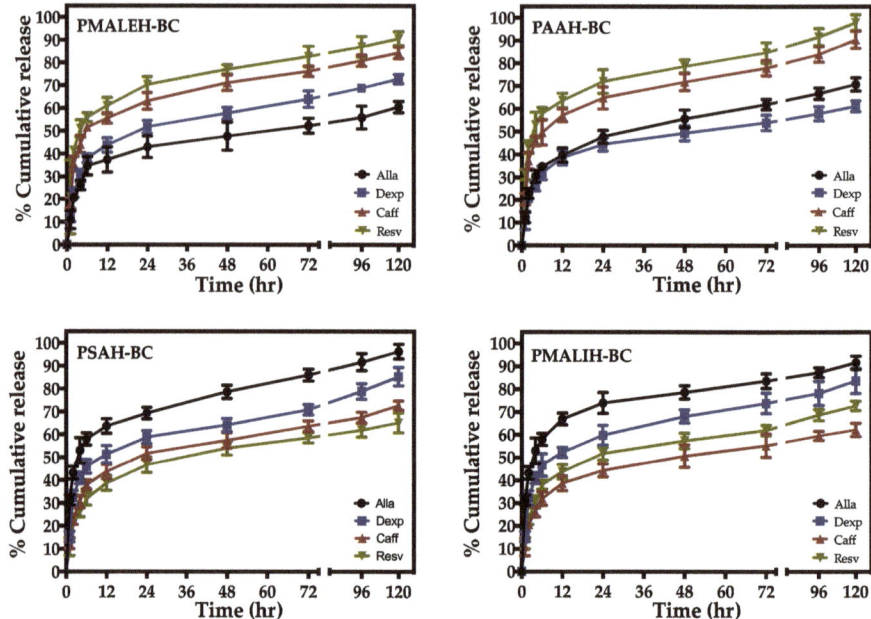

**Figure 3.** Release of BC from supramolecular PDCAH-BC in PBS at 33.5 °C; mean ± SEM, $n = 3$.

**Table 3.** The percentage of release rate during the rapid-release phase.

| Hydrogel | Rapid-Release Phase of BC (%) until 6 h | | | |
|---|---|---|---|---|
| | Allantoin | Dexpanthenol | Caffeic Acid | Resveratrol |
| PMALEH-BC | 35 ± 4.1 | 38 ± 2.0 | 52 ± 1.2 | 56 ± 2.0 |
| PAAH-BC | 35 ± 1.7 | 32 ± 3.0 | 50 ± 5.5 | 58 ± 2.6 |
| PMALIH-BC | 58 ± 2.6 | 47 ± 4.9 | 32 ± 3.0 | 38 ± 2.0 |
| PSAH-BC | 58 ± 2.6 | 46 ± 3.0 | 38 ± 2.0 | 32 ± 3.0 |

**Table 4.** The average release rate during the rapid-release phase.

| Hydrogel | Rapid-Release Phase | | | |
|---|---|---|---|---|
| | Allantoin | | Dexpanthenol | |
| | mg/6 h | mg/h | mg/6 h | mg/h |
| PMALEH-BC | 7.0 ± 0.82 | 1.17 ± 0.14 | 7.6 ± 0.4 | 1.27 ± 0.07 |
| PAAH-BC | 6.4 ± 0.34 | 1.07 ± 0.06 | 6.4 ± 0.6 | 1.07 ± 0.10 |
| PMALIH-BC | 9.4 ± 0.52 | 1.57 ± 0.09 | 9.4 ± 1.0 | 1.57 ± 0.16 |
| PSAH-BC | 9.2 ± 0.52 | 1.53 ± 0.09 | 9.2 ± 0.6 | 1.53 ± 0.10 |
| Hydrogel | Caffeic Acid | | Caffeic acid | |
| | mg/6 h | mg/h | mg/6 h | mg/h |
| PMALEH-BC | 10.4 ± 0.24 | 1.73 ± 0.04 | 11.2 ± 0.4 | 1.87 ± 0.07 |
| PAAH-BC | 10.0 ± 1.1 | 1.67 ± 0.18 | 11.6 ± 0.52 | 1.93 ± 0.09 |
| PMALIH-BC | 6.4 ± 0.60 | 1.07 ± 0.10 | 7.6 ± 0.40 | 1.27 ± 0.07 |
| PSAH-BC | 7.6 ± 0.40 | 1.27 ± 0.07 | 6.4 ± 0.60 | 1.07 ± 0.10 |

Table 5. The average release rate during the slow-release phase.

| Hydrogel | Slow-Release Phase | | | |
|---|---|---|---|---|
| | Allantoin | | Dexpanthenol | |
| | mg/114 h | mg/h | mg/114 h | mg/h |
| PMALEH-BC | 13.0 ± 0.50 | 0.11 ± 0.00 | 12.4 ± 0.40 | 0.11 ± 0.00 |
| PAAH-BC | 13.0 ± 0.58 | 0.11 ± 0.01 | 13.6 ± 0.48 | 0.12 ± 0.00 |
| PMALIH-BC | 8.4 ± 0.58 | 0.07 ± 0.01 | 10.6 ± 1.10 | 0.09 ± 0.01 |
| PSAH-BC | 8.4 ± 0.64 | 0.07 ± 0.01 | 10.8 ± 0.80 | 0.09 ± 0.01 |
| Hydrogel | Caffeic Acid | | Resveratrol | |
| | mg/114 h | mg/h | mg/114 h | mg/h |
| PMALEH-BC | 9.6 ± 0.50 | 0.08 ± 0.00 | 8.8 ± 0.62 | 0.08 ± 0.01 |
| PAAH-BC | 10.0 ± 0.76 | 0.09 ± 0.01 | 8.4 ± 0.70 | 0.07 ± 0.01 |
| PMALIH-BC | 13.6 ± 0.58 | 0.12 ± 0.01 | 12.4 ± 0.40 | 0.11 ± 0.00 |
| PSAH-BC | 12.4 ± 0.4 | 0.11 ± 0.00 | 13.6 ± 0.88 | 0.12 ± 0.01 |

The release patterns of the BC from the formulations were dependent on the crosslinker type. For instance, the higher release of allantoin and dexpanthenol occurred in PMALIH and PSAH. In contrast, the higher release of caffeic and resveratrol was produced in PMALEH and PAAH. Perhaps the crosslinker nature plays a key role in the release pattern of each biomolecule. Characteristics of crosslinkers, such as functional groups in its structure, polarity and ability to form hydrogen bonds, among others, affect the release patterns.

*3.5. FTIR Analysis*

IR spectra of PDCAH using different dicarboxylic acids such as succinic, malic, aspartic, and maleic acid are presented in Figure S1a–d, respectively. We can notice that all PDCAH spectra (PSAH, PMALIH, PAAH, and PMALEH) have most of PVA characteristic IR absorption bands (the PVA spectra, not shown here). In all spectra, we can find these representative bands that appear at around 3400 cm$^{-1}$, between 2840 and 3000 cm$^{-1}$, over 1688 cm$^{-1}$. Signals between 1150 and 1085 cm$^{-1}$ are attributed to a hydroxyl group (νOH), alkyl groups (νCH$_2$), carbonyl groups (νC=O) and the ester group (C–O–C), respectively. The last two signals indicate that the crosslinking of PVA was due to the ester linkage formed between PVA and the different dicarboxylic acids used, as shown in Figure S1. Other important absorption bands are recorded in the PSAH, PMALIH, PAAH, and PMALEH spectra and prove the presence of the succinic, malic, aspartic, and maleic acids in their structures. For example, the peak at 1627 cm$^{-1}$ present in the PMALEH spectra is a clear indication of the existence of –CO–CH=CH– stretching. On the other hand, in the PAAH spectra, we found two signals at 1630 and 1419 cm$^{-1}$ characteristic to the CO–NH group from amide. In Table S2 (See Supplementary Materials), we summarized the most characteristic bands with their assigned PVA and PVA cross-linked with succinic, malic, aspartic, and maleic acids. Finally, it is clear that the spectral changes obtained in the above analysis are evidence of cross-linking reactions between the hydroxyl group of PVA and the carboxylic groups of succinic, malic, aspartic, and maleic acids.

*3.6. Mechanical Analysis*

Mechanical properties of the PDCAH were summarized in Table S3. Young modulus (E) specifies the stiffness or rigidity of the film; tensile strength (σ) indicates the tensile strength of the film up to breaking; and the elongation at break (ε) describes the flexibility or extensibility of the films up to breaking [47].

Mechanical analysis results showed that the highest tensile strength was obtained with PSAH (19.3 MPa), corresponding to PVA crosslinked with succinic acid, which was two times higher than PAAH, 1.5-times higher than PMALEH and 1.1 times higher than PMALIH. All these values are lower than what has been previously reported for pure PVA films [48], indicating that chemical crosslinking affects the mechanical properties. Based on these results, the changes in elongation at break and tensile strength appear to follow the same trend. Of the samples measured, sample two had the lowest recorded $\varepsilon B$ value (37.8%) followed by sample one (183.9%). The mechanical properties recorded in the case of sample one may be due to the film porosity observed by SEM analysis (Figure 4). By contrast, the change in the Young modulus followed a trend opposite to that of the other mechanical properties. The results indicate that the film from PAAH (176.4 MPa) possessed a rigid structure, low elasticity, and the lowest mechanical properties. This could be related to the highly fibrous and disorganized structure observed in SEM images (Figure 4) for this sample. A similar trend had been previously described for PVA films loaded with natural fibers [48]. From these results, it is clear that the nature of the crosslinker and possibly the degree of crosslinking could alter the mechanical properties of the films.

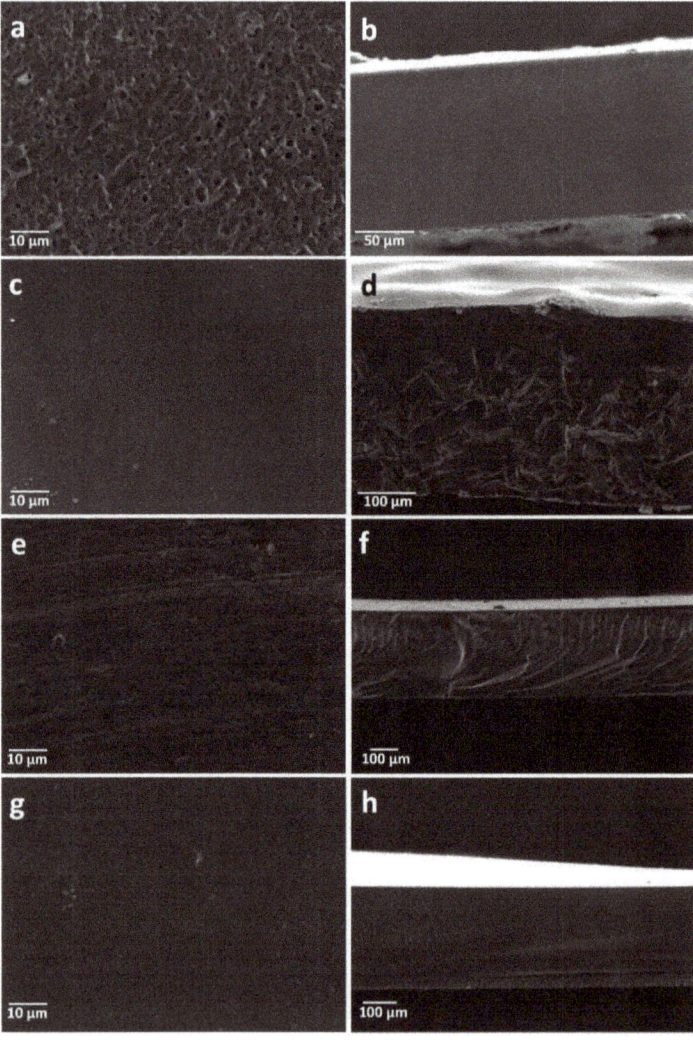

**Figure 4.** SEM micrographics from the surface and side view of PDCAH. (**a–b**) PMALEH; (**c–d**) PAAH; (**e–f**) PSAH; (**g–h**) PMALIH.

## 3.7. SEM Analysis

In Figure 4, the results from the SEM analysis of PDCAH are shown. For PMALEH (see Figure 4a), a rough and porous surface is observed, which contrasts with the film inner view (see Figure 4b) that shows a non-porous and compact structure. In the case of PAAH and PMALIH (see Figure 4c,g), similar smooth surfaces are observed, but for sample PSAH, a rough and fibrous surface is displayed. On the other side, a highly fibrous and disorganized structure is observed at the bottom of the film side view from PAAH (see Figure 4d), changing to a more compact structure at the upper part. For samples PSAH and PMALIH (see Figure 4f,h), fibrous and compact structures are presented in the film side view, which is similar to that observed in PMALEH. It seems that film morphology would be highly influenced by the crosslinker chemical structure.

## 3.8. In Vivo Wound Healing Studies

Figure 5 displays the images of the skin wound taken on day 0 and day 14, after treating with PAAH-BC, PSAH-BC, PMALIH-BC and PMALEH-BC and controls. After the 14th day, the PDCAH-BC treated wounds showed excellent results in all groups compared with the controls. The first characteristic detected was the growth of the new epidermis, which reduced towards the wound center in all treated wound lesions, resulting in a reduced area of the wounds. While the four formulations present good wound healing activity, there are some differences between them. For instance, PMALEH-BC achieved complete healing, while PAAH-BC, PSAH-BC and PMALEH-BC had a wound-healing ratio of 98 and 95%, 90% respectively. The wound healing process in the four proposed formulations was better than in the commercial and negative control (with a wound healing ratio of 85 and 40%, respectively). Such an excellent wound-healing effect of PDCAH-BC could be attributed to the synergistic effects among the bioactive compounds and their sustained release over the wound. On the one hand, allantoin has been reported to have numerous properties associated with wound healing, among them: hydrating and removing necrotic tissue, stimulating the cell mitosis as well as promoting epithelial stimulation, analgesic action and keratolytic activity [49]. On the other hand, the antioxidant agents such as resveratrol, dexpanthenol and caffeic acid have been reported with multiple activities, including anti-inflammatory and anti-bacterial activity [50–54]. Therefore, these compounds have shown positive effects for stimulating skin regeneration and promoting wound healing. Particularly, the treatment of the PDCAH-BC led to improved premature healing of the wounds. Hypothetically, the combinatorial therapy proposed in this work, where the bioactive compounds act on several therapeutic targets of the wound healing could be the key in the obtained results.

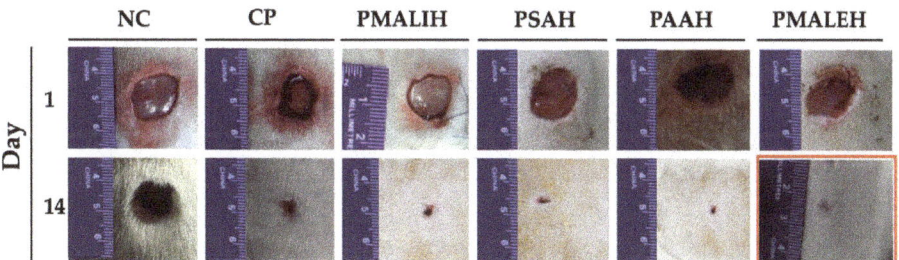

**Figure 5.** In vivo assessments of the film dressings for wound healing. NC: PSAH (Negative control, film dressing without BC); CP: Madecassol™ (Commercial Product); PMALEH, PSAH, PAAH and PMALIH as proposed film dressings.

## 3.9. Histological Analysis

Wound healing is a complex process which is involved of the following overlapping but well-defined stages: hemostasis, inflammation, migration, proliferation and remodeling [55,56]. Hematoxylin and eosin (H&E), Masson's trichrome and Giemsa staining were utilized to evaluate the wound healing progress. The histology analysis of wounds covered with PDCAH-BC and controls on the 14th postoperative day is shown in Figure 6. In the Control a, the histological analysis shows a limited organization of the area under repair and the absence of reepithelialization was observed. On the other hand, no structural epithelium organization was observed and there was an absence of hair follicle formation. An increase in the cellular content such as fibroblasts, inflammatory and endothelial cells was observed, which is interpreted as a smaller organization of the connective tissue. A lot of blood vessels were observed, showing deep angiogenesis in the repair area. Moreover, moderate hemorrhage composed of extravasated red blood cells towards the deep area of the sample was observed. In addition, some incipient hair follicles at the edges of the healing area were detected (Figure 6a). In contrast, in sample c (PMALIH-BC), stratified epithelium and re-epithelialization in the repair zone was observed. The connective tissue presented a limited organization with high cellularity and vascularization. In sample d (PAAH-BC), the repair zone with a moderated organization and the connective tissue were detected. Moreover, high cellularity and vascularization were observed. In sample e (PSAH-BC), the repair zone with a moderated organization of connective tissue, vascularization and deep cellularity was revelated. The presence of hair follicles was evidenced in the deep area of scarring.

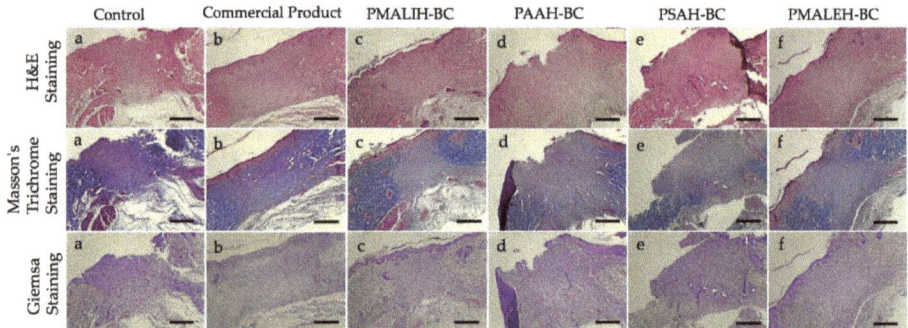

**Figure 6.** Histological images of H&E, Masson's trichrome and GIEMSA stained sections after 14 days of wound healing for each sample from "a" to "f" respectively. Control: PDCAH (**a**); Commercial Product: Madecassol® (**b**); PMALIH-BC (**c**); PAAH-BC (**d**); PSAH-BC (**e**) and PMALEH-BC (**f**). Scale bars is 100 μm.

In general, in all PDCAH-BC (Figure 6c–f) a better and faster reepithelialization process and retraction of the wound healing area was exhibited. A better organization of the granulation tissue in relation to control (a) was detected. Also, there was a greater delimitation of the scar area. In all PDCAH-BC, the epithelium showed signs of better structural organization, which included defined basal, spiny and superficial strata, as well as the beginning of granular stratum formation at the lateral edges of the repair area. The basal stratum of the epithelium had normotypic hyperchromatic cells compatible with proliferative activity in the reepithelialization zone. A similar characteristic in samples from PSAH-BC, PAAH-BC, PMALIH-BC and PMALEH-BC was observed. An apparent technical defect due to the detachment of scar tissue in samples d and e (defect during sample processing) was observed. However, the microscopical evidence suggests that the histological aspect of the wound healing process is better in sample from PMALEH-BC than PSAH-BC, PAAH-BC and PMALIH-BC as shown in Figure 6. In the case of the commercial product, similar features were detected than PMALEH, however, this formulation was daily applied until day 14 unlike the prepared formulations in this work applied only one time.

These results are in concordance with the closure evaluations (Figure 5) in which PMALEH-BC presented 100% wound closure. Moreover, these results could be supported by the release profile of PMALEH (Figure 3), in which the fastest release was produced with caffeic acid and resveratrol (with radical scavenging effects in hemostasis and inflammation [57]). Subsequently, PMALEH-BC allowed the slow release (later) of allantoin and dexpanthenol so that they could carry out their in later stages of wound healing (proliferation and remodeling [49,50]).

*3.10. PDCAH Cytotoxicity Studies*

This study was performed to quantify the cytotoxicity of the prepared formulations on fibroblast cells. The cytotoxicity of the sterilized PMALEH, PAAH, PSAH and PMALIH was analyzed by a cell viability assay using L929 fibroblast cells after 24 h. Figure 7 displayed fibroblast cell viability exposed to three different concentrations of prepared formulations (a concentration range of 500–2500 µg·mL$^{-1}$ for each PDCAH analyzed). As indicated in Figure 7, at 500 µg·mL$^{-1}$ the cell viability is close to 100% for all four cases. On the other hand, it is observed that when significantly increasing PDCAH concentration, the fibroblast cell viability only declines slightly. In other words, among the concentration range of 1500 and 2500 µg·mL$^{-1}$, the cell viability decreases between 95 and 88% for the all four prepared formulations, respectively. These results confirm that the PDCAH have minimum toxicity over the fibroblast cell model. These performed assays allow concluding that these proposed formulations could be biocompatible for medical applications. Therefore, PDCAH could be considered as safe formulations for sustained release of bioactive compounds with wound healing properties.

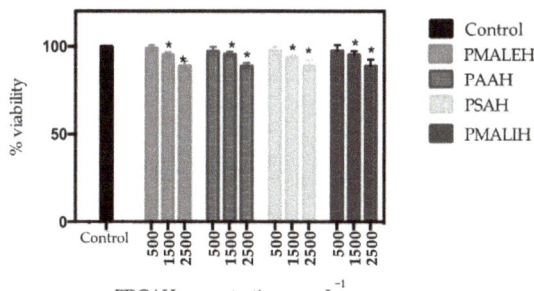

**Figure 7.** Percentage of cell viability obtained from the MTT assay of the L929 fibroblast cells with respect to a negative control (without PDCAH). $n = 3$, * = $p < 0.05$ with respect to the control.

## 4. Conclusions

In this article, an *in-silico* strategy has been implemented to quickly evaluate the energy contribution of each compound (allantoin, dexpanthenol, caffeic acid and resveratrol) in the PVAnp-compound binding affinity. According to the results obtained from ΔE, the pores generated between PVA and the succinic, malic, maleic and aspartic acids have better ΔE for the 4 compounds of interest simultaneously. Therefore, hydrogels PVA cross-linked with these acids were the candidates to be evaluated experimentally.

Thus, based on the in-silico obtained results, novel film dressings based on hydrogels with unique properties were successfully prepared. Starting with the rational design of these formulations, this work concluded with in vivo wound healing studies that yielded promising results. Specifically, a series of hydrogels loaded with bioactive compounds with wound healing activity such as allantoin, dexpanthenol, caffeic acid, and resveratrol was developed. Moreover, an enhanced wound healing process in a full-thickness skin defect model with these formulations was demonstrated. The in vitro release studies exhibited that it is possible to carry out a combinatorial and coordinated sustained release of all four bioactive compounds, demonstrating an excellent strategy to achieve wound healing. These formulations have been designed by simply conjugating PVA chains and maleic, malic, aspartic and succinic acids as crosslinking agents.

The release profile of allantoin, dexpanthenol, caffeic acid and resveratrol exhibited some differences for each PDCAH. This difference seems to be governed by the affinity of the bioactive compound type and the crosslinking agent type (intermolecular interactions). The swelling index results concluded that these formulations based on hydrogels are stimuli-responsive to pH, time and crosslinking agent type. Moreover, in a great part of formulations, a rough, porous surface and good mechanical properties were observed. On the other hand, all four prepared film dressings showed good biocompatibility with L929 mouse connective tissue fibroblasts. The results revealed a viability of more than 88%.

In vivo studies, the PDCAH-BC treated wounds showed excellent results compared with the controls. Macroscopically, the growth of the new epidermis towards the wound center in all treated wound lesions was detected for all four cases (PMALIH-BC, PAAH-BC, PSAH-BC, and PMALEH-BC), resulting in a reduced area of the wounds. While the four formulations present good wound healing activity, there are some differences between them; this may be due to the nature of the crosslinking agent in each case. The microscopic evidence suggests that the histological aspect of the wound healing process is in concordance with the wound closure results. In conclusion, novel film dressings with simultaneous and sustained-release properties have been obtained and could be excellent candidates for skin wound healing.

**Supplementary Materials:** The following are available online at http://www.mdpi.com/1999-4923/11/9/447/s1, Table S1: Design of PVA hydrogel nanopores (Hnp) crosslinked with different dicarboxylic acids. Figure S1: FTIR spectra of PMALEH, PAAH, PSAH and PMALIH. Table S2: Vibration modes and band frequencies in PVA and PDCAH. Table S3. Mechanical parameters of PDCAH.

**Author Contributions:** E.F.D.-L., F.Á.-S., A.M., G.C, O.V, G.C.-B, W.D., B.V., S.V, and S.P. contributed to the conceptualization, methodology, validation, formal analysis and investigation. E.F.D.-L. and F.Á.-S contributed for the writing—original draft preparation. E.F.D.-L. and A.M. contributed for supervision, writing—review and editing, project administration, resources and funding acquisition.

**Funding:** This research work was financially supported by FONDECYT (Chile) through project No. 11170155 (E.F.D.-L.), project No. 11170008 (O.V.), project No. 11180059 (A.M.), project CONICYT PIA/APOYO CCTE AFB170007 (G.C.) and project No. 3170909 (F.A.-S.).

**Acknowledgments:** We thank Rachael Jiménez-Lange from the Academic Writing Center in the Programa de Idiomas at the Universidad de Talca.

**Conflicts of Interest:** The authors declare no conflict of interest.

### References

1. Kondo, T.; Ishida, Y. Molecular pathology of wound healing. *Forensic Sci. Int.* **2010**, *203*, 93–98. [CrossRef]
2. Lee, J.H.; Kim, H.L.; Lee, M.H.; You, K.E.; Kwon, B.J.; Seo, H.J.; Park, J.C. Asiaticoside enhances normal human skin cell migration, attachment and growth in vitro wound healing model. *Phytomedicine* **2012**, *19*, 1223–1227. [CrossRef] [PubMed]
3. Araújo, L.U.; Grabe-Guimarães, A.; Mosqueira, V.C.F.; Carneiro, C.M.; Silva-Barcellos, N.M. Profile of wound healing process induced by allantoin. *Acta Cir. Bras.* **2010**, *25*, 460–466. [CrossRef] [PubMed]
4. Croft, C.; Tarin, D. Ultrastructural studies of wound healing in mouse skin. *J. Anat.* **1970**, *106*, 63–77.
5. Schultz, G.S.; Sibbald, R.G.; Falanga, V.; Ayello, E.A.; Dowsett, C.; Harding, K.; Romanelli, M.; Stacey, M.C.; Teot, L.; Vanscheidt, W. Wound bed preparation: A systematic approach to wound management. *Wound Repair Regen.* **2003**, *11*, S1–S28. [CrossRef]
6. Eming, S.A.; Martin, P.; Tomic-Canic, M. Wound repair and regeneration: Mechanisms, signaling, and translation. *Sci. Transl. Med.* **2014**, *6*, 265sr6. [CrossRef]
7. Kamoun, E.; Kenawy, E.; Chen, X. A review on polymeric hydrogel membranes for wound dressing applications: PVA-based hydrogel dressings. *J. Adv. Res.* **2017**, *8*, 217–233. [CrossRef]
8. Caló, E.; Khutoryanskiy, V.V. Biomedical applications of hydrogels: A review of patents and commercial products. *Eur. Polym. J.* **2015**, *65*, 252–267. [CrossRef]
9. Kim, S.J.; Lee, K.J.; Kim, S.I. Thermo-sensitive Swelling Behavior of Poly(2-Ethyl-2-oxazoline)/Poly(Vinyl Alcohol) Interpenetrating Polymer Network Hydrogels. *J. Macromol. Sci. A* **2004**, *41*, 267–274. [CrossRef]
10. Yu, L.; Ding, J. Injectable hydrogels as unique biomedical materials. *Chem. Soc. Rev.* **2008**, *37*, 1473–1481. [CrossRef]
11. Chai, Q.; Jiao, Y.; Yu, X. Hydrogels for Biomedical Applications: Their Characteristics and the Mechanisms behind Them. *Gels* **2017**, *3*, 6. [CrossRef] [PubMed]

12. Rodríguez Nuñez, Y.A.; Castro, R.I.; Arenas, F.A.; López-Cabaña, Z.E.; Carreño, G.; Carrasco-Sánchez, V.; Marican, A.; Villaseñor, J.; Vargas, E.; Santos, L.S.; et al. Preparation of Hydrogel/Silver Nanohybrids Mediated by Tunable-Size Silver Nanoparticles for Potential Antibacterial Applications. *Polymers* **2019**, *11*, 716. [CrossRef] [PubMed]
13. Hoare, T.R.; Kohane, D.S. Hydrogels in drug delivery: Progress and challenges. *Polymer* **2008**, *49*, 1993–2007. [CrossRef]
14. Grijalvo, S.; Mayr, J.; Eritja, R.; Díaz, D.D. Biodegradable liposome-encapsulated hydrogels for biomedical applications: A marriage of convenience. *Biomater. Sci. UK* **2016**, *4*, 555–574. [CrossRef] [PubMed]
15. Ribeiro, M.P.; Espiga, A.; Silva, D.; Baptista, P.; Henriques, J.; Ferreira, C.; Silva, J.C.; Borges, J.P.; Pires, E.; Chaves, P.; et al. Development of a new chitosan hydrogel for wound dressing. *Wound Repair Regen.* **2009**, *17*, 817–824. [CrossRef] [PubMed]
16. Oliveira, R.N.; McGuinness, G.B.; Ramos, M.E.; Kajiyama, C.E.; Thiré, R.M. Properties of PVA Hydrogel Wound-Care Dressings Containing UK Propolis. *Macromol. Symp.* **2016**, *368*, 122–127. [CrossRef]
17. León, K.; Santiago, J. Propiedades antimicrobianas de películas de quitosano-alcohol polivinílico embebidas en extracto de sangre de grado. *Rev. Soc. Quim. Perú* **2007**, *73*, 158–165.
18. Marican, A.; Avila-Salas, F.; Valdés, O.; Wehinger, S.; Villaseñor, J.; Fuentealba, N.; Arenas-Salinas, M.; Argandoña, Y.; Carrasco-Sánchez, V.; Durán-Lara, E.F. Rational Design, Synthesis and Evaluation of γ-CD-Containing Cross-Linked Polyvinyl Alcohol Hydrogel as a Prednisone Delivery Platform. *Pharmaceutics* **2018**, *10*, 30. [CrossRef]
19. Schanuel, F.S.; Santos, K.S.R.; Monte-Alto-Costa, A.; de Oliveira, M.G. Combined nitric oxide-releasing poly (vinyl alcohol) film/F127 hydrogel for accelerating wound healing. *Colloids Surf. B* **2015**, *130*, 182–191. [CrossRef]
20. Chavda, H.V.; Patel, C.N. Effect of crosslinker concentration on characteristics of superporous hydrogel. *Int. J. Pharm. Investig.* **2011**, *1*, 17–21. [CrossRef]
21. Thompson, A.; Nguyen, S.; Nave, F. Characterization of PVA-IDA Hydrogel Crosslinked with 1.25%, 2.5% and 5% Glutaraldehyde. *J. Chem.* **2013**, *1*, 1–7. [CrossRef]
22. Orienti, I.; Treré, R.; Luppi, B.; Bigucci, F.; Cerchiara, T.; Zuccari, G.; Zecchi, V. Hydrogels Formed by Crosslinked Poly (vinyl alcohol) as Sustained Drug Delivery Systems. *Arch. Pharm.* **2002**, *2*, 89–93. [CrossRef]
23. Vrana, N.E.; Liu, Y.; McGuiness, G.B.; Cahill, P.A. Characterization of Poly (vinyl alcohol)/Chitosan Hydrogels as Vascular Tissue Engineering Scaffolds. *Macromol. Symp.* **2008**, *269*, 106–110. [CrossRef]
24. Peng, Z.; Li, Z.; Zhang, F.; Peng, X. Preparation and Properties of Polyvinyl Alcohol/Collagen Hydrogel. *J. Macromol. Sci. B* **2012**, *51*, 1934–1941. [CrossRef]
25. Chang, C.; Lue, A.; Zhang, L. Effects of Crosslinking Methods on Structure and Properties of Cellulose/PVA Hydrogels. *Macromol. Chem. Phys.* **2008**, *209*, 1266–1273. [CrossRef]
26. Liu, Y.; Vrana, N.E.; Cahill, P.A.; McGuinness, G.B. Physically Crosslinked Composite Hydrogels of PVA With Natural Macromolecules: Structure, Mechanical Properties, and Endothelial Cell Compatibility. *J. Biomed. Mater. Res. B* **2009**, *90*, 492–502. [CrossRef] [PubMed]
27. Reis, E.F.; Campos, F.S.; Lage, A.P.; Leite, R.C.; Heneine, L.G.; Vasconcelos, W.L.; Lobato, Z.I.; Mansur, H.S. Synthesis and Characterization of Poly(Vinyl Alcohol) Hydrogels and Hybrids for rMPB70 Protein Adsorption. *Mater. Res.* **2006**, *9*, 185–191. [CrossRef]
28. Roberts, J.; Farrugia, B.; Green, R.; Rnjak-Kovacina, J.; Martens, P. In situ formation of poly (vinyl alcohol)–heparin hydrogels for mild encapsulation and prolonged release of basic fibroblast growth factor and vascular endothelial growth factor. *J. Tissue Eng.* **2016**, *7*, 1–10. [CrossRef]
29. Singh, B.; Pal, L. Sterculia crosslinked PVA and PVA-poly(AAm) hydrogel wound dressings for slow drug delivery: Mechanical, mucoadhesive, biocompatible and permeability properties. *J. Mech. Behav. Biomed.* **2012**, *9*, 9–21. [CrossRef]
30. Buwalda, S.J.; Vermonden, T.; Hennink, W.E. Hydrogels for Therapeutic Delivery: Current Developments and Future Directions. *Biomacromolecules* **2017**, *18*, 316–330. [CrossRef]
31. Pirinen, S.; Karvinen, J.; Tiitu, V.; Suvanto, M.; Pakkanen, T.T. Control of swelling properties of polyvinyl alcohol/hyaluronic acid hydrogels for the encapsulation of chondrocyte cells. *J. Appl. Polym. Sci.* **2015**, *132*, 28. [CrossRef]
32. Valderruten, N.E.; Valverde, J.D.; Zuluaga, F.; Durántez-Ruiz, E. Synthesis and characterization of chitosan hydrogels cross-linked with dicarboxylic acids. *React. Funct. Polym.* **2015**, *84*, 21–28. [CrossRef]

33. US Food and Drug Administration Guidance for Industry: Co-development of two or more new investigational drugs for use in combination. June 2013. Available online: https://www.fda.gov/regulatory-information/search-fda-guidance-documents/codevelopment-two-or-more-new-investigational-drugs-use-combination (accessed on 28 July 2019).
34. Wu, M.; Sirota, M.; Butte, A.J.; Chen, B. Characteristics of drug combination therapy in oncology by analyzing clinical trial data on clinicaltrials.gov. *Pac. Symp. Biocomput.* **2015**, 68–79.
35. Bouhadir, K.H.; Alsberg, E.; Mooney, D.J. Hydrogels for combination delivery of antineoplastic agents. *Biomaterials* **2001**, *22*, 2625–2633. [CrossRef]
36. Singh, B.; Dhimana, A. Designing bio-mimetic moxifloxacin loaded hydrogel wound dressing to improve antioxidant and pharmacology properties. *RSC Adv.* **2015**, *5*, 44666–44678. [CrossRef]
37. Hassan, A.; Niazi, M.B.K.; Hussain, A.; Farrukh, S.; Ahmad, T. Development of anti-bacterial PVA/starch based hydrogel membrane for wound dressing. *J. Polym. Environ.* **2018**, *26*, 235–243. [CrossRef]
38. ChemAxon Ltd. *MarvinSketch Program Version 19.01 (For OSX)*; ChemAxon Ltd.: Budapest, Hungary, 2019; Available online: https://chemaxon.com/products/marvin (accessed on 20 June 2018).
39. Frisch, M.J.; Trucks, G.W.; Schlegel, H.B.; Scuseria, G.E.; Robb, M.A.; Cheeseman, J.R.; Montgomery, J.A.J.; Vreven, T.; Kudin, K.; Burant, J. *Gaussian 16, Revision, A.03*, Gaussian Inc.: Wallingford, CT, USA, 2016.
40. Fan, C.F.; Olafson, B.D.; Blanco, M. Application of molecular simulation to derive phase diagrams of binary mixtures. *Macromolecules* **1992**, *25*, 3667–3676. [CrossRef]
41. Avila-Salas, F.; Sandoval, C.; Caballero, J.; Guiñez-Molinos, S.; Santos, L.S.; Cachau, R.E.; González-Nilo, F.D. Study of Interaction Energies Between the PAMAM Dendrimer and Nonsteroidal Anti-inflammatory Drug Using a Distributed Computational Strategy and Experimental Analysis by ESI-MS/MS. *J. Phys. Chem. B* **2012**, *116*, 2031–2039. [CrossRef]
42. Murakami, K.; Aoki, H.; Nakamura, S.; Nakamura, S.I.; Takikawa, M.; Hanzawa, M.; Kishimoto, S.; Hattori, H.; Tanaka, Y.; Kiyosawa, T.; et al. Hydrogel blends of chitin/chitosan, fucoidan and alginate as healing-impaired wound dressings. *Biomaterials* **2010**, *31*, 83–90. [CrossRef]
43. Mosmann, T. Rapid Colorimetric Assay for Cellular Growth and Survival: Application to Proliferation and Cytotoxicity Assays. *J. Immunol. Methods* **1984**, *65*, 55–63. [CrossRef]
44. Gupta, N.V.; Shivakumar, H.G. Investigation of Swelling Behavior and Mechanical Properties of a pH-Sensitive Superporous Hydrogel Composite. *Iran. J Pharm. Res.* **2012**, *11*, 481–493. [PubMed]
45. Avila-Salas, F.; Rodriguez Nuñez, Y.A.; Marican, A.; Castro, R.I.; Villaseñor, J.; Santos, L.S.; Wehinger, S.; Durán-Lara, E.F. Rational Development of a Novel Hydrogel as a pH-Sensitive Controlled Release System for Nifedipine. *Polymers* **2018**, *10*, 806. [CrossRef] [PubMed]
46. Kristin, E.; Curtis, W.F. Protein diffusion in photopolymerized poly(ethylene glycol) hydrogel networks. *Biomed. Mater.* **2011**, *6*, 055006.
47. Masri, C.; Chagnon, G.; Favier, D. Influence of processing parameters on the macroscopic mechanical behavior of PVA hydrogels. *Mater. Sci. Eng. C* **2017**, *75*, 769–776. [CrossRef] [PubMed]
48. Chiellini, E.; Cinelli, P.; Imam, S.H.; Mao, L. Composite films based on biorelated agro-industrial waste and poly(vinyl alcohol). preparation and mechanical pxroperties characterization. *Biomacromolecules* **2001**, *2*, 1029–1037. [CrossRef] [PubMed]
49. Özçelik, B.; Kartal, M.; Orhan, I. Cytotoxicity, antiviral and antimicrobial activities of alkaloids, flavonoids, and phenolic acids. *Pharm. Biol.* **2011**, *49*, 396–402. [CrossRef] [PubMed]
50. Heise, R.; Skazik, C.; Marquardt, Y.; Czaja, K.; Sebastian, K.; Kurschat, P.; Gan, L.; Denecke, B.; Ekanayake-Bohlig, S.; Wilhelm, K.-P.; et al. Dexpanthenol modulates gene expression in skin wound healing in vivo. *Skin Pharmacol. Physiol.* **2012**, *25*, 241–248. [CrossRef] [PubMed]
51. Liu, F.C.; Tsai, Y.F.; Tsai, H.I.; Yu, H.P. Anti-inflammatory and organ-protective effects of resveratrol in trauma-hemorrhagic injury. *Mediat. Inflamm.* **2015**, *2015*, 1–9. [CrossRef] [PubMed]
52. Kramer, A.; Assadian, O.; Koburger-Janssen, T. Antimicrobial efficacy of the combination of chlorhexidine digluconate and dexpanthenol. *GMS Hyg. Infect. Control* **2016**, *11*, 1–6.
53. Natarajan, K.; Singh, S.; Burke, T.R.; Grunberger, D.; Aggarwal, B.B. Caffeic acid phenethyl ester is a potent and specific inhibitor of activation of nuclear transcription factor NF-kB. *Proc. Natl. Acad. Sci. USA* **1996**, *93*, 9090–9095. [CrossRef]
54. Yaman, I.; Derici, H.; Kara, C.; Kamer, E.; Diniz, G.; Ortac, R.; Sayin, O. Effects of resveratrol on incisional wound healing in rats. *Surge. Today* **2013**, *43*, 1433–1438. [CrossRef] [PubMed]

55. Qu, J.; Zhao, X.; Liang, Y.; Zhang, T.; Ma, P.X.; Guo, B. Antibacterial adhesive injectable hydrogels with rapid self-healing, extensibility and compressibility as wound dressing for joints skin wound healing. *Biomaterials* **2018**, *183*, 185–199. [CrossRef] [PubMed]
56. Zhao, X.; Wu, H.; Guo, B.; Dong, R.; Qiu, Y.; Ma, P.X. Antibacterial anti-oxidant electroactive injectable hydrogel as self-healing wound dressing with hemostasis and adhesiveness for cutaneous wound healing. *Biomaterials* **2017**, *122*, 34–47. [CrossRef] [PubMed]
57. Kadoma, Y.; Fujisawa, S. Radical-scavenging activity of dietary phytophenols in combination with co-antioxidants using the induction period method. *Molecules* **2011**, *16*, 10457–10470. [CrossRef] [PubMed]

© 2019 by the authors. Licensee MDPI, Basel, Switzerland. This article is an open access article distributed under the terms and conditions of the Creative Commons Attribution (CC BY) license (http://creativecommons.org/licenses/by/4.0/).

*Article*

# Use of nPSi-βCD Composite Microparticles for the Controlled Release of Caffeic Acid and Pinocembrin, Two Main Polyphenolic Compounds Found in a Chilean Propolis

Dina Guzmán-Oyarzo [1], Tanya Plaza [2], Gonzalo Recio-Sánchez [2,3], Dulcineia S. P. Abdalla [4], Luis A. Salazar [1,*] and Jacobo Hernández-Montelongo [2,3,*]

[1] Center of Molecular Biology and Pharmacogenetics, Scientific and Technological Bioresource Nucleus (BIOREN), Universidad de La Frontera, Avenida Francisco Salazar 01145, Temuco 4811230, Chile; dina.guzman.o@gmail.com

[2] Bioproducts and Advanced Materials Research Center (BioMA), Faculty of Engineering, Universidad Católica de Temuco, Avenida Rudecindo Ortega 02950, Temuco 4813302, Chile; tplaza.neira@gmail.com (T.P.); grecio@uct.cl (G.R.-S.)

[3] Department of Physical and Mathematical Sciences, Faculty of Engineering, Universidad Católica de Temuco, Temuco 4813302, Chile

[4] Department of Clinical and Toxicological Analyses, Faculty of Pharmaceutical Sciences, Universidade de São Paulo, Avenida Professor Lineu Prestes 580, CEP 05508-000 São Paulo, SP, Brazil; dspabdalla@gmail.com

* Correspondence: luis.salazar@ufrontera.cl (L.A.S.); jacobo.hernandez@uct.cl (J.H.-M.); Tel.: +56-452-596724 (L.A.S.); +56-452-553947 (J.H.-M.)

Received: 13 May 2019; Accepted: 17 June 2019; Published: 19 June 2019

**Abstract:** Propolis is widely recognized for its various therapeutic properties. These are attributed to its rich composition in polyphenols, which exhibit multiple biological properties (e.g., antioxidant, anti-inflammatory, anti-angiogenic). Despite its multiple benefits, oral administration of polyphenols results in low bioavailability at the action site. An alternative to face this problem is the use of biomaterials at nano-micro scale due to its high versatility as carriers and delivery systems of various drugs and biomolecules. The aim of this work is to determine if nPSi-βCD microparticles are a suitable material for the load and controlled release of caffeic acid (CA) and pinocembrin (Pin), two of the main components of a Chilean propolis with anti-atherogenic and anti-angiogenic activity. Polyphenols and nPSi-βCD microparticles cytocompatibility studies were carried out with human umbilical vein endothelial cells (HUVECs). Results from physicochemical characterization demonstrated nPSi-βCD microparticles successfully retained and controlled release CA and Pin. Furthermore, nPSi-βCD microparticles presented cytocompatibility with HUVECs culture at concentrations of 0.25 mg/mL. These results suggest that nPSi-βCD microparticles could safely be used as an alternate oral delivery system to improve controlled release and bioavailability of CA or Pin—and eventually other polyphenols—thus enhancing its therapeutic effect for the treatment of different diseases.

**Keywords:** controlled release; nanoporous silicon; βCD polymer; caffeic acid; pinocembrin; polyphenols; HUVECs

## 1. Introduction

Since ancient times, the use of natural compounds has been of great importance for medicine mainly in the prevention and treatment of different pathologies [1,2]. That is why they represent the main source of used compounds in the discovery and/or development of new drugs [3]. An example of natural compounds with bioactive potential is propolis, which is a resinous compound produced by bees from plants exudates. Studies both in vitro and in vivo have identified a wide variety of biological activities for propolis: antibacterial [4], antifungal [5], antioxidant [6], anti-inflammatory [7], anti-carcinogenic [8] and anti-angiogenic [9]. These activities are attributed to its polyphenols rich composition, molecules that present different biological properties: relaxing [10], antioxidant [11], antithrombotic [12], antiangiogenic [13], anti-inflammatory [14], anti-carcinogenic [15], among others. Biochemically, polyphenols are secondary metabolites exclusively synthesized by plants and their entire structure is based in one or more hydroxyl groups attached to an aromatic ring (benzene) [16]. Since the role of polyphenols in plants is related to growth, development and defense, they are found in leaves, fruits and seeds, as well as in a wide range of food of plant origin (vegetables, tea, cocoa, wine, etc.) [17]. Concerning to the presence and abundance of polyphenols in propolis, they are very variable due to their close dependence with the botanical origin of plants, climate, geographical location, year and time of collection [3,18,19]. Examples of this important dependence are the studies of three Brazilians, one Polish and one Chilean propolis. For the Brazilians propolis, Daleprane et al. [9] reported that artepellin C, pinocembrin and kampferol were the main components of green propolis; 3-hydroxy-8,9-dimethoxypterocarpane, medicarpine and daidezein were the main components of red propolis; and pinocembrin, phenyl ester of caffeic acid, quercetin and galangin were the main components of brown propolis. For the Polish propolis, Szliszka et al. [20] detected that was mainly composed by the flavonoids pinobanksin, chrysin and methoxyflavanone; and the phenolics acids coumarin, ferulic and caffeic. Finally, for the Chilean propolis with anti-atherogenic and anti-angiogenic activity [21], the main polyphenols detected in the ethanolic extract were caffeic acid (a phenolic acid) and pinocembrin (a flavonone) [22].

Daily intake of polyphenols has multiple health benefits [23] because they reduce the risk of developing non-communicable diseases such as diabetes [11], cancer [24] and cardiovascular diseases [25]. In vivo studies reported that supplementation of the diet with persimmon extract rich in polyphenols maintains plasma lipid levels in hypercholesterolemic mice [26]; whereas the use of a mixture of resveratrol, CA and catechin significantly reduces the atheroma plaque in ApoE knockout mice [27]. Although the consumption of polyphenols contributes to the prevention of diseases, its oral administration without compound protection translates into a low efficiency at the action site. This is due to several factors such as concentration, binding site, chemical structure, stability in the gastrointestinal environment and aqueous solubility, which, in general, have a negative impact on absorption levels, metabolization degree, distribution throughout the body, life span and compound excretion [2,28]. Finally, the pharmacokinetics of polyphenols is also influenced by age, health status, intestinal microbiota and diet of patients, as well as by their oral antibiotic treatments [29,30]. All of the above is translated into different reports of low bioavailability of polyphenols, for example, 0.56–4.54 nmol/L for anthocyanins [31], 0.46–1.28 µmol/L for flavonones [32], and 37–60 nmol/L for phenolic acids [33].

Due to the low bioavailability of polyphenols after oral intake, several strategies have been developed to improve the bioavailability and bioactivity of these compounds. One of them is the use of microparticles based on biomaterials whose main function is to protect and transport the entire biomolecule [34]. Concerning this, nanoporous silicon (nPSi), is an excellent biomaterial that has been successfully used for the controlled release of different drugs and biomolecules, due to its large surface area, porous structure, biocompatibility, biodegradability, bioresorbability and resistance to low pH [35–37]. Moreover, because of the versatility of its surface chemistry, different functionalization strategies routes have been explored in order to enhance the load and controlled release of drugs [38]. A refined technique is to embed polymers into their nanopores to form composites [39]. In this regard, β-cyclodextrin (βCD), which is a biocompatible and aqueous soluble molecule, has been successfully used in drug delivery applications. The wide application of βCD in this field is related to the possibility to form the "host-guest" complexation (βCD/drug) [40]; drugs are encapsulated into its lipophilic cavity structure, whereas its outer hydrophilic surface can be crosslinked with other molecules (i.e., citric acid), yielding a 3D-polymer network suitable for drug delivery applications. Therefore by combining a flexible and soft βCD polymer within the highly porous inorganic matrix of nPSi as substrate, both stability and control of drugs release can be improved, increasing their therapeutic potential by reducing their degradation before they reach the target tissues [40]. Based on this, we hypothesize that nPSi-βCD composite is a safe alternative system for oral administration of CA and Pin since it has no toxic effects on human cells. The aim of this work is to determine if nPSi-βCD microparticles are a suitable and safe material for the load and controlled release of caffeic acid (CA) and pinocembrin (Pin), two of the main components of a Chilean propolis with anti-atherogenic and anti-angiogenic activity. This study includes the synthesis and physicochemical characterizations of nPSi-βCD microparticles loaded or not with CA or Pin, their respective release profiles and the corresponding cytocompatibility tests for each polyphenol and composite.

## 2. Materials and Methods

### 2.1. Materials

Caffeic acid (CA, $M_W \approx 180.16$ g/mol), pinocembrin (Pin, $M_W \approx 256.25$ g/mol), chitosan (Chi, 75–85% deacetylated, low $M_W \approx 5 \times 10^4$ g/mol), β-cyclodextrin (βCD, $M_W \approx 1134.98$ g/mol), citric acid ($M_W \approx 210.14$ g/mol), g $NaH_2PO_2 \cdot H_2O$ ($M_W \approx 105.99$ g/mol) and phosphate buffer solution (PBS) 0.01 M (0.138 M NaCl, 0.0027 M KCl, pH = 7.4 at 25 °C) were purchased from MiliporeSigma, St. Louis, MO, USA. Acetone ($C_3H_6O$), dimethyl sulfoxide (DMSO, $C_2H_6OS$), isopropanol ($C_3H_7OH$), ethanol (EtOH, $C_2H_5OH$), glacial acetic acid ($CH_3COOH$), hydrogen peroxide ($H_2O_2$), sodium hydroxide (NaOH), hydrochloric acid (HCl) and hydrofluoric acid (HF) were acquired from Merck, Darmstadt, Germany. All chemicals were used without further purification, and solutions were prepared using Milli-Q water with resistivity of 18.2 M·Ω·cm (pH ~7.6, otherwise mentioned). Silicon (Si) wafers ($p^+$ type, boron-doped, orientation <100> resistivity of 0.001–0.005 Ω·cm) were purchased from University Wafer, South Boston, MA, USA. Fetal bovine serum (FBS), L-Glutamine, penicillin-streptomycin solution and D-PBS were purchased from Corning, Manassas, VA, USA. CellTiter-FluorTM assay and the CellTiter 96® AQueous One Solution cell proliferation assay (MTS) were acquired from Promega, Madison, WI, USA.

### 2.2. Sample Preparation

Si wafers were cleaned by ultrasonication in acetone, isopropanol and distilled water, for a period of 15 min in each solvent. Acetone removed greasy and oily substances; isopropanol was necessary to rinse acetone off, and distilled water removed any isopropanol residues. Then, nPSi layers were fabricated by electrochemical etching from the cleaned Si wafers in HF (48%):EtOH (1:2) solution under controlled formation conditions: etching time of 30 min and current density of 80 mA·cm$^{-2}$. Afterward, an electropolishing pulse was applied to get free-standing nPSi layers. For that, the applied current density was enhanced to 150 mA/cm$^2$ during 2 s. nPSi free-standing layers were scraped with a diamond tip to obtain microparticles. They were milled, collected in EtOH and subjected to 10 min ultrasound agitation for homogenization. Finally, the obtained nPSi particles were chemically oxidized by $H_2O_2$ (30%, v/v) for 12 h in orbital agitation and rinsed with EtOH (Figure 1A).

Oxidized nPSi microparticles were the substrate to synthetize the composite according to the protocol of Hernandez-Montelongo et al. [41] (Figure 1B). nPSi microparticles were immersed in a Chi solution for 15 min and after rinsed with EtOH (nPSi-CHI). The Chi solution (1% w/v) was previously prepared with Chi powder in 100 mM glacial acetic acid, then, the pH value was adjusted at 4 with a 0.1 M HCl and/or NaOH solution. For the composites (nPSi-βCD) synthesis, a monomer solution was prepared with 10 g βCD, 3 g $NaH_2PO_2 \cdot H_2O$ as catalyst, and 10 g citric acid in 100 mL of distilled water. Then, nPSi-Chi was immersed in this solution for 15 min while stirring. Samples were dried, first at room temperature, and later at 90 °C for 1 h in each case. The βCD–citric acid in situ polymerization in nPSi-CHI was carried out at 140 °C for 25 min. Afterward, samples were rinsed with EtOH, dried at 90 °C for 1 h and milled for homogenization.

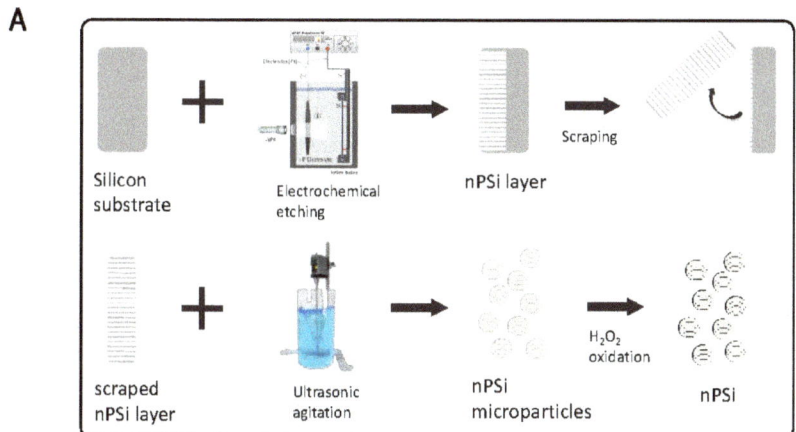
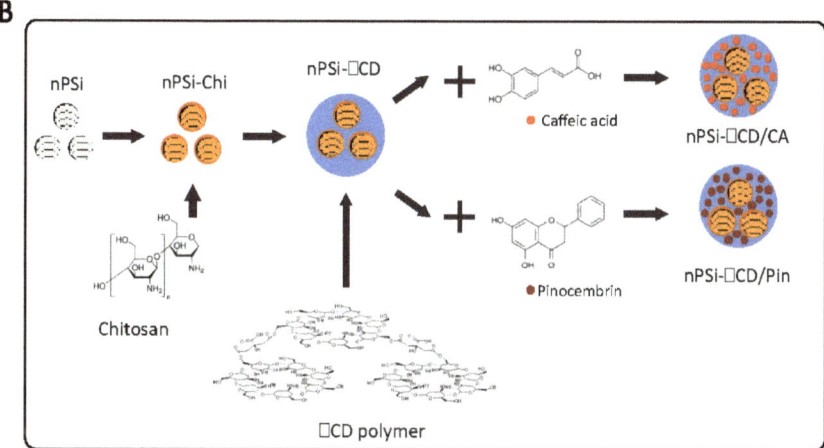

**Figure 1.** Experimental scheme for the synthesis of: (**A**) nPSi microparticles, and (**B**) nPSi-βCD composite microparticles. nPSi: nanoporous silicon, βCD: β-ciclodextrin polymer.

## 2.3. Physicochemical Characterization

The zeta potential of samples was measured by a ZetaSizer Nano–ZS (Malvern Ltd., Royston, UK) in distilled water. Attenuated total reflectance Fourier-transform infrared spectroscopy (ATR-FTIR) was used for chemical analyses of the microparticles. An FTIR spectrometer (CARY 630 FTIR Agilent Technologies, Santa Clara, CA, USA) was used in a range between 4000 and 600 cm$^{-1}$ with a resolution of 1 cm$^{-1}$ (NS = 4). The obtained spectra were mathematically processed by data smoothing and spectral normalized. The morphology of the samples was investigated by a variable pressure scanning electron microscope (VP-SEM, SU-3500 Hitachi, Tokyo, Japan) using an acceleration voltage of 5 kV. The size distribution of samples was presented as histograms; data was obtained from the SEM images that were processed using freely available ImageJ software, version 1.52k, National Institutes of Health, Bethesda, Maryland, USA. The atomic percentage was obtained by energy-dispersive X-ray analysis (EDX) with an INCA X-sight from Oxford Instruments within the VP-SEM equipment. Thermogravimetric analyses (TGA) were conducted in a $N_2$ atmosphere at a heating rate of 10 °C/min

(DTG-60H Shimadzu, Tokyo, Japan). Porosity of nPSi samples was obtained by gravimetric analysis according to the following equation:

$$\%P = (m_1 - m_2/m_1 - m_3) \times 100 \tag{1}$$

where $m_1$ is the mass of Si wafer before electrochemical etching, $m_2$ is the mass of sample just after anodization and $m_3$ is the mass of sample after a rapid dissolution of the whole porous layer in a 3% KOH solution.

### 2.4. Polyphenols Loading

Polyphenols, CA and Pin, were reconstituted with 100% DMSO (200 µM) and stored at −20 °C until required. Five mg of samples were loaded with CA and Pin using 1 mL of concentrated aqueous solution of each polyphenol (2 mM) and placed in a horizontal shaker incubator (NB-2005LN Biotek, Winooski, VT, USA) for 12 h at 50 RPM and room temperature. After polyphenol loading, samples were rinsed to remove the unentrapped molecules, they were dried at room temperature and milled for homogenization. To determine the maximum polyphenol loading, samples were hydrolyzed in 0.1 M NaOH solutions then they were analyzed by UV–visible spectrometry (UVmini-1240 spectrometer Shimadzu, Tokyo, Japan). CA and Pin were detected at 310 and 322 nm, respectively. Polyphenol entrapment efficiency (%PEE) and polyphenol loading efficiency (%PLE) and were calculated from Equations (2) and (3), respectively [42]:

$$\%PEE = (m_{p\_m}/m_{p\_i}) \times 100 \tag{2}$$

$$\%PLE = (m_{p\_m}/m_m) \times 100 \tag{3}$$

where $m_{p\_m}$ is the mass of polyphenol in microparticles, $m_{p\_i}$ is the mass of polyphenol fed initially and $m_m$ is the mass of microparticles.

### 2.5. Polyphenols Release Profiles

Polyphenols release data were collected at different times using 5 mg of charged samples in 3 mL of PBS solution (37 °C) as release medium in agitation at 100 RPM. All experiments were conducted in triplicate and nPSi samples were used as controls in these kinetic experiments.

In order to determine the mechanism of drug release, three models were fitted to the release profiles: First order, Higuchi and Korsmeyer–Peppas models. The first order equation is [43]:

$$\ln M_t - \ln M_0 = k_1 t \tag{4}$$

where $M_t$ is the absolute cumulative amount of drug released at time point $t$, $M_0$ is the initial amount of drug in the solution, and $k_1$ is the first order release kinetic constant. The Higuchi equation is [44]:

$$M_t = k_H t^{1/2} \tag{5}$$

where $M_t$ is the absolute cumulative amount of drug released at time point $t$, and $k_H$ is Higuchi release kinetic constant. The Korsmeyer–Peppas semiempirical model is given by [45]:

$$M_t/M_\infty = k_{KP} t^n \tag{6}$$

where $M_t/M_\infty$ is the fractional drug release, $t$ is the release time, $k_{KP}$ is the Korsmeyer–Peppas release kinetic constant and $n$ is an exponent which characterizes the mechanism of release. The fitting of models was conducted with SigmaPlot v14.0, Systat Software, Inc., San Jose, USA.

## 2.6. Cytotoxicity Assays

### 2.6.1. Cell Culture

For cell culture, human umbilical vein endothelial cells (HUVECs) were obtained from the Cell Applications Inc (San Diego, CA, USA), and maintained in Endothelial Cell Growth Medium (Cell Applications, San Diego, CA, USA) supplement with 10% FBS, 1% L-Glutamine and 1% penicillin-streptomycin solution. The cell culture was routinely grown under specific conditions in a humidified atmosphere incubator of 95% air and 5% $CO_2$ at 37 °C. Cells were used at no more than seven passages.

### 2.6.2. Polyphenols Cytotoxicity

For the in vitro viability assays, CellTiter 96® Aqueous One Solution Cell Proliferation Assay (MTS) Promega (Madison, WI, USA) was used to determine the toxic effect of CA and Pin on HUVECs viability. The MTS assay is based on the conversion of a tetrazolium salt into a colored aqueous soluble formazan product by mitochondrial activity of viable cells at 37 °C. The amount of formazan produced by dehydrogenase enzymes is directly proportional to the number of living cells in culture. The viability assays were performed according to the manufacturer's protocols. HUVECs were briefly placed into 96-well plates ($2.5 \times 10^3$ cells/per well) in 100 µL and incubated at 37 °C. Then, cells were exposed to increase concentrations up to 2000 µM of polyphenols. The compound was prepared in dimethylsulfoxide (DMSO). After 24 h of incubation, the medium was removed and 20 µL MTS reagent was added to the wells, followed by a 4-h incubation at 37 °C. The absorbance was determined by a microplate reader (NanoQuant, Infinite® M200PRO–Tecan, Redwood, CA, USA) at 490 nm. Results were expressed as the percentage of viability relative to the control. The cell viability was calculated as follows: cell viability (%) = (OD of treatment group/OD of control group) × 100. Dose-dependent viability curves were determined using the cell viability trends.

### 2.6.3. nPSi-βCD Composite Cytotoxicity

To determine the effect of the composite (nPSi-βCD) on HUVEC cell viability, a CellTiter-Fluor™ Cell Viability assay (Promega, Madison, WI, USA) was used. This assay measures a conserved and constitutive protease activity within live cells using a fluorogenic peptide substrate (glycyl-phenylalanyl-aminofluorocoumarin; GF-AFC). The substrate enters intact cells where it is cleaved by the live-cell protease activity to generate a fluorescent signal proportional to the number of living cells. $1 \times 10^5$ cells were exposed to different concentrations of composite and were photographed using a confocal laser microscope (CLSM, FV1000 Olympus, Tokyo, Japan) with excitation and emission wavelengths of 390 nm and 505 nm, respectively. The assay was performed according to the manufacturer's protocols. The fluorescence intensity analysis was performed with Olympus Fluoview (FV10 v2.0c) software (Olympus Corporation, Tokyo, Japan). Data was analyzed statistically by analysis of variance (ANOVA) using Kruskal-Wallis test, and post-hoc test were also conducted using Dunn's multiple comparisons. The level of significance was $p < 0.05$ and the results were expressed as the arithmetic mean of three biological replicates with its corresponding standard deviation. The statistical analysis was performed using the GraphPad Prism v7.0c (GraphPad Software, San Diego, CA, USA).

## 3. Results and Discussion

As the synthesis of composite microparticles was obtained by electrostatic attraction of oppositely charges, zeta potential analysis was performed (Figure 2A). This technique provides the net electrical charge of the microparticles generated by their functional groups. In the case of nPSi, its negative zeta potential value (−29.06 ± 0.06 mV) would correspond to the negatively charged silanol groups produced by the chemical oxidation with $H_2O_2$ [46]. nPSi-Chi showed positive values (16.5 ± 0.6 mV) because the grafting with chitosan would generate a rich aminated surface [47]. On the other hand, the sharp negative zeta potential of nPSi-βCD (−39.8 ± 1.73 mV) was according to βCD value

(−28.2 ± 9 mV), which is generated by the hydrophilic outer surface cavity (C–OH groups) of βCD molecules [48].

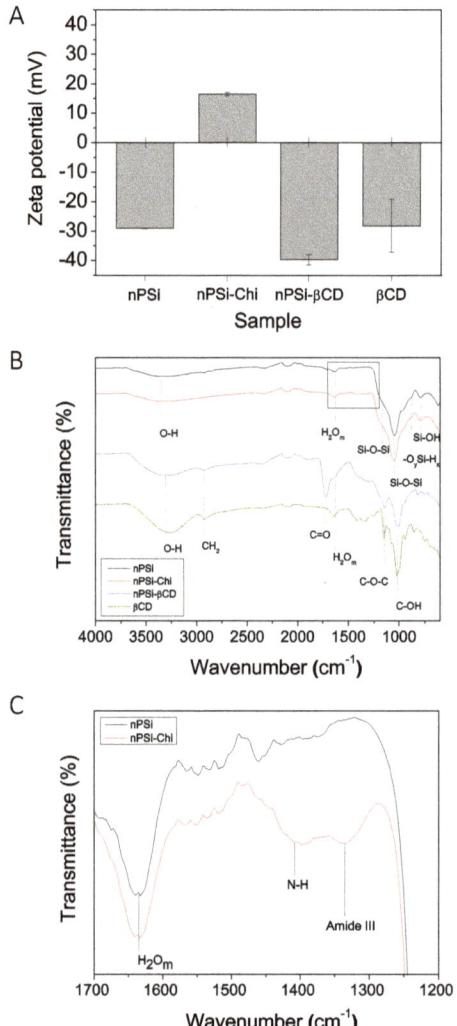

**Figure 2.** Monitoring of the synthesis process of nPSi-βCD composite microparticles via: (**A**) Zeta potential and (**B**) Attenuated total reflectance Fourier transform infrared spectroscopy analysis (ATR-FTIR) (**C**) Zoom in of ATR-FTIR spectra.

In that sense, ATR-FTIR analysis was performed to determine the chemical changes of nPSi microparticles during the cascade synthesis processes (Figure 2B). The spectrum of nPSi showed a sharp transmittance peak at 1050 cm$^{-1}$ with a shoulder at 1170 cm$^{-1}$, which both correspond to Si–O–Si stretching mode [49]. Besides, weak bands at 880 and 795 cm$^{-1}$ related to $-O_y$Si-$H_x$ and SiOH, respectively, and the O–H stretching band from SiOH and adsorbed $H_2O$ at 3350 cm$^{-1}$ were detected [49]. Moreover, molecular water ($H_2O_m$) absorbance band was observed at 1630 cm$^{-1}$ [50]. These detected functional groups are in agreement with the chemical oxidation of nPSi via $H_2O_2$. On the other hand, the spectrum of nPSi-Chi presented the same functional groups as nPSi plus weak bands of N–H and amide III detected at 1408 and 1320–1346 cm$^{-1}$ [51–53], respectively (Figure 2C). Those bands are related to the polyamino-saccharide chains of Chi, which were used to link the βCD polymer with nPSi microparticles. Regarding the spectrum of nPSi-βCD, bands corresponded to the spectrum of native βCD were observed: C–OH stretching (1021 cm$^{-1}$) [49], C–O–C stretching (1150 cm$^{-1}$) [13], H$_2$O$_m$ (1630 cm$^{-1}$) [50], CH2 asymmetric stretching (2930 cm$^{-1}$) and O-H stretching from hydroxyl groups (3300 cm$^{-1}$) [49]. nPSi-βCD barely showed an extra band than βCD at 1721 cm$^{-1}$ which correspond to C=O groups generated during the polymerization achieved between βCD and citric acid [49].

SEM images were produced (Figure 3A) to analyze the size and morphology of samples at the different stages of synthesis. Moreover, the obtained distribution size from these images is shown in Figure 3B. nPSi and nPSi-Chi presented irregular shapes with an average size of 2.0–2.5 μm, and both kind of microparticles showed rougher surface due to their columnar pores of ~50 nm width. In addition, gravimetric analysis presented an average porosity of 75 ± 5%. In the case of the nPSi-βCD sample, the microparticle shapes were also irregular with a higher size around of 14.0 μm, and their faces exhibited a softer appearance. In fact, folds produced by the polymerization could be also observed. The increase in particle size may have been because the small particles agglomerated during the polymerization forming higher particles. Similar size distribution of this kind of particles for oral drug delivery system has been previously reported by Salonen et al. [54]. On the other hand, EDX analysis was performed on each sample (Figure 3C1) and the atomic percentage and C/Si ratio were obtained (Figure 3C2). nPSi mainly exhibited Si and O signal due to the oxidation performed by H$_2$O$_2$. Although the C/Si ratio was 0.76 ± 0.2, C signal was considerable high (~18%). This was most probably due to contamination when handling. In the case of nPSi-Chi, due to the previous characterization (Zeta potential and ATR-FTIR), N signal from amines groups of the incorporated chitosan was expected to be identified but it was not. This can be explained because incorporated chitosan was most likely a superficial layer and N signal was not strong enough to be detected by EDX technique. However, it is possible to observe that the C/Si ratio increased twice up to 1.5 ± 0.3 due to the polymer grafting [55]. Regarding the nPSi-βCD sample, a high increase of C signal was identified: the C/Si was raised up to 17.4 ± 8 due to the in situ polymerization of βCD and citric acid. In fact, Na and P traces from the catalyst were also detected.

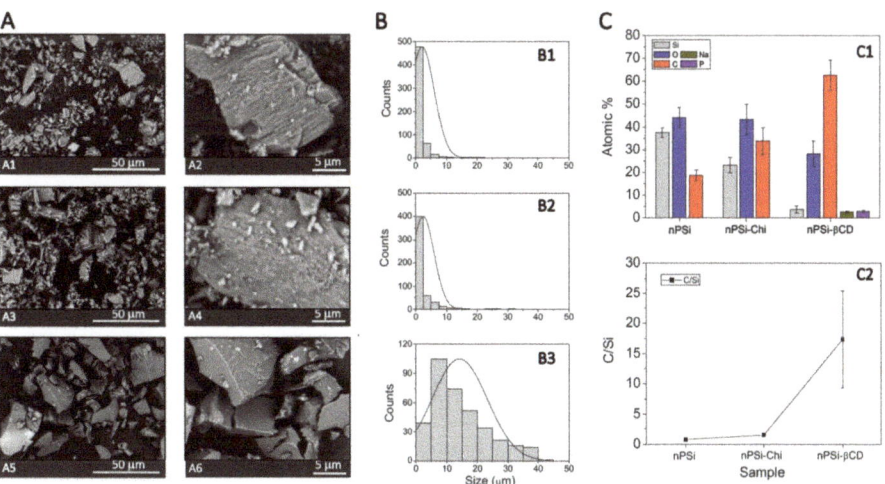

**Figure 3.** (**A**) Scanning electron microscope images (SEM) of samples at the different stages of synthesis: (**A1,A2**) for nPSi, (**A3,A4**) for nPSi-Chi, (**A5,A6**) for nPSi-βCD. (**B**) Histograms of particle size distribution: (**B1**) for nPSi, (**B2**) for nPSi-Chi, (**B3**) for nPSi-βCD. (**C**) Atomic % and C/Si ratio of samples obtained from energy-dispersive X-ray analysis (EDX), (**C1,C2**), respectively.

In order to evaluate changes in nPSi-βCD composite microparticles with the loaded polyphenols, same set of previous physicochemical characterization was carried out. Figure 4A shows how zeta potential of nPSi-βCD was reduced after the addition of both polyphenols: nPSi-βCD/CA and nPSi-βCD/Pin presented 5.5 and 10.4 mV lower zeta potential values than nPSi-βCD, respectively. Regarding the ATR-FTIR analysis (Figure 4B), nPSi-βCD/CA and nPSi-βCD/Pin spectra exhibited the same functional groups than nPSi-βCD spectrum. Nevertheless, both nPSi-βCD/CA and nPSi-βCD/Pin exhibited two extra bands related to the bending modes of CH, which are associated to incorporation of both polyphenols (Figure 4C): β(CH) and γ(CH) at 1187 and 940 cm$^{-1}$, respectively. β denotes in-plane bending modes and γ designates out-of-plane bending modes [56].

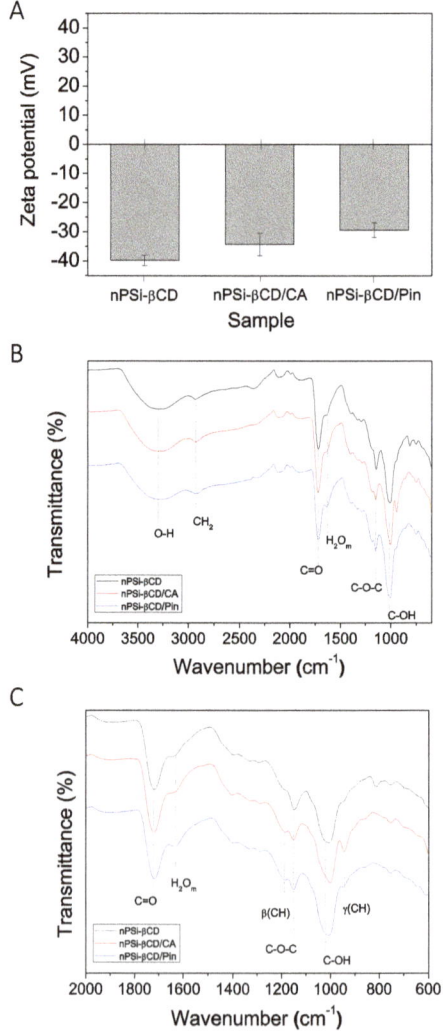

**Figure 4.** Monitoring of the polyphenols loading on PSi-βCD composite microparticles via: (**A**) Zeta potential and (**B**) Attenuated total reflectance Fourier transform infrared spectroscopy analysis (ATR-FTIR). (**C**) Zoom in of ATR-FTIR spectra.

Regarding morphology, which was observed by SEM images (Figure 5A), nPSi-βCD/CA and nPSi-βCD/Pin did not present significantly changes in comparison with nPSi-βCD; the surface of both nPSi-βCD/CA and nPSi-βCD/Pin microparticles exhibited a softer appearance and some folds produced by the polymerization. However, the size of the loaded microparticles was higher than nPSi-βCD (Figure 5B): ~19 and ~22 μm, respectively. As polyphenols are highly hydrophobics, they could tend to agglomerate smaller particles. As Pin is more hydrophobic than CA, this could generate more aggregation, and therefore, higher microparticles. On the other hand, EDX analysis were also performed on the samples (Figure 5C1), atomic percentage and C/Si ratio were obtained too (Figure 5C2). nPSi-βCD/CA and nPSi-βCD/Pin did not present any more Na and P traces, and the C/Si ratio considerably increased in comparison with nPSi-βCD: nPSi-βCD showed a C/Si ratio of 17.4 ± 8, and nPSi-βCD/CA and nPSi-βCD/Pin presented 88.75 ± 22.2 and 105.7 ± 30.5, respectively.

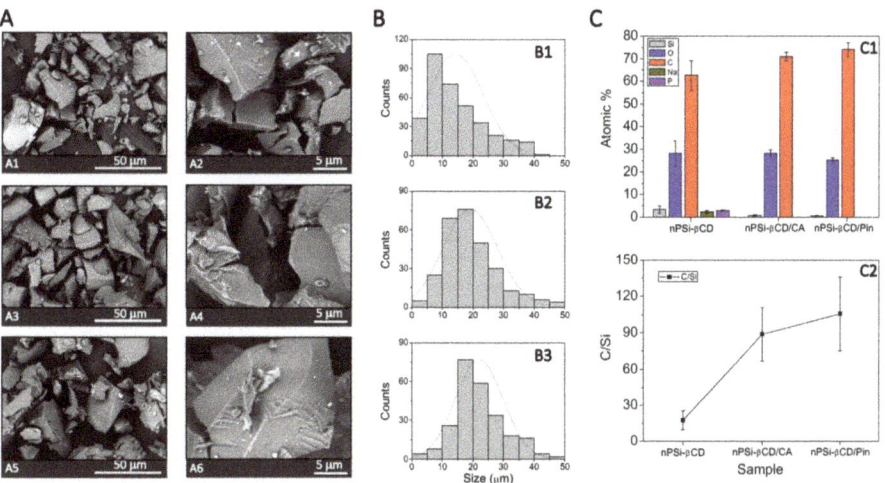

**Figure 5.** (**A**) Scanning electron microscope images (SEM) of nPSi-βCD loading with polyphenols: (**A1,A2**) for nPSi-βCD, (**A3,A4**) for nPSi-βCD/CA, (**A5,A6**) for nPSi-βCD/Pin. (**B**) Histograms of particle size distribution: (**B1**) for nPSi-βCD, (**B2**) for nPSi-βCD/CA, B3 for nPSi-βCD/Pin. (**C**) Atomic % and C/Si ratio of samples obtained from energy-dispersive X-ray analysis (EDX), (**C1,C2**), respectively.

In order to determine the functionalization degree of Chi and βCD polymer integrated onto nPSi substrates, TGA analyses were performed (Figure 6A). The plot illustrates the percent mass as a function of samples temperature under a nitrogen purge. As expected, nPSi sample practically did not present degradation, but nPSi-Chi showed a slight decomposition of around 3%, this is in accordance previous characterization that suggests chitosan grafting was just superficially. Moreover, the thermogravimetric analysis of native βCD was monitored as reference. The βCD decomposition was clearly appreciable; the first stage with was at 100 °C corresponds to the level of absorbed water (~10.5%). The second stage, which started at 310 °C and finished at 350 °C, is related to the melting, decomposition and turning into char of the glucose units of the βCD molecules [57]. In the case of nPSi-βCD, the phenomenon was gradual, due to the stronger 3D structure net of βCD polymer, but similar to the native βCD reference. Considering the residual weight at 600 °C, it is possible to ponder that nPSi-βCD was composed by 32% nPSi, 62% βCD polymer, 3% Chi and 3% humidity. The high percent of βCD polymer (62%) in composite composition can be explained, in addition to the electrostatic interactions with polymers showed by zeta potential, with porosity of samples which also worked as an anchor holding the polymer film.

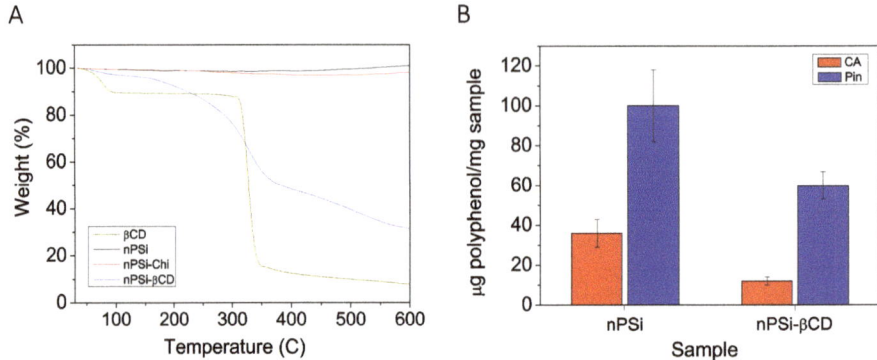

**Figure 6.** (**A**) Thermogravimetric analysis (TGA) of microparticles at different stages of synthesis, and (**B**) Polyphenols capacity loading of nPSi-βCD composite microparticles.

%PEE and %PLE of samples were determined by UV-Vis spectroscopy at 310 and 322 nm for CA and Pin, respectively. Although nPSi (control) did not exhibit a chemical surface compatible with polyphenols, samples presented high values of %PEE 50 ± 2.0 and 97.5 ± 2.0 for CA and Pin respectively. This can be explained by the high surface area of their nanopores. In the case of nPSi-βCD, microparticles exhibited 16.6 ± 1.0 %PEE and 58.5 ± 1.5 %PEE for CA and Pin, respectively. In the same sense, nPSi samples presented higher %PLE than nPSi-βCD. Figure 6B shows the polyphenols capacity loading of both kind of microparticles. nPSi presented a load of 36 ± 7 µg CA/mg nPSi (3.6 ± 0.7% PLE) and 100 ± 18 µg Pin/mg nPSi (10.0 ± 1.8% PLE), and nPSi-βCD showed a load of 12 ± 2 µg CA/mg nPSi (1.2 ± 0.2% PLE) and 60 ± 7 µg Pin/mg nPSi (6.0 ± 0.7 %PLE). Due to previous characterization results, it is very possible that polyphenols were mainly adsorbed in the large corona of βCD polymer around the small nPSi microparticles, which were the substrate of the composite.

To evaluate the polyphenols controlled release functionality, loaded microparticles of nPSi and nPSi-βCD were immersed in PBS batches at 37 °C under stirring. The obtained polyphenols release profiles are shown in Figure 7A1,A2 for CA, and Figure 7B1,B2 for Pin. After 24 h of release, all samples presented higher values of %cumulative release. nPSi samples showed 97.6 ± 17.6 and 94.2 ± 17.0 for CA and Pin, respectively. In the case of nPSi-βCD, microparticles exhibited 93.8 ± 11.2 and 92.3 ± 11.0 for CA and Pin, respectively. Profiles presented a clear contrasting behavior between the control (nPSi) and composite (nPSi-βCD). Results visibly showed that nPSi-βCD worked much better than nPSi: both polyphenols retained into nPSi showed a fast release profile during the first minutes, in contrast with nPSi-βCD, which showed a controlled released for more than 5 h.

To attain deeper perception of the mechanisms that govern the release of polyphenols from the samples, three release models were fitted to the experimental data: first order, Higuchi and Korsmeyer-Peppas models (Table 1). In the case of CA release, according to the $r^2$ obtained values for nPSi, it presented better adjustment with the first order model, where immediate-release dosage was dispersed in a single action [58]. However, for the release of CA using nPSi-βCD, CA release kinetics were described with a more accurate precision by the Korsmeyer-Peppas model. This means that the governing factor of CA release was not the dissolution from samples, but a Fickian diffusion process [41]. Moreover, in that sense, since the release exponent $n$ from the Korsmeyer-Peppas model was smaller than 0.5, only diffusive release can be suggested. Therefore, erosion process could be insignificant [41]. For the case of Pin, both release profiles from nPSi and nPSi-βCD microparticles, were better adjusted to the Korsmeyer-Peppas model. However, nPSi presented an $n$ value closed to zero. Regarding nPSi-βCD, it showed an $n > 0.5$, which suggest that besides Fickian diffusion, erosion process could also be contributing in the Pin release [41].

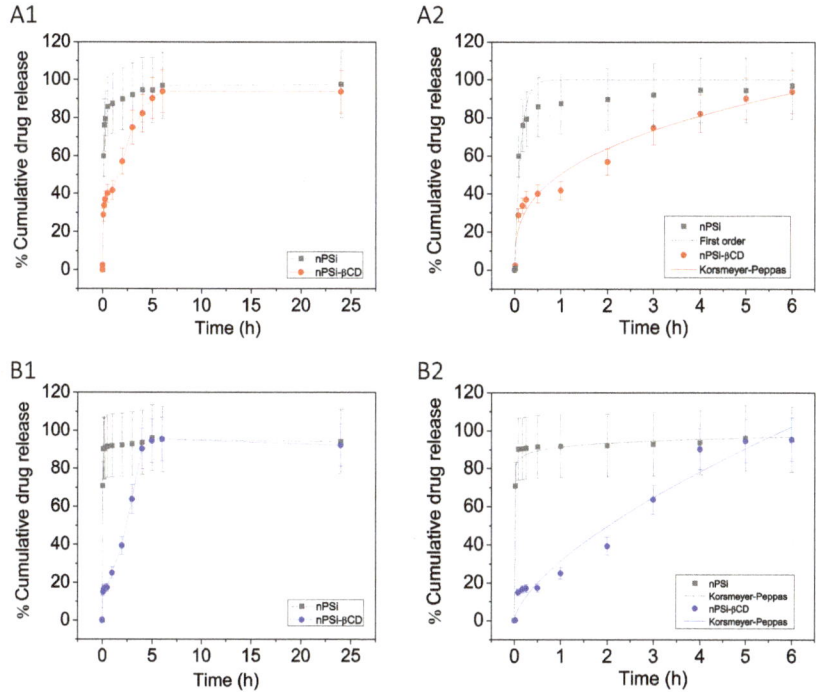

**Figure 7.** Polyphenols release profiles in phosphate-buffered saline (PBS) at 37 °C: (**A1,A2**) CA and (**B1,B2**) Pin.

**Table 1.** *In vitro* release kinetics of caffeic acid and pinocembrin in PBS at 37 °C.

| Polyphenol | Sample | First order $lnM_t - lnM_0 = k_1 t$ | | Higuchi $M_t = k_H t^{1/2}$ | | Korsmeyer-Peppas $\frac{M_t}{M_\infty} = k_{KP} t^n$ | | |
|---|---|---|---|---|---|---|---|---|
| | | $k_1 (\%h^{-1})$ | $r^2_{adj}$ | $k_H (\%h^{-1/2})$ | $r^2_{adj}$ | $k_{KP} (h^{-n})$ | $n$ | $r^2_{adj}$ |
| Caffeic acid | nPSi | 8.1047 | 0.9355 | 52.7074 | −0.0040 | 79.7830 | 0.197 | 0.7832 |
| | nPSi-βCD | 0.5694 | 0.7982 | 41.6744 | 0.9074 | 50.4833 | 0.3419 | 0.9649 |
| Pinocembrin | nPSi | 76.3252 | 0.9271 | 54.5954 | −2.1444 | 91.4797 | 0.0317 | 0.9774 |
| | nPSi-βCD | 0.3892 | 0.9433 | 38.5733 | 0.9451 | 31.3890 | 0.6584 | 0.9654 |

To evaluate the cytotoxic effect of CA and Pin, human umbilical vein endothelial cells (HUVECs) cultures were performed. HUVECs are a classic model to study endothelial functions, such as angiogenesis. Angiogenesis is the formation of new blood vessels from pre-existing vessels [59]. Although it is a physiological process, the abnormal growth of vessels promotes the development and/or progression of some diseases such as cardiovascular diseases. Regarding this, the MTS test was used to study the impact of CA and Pin on the viability of HUVECs. Results showed that viability gradually decreased and responded in a dose-dependent manner for both polyphenols. In the case of CA, cell viability was slightly reduced from 100% to 80% for concentrations from 2 to 200 µM ($p > 0.05$), while surviving cells were ≤70% for concentrations ≥500 µM. ($p < 0.01$) (Figure 8A). Regarding to the effect of Pin (Figure 8B), the viability was higher than 80% from 2 to 100 µM, but it decreased to 50% at 200 µM ($p > 0.05$). Moreover, cell viability was reduced to less than 20% at concentrations ≥ 500 µM ($p < 0.01$). According to this, concentrations up to 200 and 100 µM for CA and Pin, respectively, maintained cell viability ≥80%, that is to say, they did not generate cellular cytotoxicity.

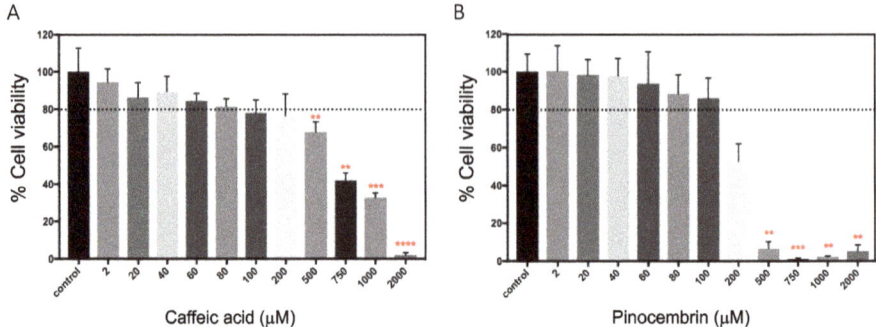

**Figure 8.** Viability of HUVECs treated with polyphenols. Cells were exposed to different concentrations of polyphenols for 24 h and cellular viability was measured by tetrazolium salt (MTS) assay. (**A**) Cells treated with 2–2000 mM of caffeic acid, and (**B**) Cells treated with 2–2000 mM of pinocembrin. Unexposed cells to nPSi-βCD microparticles were used as a control. The dashed line indicates the cell viability of 80%. All results are presented as the mean ± standard deviation (SD). The experimental data from all relevant studies were analyzed using Analysis of Variance (ANOVA) and Kruskal–Wallis test, which indicate the statistical significance when the percentage of cells viability exposed to the different microparticle concentrations are different from the control. ** $p < 0.01$, *** $p < 0.001$, **** $p < 0.0001$, (n = 3).

On the other hand, the effect nPSi-βCD microparticles on the viability of HUVECs was also studied. Cells were cultured in the presence of composite microparticles at concentrations of 0.25, 0.50, 1.25 and 2.5 mg/mL for 6 and 24 h. In the microscopic observation (Figure 9), HUVECs exposed to the lowest concentrations (0.25 and 0.50 mg/mL) exhibited a normal flattened and thin morphology, suggesting that microparticles were well tolerated by cells. Instead, the highest concentrations (1.25 and 2.5 mg/mL), generated a large amount of rounded and suspended cells in the culture medium. This indicates that nPSi-βCD microparticles concentrations higher than 0.50 mg/mL were not well tolerated by HUVECs, affecting its cell adhesion capacity, an essential survival characteristic of this type of cells.

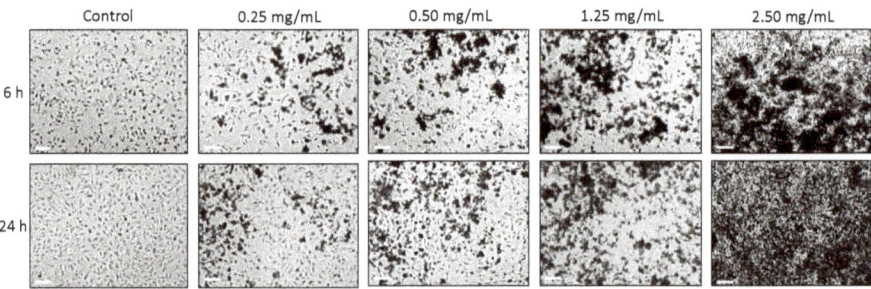

**Figure 9.** Microscopical images of the effect of nPSi-βCD microparticles exposition to HUVECs during a 6 and 24 h culture. Images were taken at 8X magnification. Scale bars: 1 mm.

Concerning the viability percentage of HUVECs exposed to different concentrations of nPSi-βCD microparticles (Figure 10B), results showed that cell viability at 0.25 mg/mL was higher than 80% but it started to decrease at concentrations equal or higher than 0.50 mg/mL (63%; $p < 0.01$). Specifically, at concentrations of 1.25 and 2.50 mg/mL, cell viability was very low reaching values as low as 20%. This increase in cell mortality from 0.25 mg/mL to ≥1.25 mg/mL of nPSi-βCD microparticles may be due to an alteration in the basic cellular functions such as the adherence capacity affected, for example, cell communication, differentiation and migration leading to cell death [60].

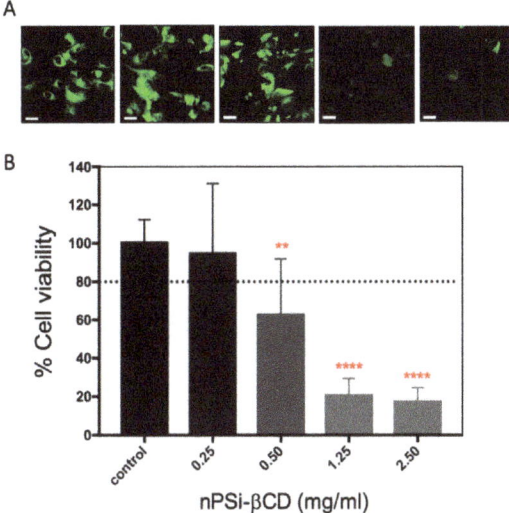

**Figure 10.** Cytotoxicity of HUVECs exposed to nPSi-βCD microparticles. The cells were exposed for 6 h and then cellular viability was evaluated with CellTiter-FluorTM assay and fluorescence intensity measuring was with confocal laser microscope (**A**) HUVECs exposed to nPSi-βCD microparticles. Scale bars: 20 μm. (**B**) Viability percentage of cells exposed to composite microparticles. Unexposed cells to nPSi-βCD microparticles were used as a control. Dashed line indicates cell viability of 80%. All results are presented as the mean ± standard deviation (SD). The experimental data from all the studies were analyzed using Analysis of Variance (ANOVA) and Kruskal–Wallis test, which indicate the statistical significance when the percentage of cells viability exposed to the different microparticles concentrations are different from the control. ** $p < 0.01$, **** $p < 0.0001$. (n = 3).

## 4. Conclusions

Physicochemical characterizations showed that nPSi-βCD microparticles were suitable to be used an alternative as carrier and controlled oral delivery system of both polyphenols, CA and Pin. The release profiles indicated that nPSi-βCD composite presented a better-controlled release of polyphenols than nPSi without βCD polymer. Moreover, nPSi-βCD samples loaded higher amount of Pin than CA, and the release of Pin was higher controlled than CA. For the CA case, a purely diffusive mechanism of release was suggested, but for the Pin, erosion process could be also contributing during the release. On the other hand, nPSi-βCD microparticles presented cytocompatibility HUVECs culture at concentrations of 0.25 mg/mL. Then, these results indicate that nPSi-βCD composite microparticles could be safely used as an alternative oral delivery system to improve controlled release and bioavailability of CA and Pin, and eventually other polyphenols with therapeutic potential.

**Author Contributions:** D.G.-O., L.A.S. and J.H.-M. conceived and designed experiments; T.P. and G.R.-S. synthetized the nPSi and nPSi-βCD microparticles; D.G.-O., T.P., G.R.-S. and J.H.-M. performed the physicochemical characterization; D.G.-O. and D.S.P.A. performed the biological experiments, D.G.-O. and J.H.-M. analyzed the data; D.G.-O., G.R.-S., L.A.S. and J.H.-M. contributed the reagent and analytical tools; G.D.-O. and J.H.-M. wrote the manuscript. All authors revised the manuscript.

**Funding:** This work was financially supported by Fondo Nacional de Desarrollo Científico Tecnológico FONDECYT–Chile (Grant numbers 11180395 and 1171765), CONICYT Scholarship (21140154) and UCT project (2017PF-JH-07).

**Conflicts of Interest:** The authors declare no conflict of interest.

## References

1. Li, A.N.; Li, S.; Zhang, Y.J.; Xu, X.R.; Chen, Y.M.; Li, H.B. Resources and biological activities of natural polyphenols. *Nutrients* **2014**, *6*, 6020–6047. [CrossRef] [PubMed]
2. Fraga, C.G.; Galleano, M.; Verstraeten, S.V.; Oteiza, P.I. Basic biochemical mechanisms behind the health benefits of polyphenols. *Mol. Asp. Med.* **2010**, *31*, 435–445. [CrossRef] [PubMed]
3. Silva-Carvalho, R.; Baltazar, F.; Almeida-Aguiar, C. Propolis: A Complex Natural Product with a Plethora of Biological Activities That Can Be Explored for Drug Development. *Evid. Based Complement. Altern. Med.* **2015**, *2015*, 206439. [CrossRef] [PubMed]
4. Grange, J.M.; Davey, R.W. Antibacterial properties of propolis (bee glue). *J. R. Soc. Med.* **1990**, *83*, 159–160. [CrossRef] [PubMed]
5. Dobrowolski, J.W.; Vohora, S.B.; Sharma, K.; Shah, S.A.; Naqvi, S.A.H.; Dandiya, P.C. Antibacterial, antifungal, antiamoebic, antiinflammatory and antipyretic studies on propolis bee products. *J. Ethnopharmacol.* **1991**, *35*, 77–82. [CrossRef]
6. Daleprane, J.B.; Abdalla, D.S. Emerging roles of propolis: Antioxidant, cardioprotective, and antiangiogenic actions. *Evid. Based Complement. Altern. Med.* **2013**, *2013*, 175135. [CrossRef] [PubMed]
7. Tan-No, K.; Nakajima, T.; Shoji, T.; Nakagawasai, O.; Niijima, F.; Ishikawa, M.; Endo, Y.; Sato, T.; Satoh, S.; Tadano, T. Anti-inflammatory effect of propolis through inhibition of nitric oxide production on carrageenin-induced mouse paw edema. *Biol. Pharm. Bull.* **2006**, *29*, 96–99. [CrossRef] [PubMed]
8. Grunberger, D.; Banerjee, R.; Eisinger, K.; Oltz, E.M.; Efros, L.; Caldwell, M.; Estevez, V.; Nakanishi, K. Preferential cytotoxicity on tumor cells by caffeic acid phenethyl ester isolated from propolis. *Experientia* **1988**, *44*, 230–232. [CrossRef] [PubMed]
9. Daleprane, J.B.; Freitas, V.S.; Pacheco, A.; Rudnicki, M.; Faine, L.A.; Dörr, F.A.; Ikegaki, M.; Salazar, L.A.; Ong, T.P.; Abdalla, D.S. Anti-atherogenic and anti-angiogenic activities of polyphenols from propolis. *J. Nutr. Biochem.* **2012**, *23*, 557–566. [CrossRef] [PubMed]
10. Woodman, O.L.; Chan, E.C. Vascular and anti-oxidant actions of flavonols and flavones. *Clin. Exp. Pharmacol. Physiol.* **2004**, *31*, 786–790. [CrossRef]
11. Sedlak, L.; Wojnar, W.; Zych, M.; Wyględowska-Promieńska, D.; Mrukwa-Kominek, E.; Kaczmarczyk-Sedlak, I. Effect of Resveratrol, a Dietary-Derived Polyphenol, on the Oxidative Stress and Polyol Pathway in the Lens of Rats with Streptozotocin-Induced Diabetes. *Nutrients* **2018**, *10*, 1423. [CrossRef] [PubMed]
12. Bijak, M.; Ziewiecki, R.; Saluk, J.; Ponczek, M.; Pawlaczyk, I.; Krotkiewski, H.; Wachowicz, B.; Nowak, P. Thrombin inhibitory activity of some polyphenolic compounds. *Med. Chem. Res.* **2014**, *23*, 2324–2337. [CrossRef] [PubMed]
13. Paeng, S.H.; Jung, W.K.; Park, W.S.; Lee, D.S.; Kim, G.Y.; Choi, Y.H.; Seo, S.K.; Jang, W.H.; Choi, J.S.; Lee, Y.M.; et al. Caffeic acid phenethyl ester reduces the secretion of vascular endothelial growth factor through the inhibition of the ROS, PI3K and HIF-1$\alpha$ signaling pathways in human retinal pigment epithelial cells under hypoxic conditions. *Int. J. Mol. Med.* **2015**, *35*, 1419–1426. [CrossRef] [PubMed]
14. Gao, Y.; Liu, F.; Fang, L.; Cai, R.; Zong, C.; Qi, Y. Genkwanin inhibits proinflammatory mediators mainly through the regulation of miR-101/MKP-1/MAPK pathway in LPS-activated macrophages. *PLoS ONE* **2014**, *9*, e96741. [CrossRef] [PubMed]
15. Turrini, E.; Ferruzzi, L.; Fimognari, C. Potential Effects of Pomegranate Polyphenols in Cancer Prevention and Therapy. *Oxid. Med. Cell. Longev.* **2015**, *2015*, 938475. [CrossRef] [PubMed]
16. Vermerris, W.; Nicholson, R. *Phenolic Compound Biochemistry*; Springer: Dordrecht, The Netherlands, 2006; ISBN 9781402051630.
17. Manach, C.; Scalbert, A.; Morand, C.; Rémésy, C.; Jiménez, L. Polyphenols: Food sources and bioavailability. *Am. J. Clin. Nutr.* **2004**, *79*, 727–747. [CrossRef]
18. Barrientos, L.; Herrera, C.L.; Montenegro, G.; Ortega, X.; Veloz, J.; Alvear, M.; Cuevas, A.; Saavedra, N.; Salazar, L.A. Chemical and botanical characterization of Chilean propolis and biological activity on cariogenic bacteria Streptococcus mutans and Streptococcus sobrinus. *Braz. J. Microbiol.* **2013**, *44*, 577–585. [CrossRef]
19. Bankova, V. Chemical diversity of propolis and the problem of standardization. *J. Ethnopharmacol.* **2005**, *100*, 114–117. [CrossRef]

20. Szliszka, E.; Sokół-Łętowska, A.; Kucharska, A.Z.; Jaworska, D.; Czuba, Z.P.; Król, W. Ethanolic Extract of Polish Propolis: Chemical Composition and TRAIL-R2 Death Receptor Targeting Apoptotic Activity against Prostate Cancer Cells. *Evid. Based Complement. Altern. Med.* **2013**, *2013*, 757628. [CrossRef]
21. Cuevas, A.; Saavedra, N.; Rudnicki, M.; Abdalla, D.S.; Salazar, L.A. ERK1/2 and HIF1α Are Involved in Antiangiogenic Effect of Polyphenols-Enriched Fraction from Chilean Propolis. *Evid. Based Complement. Altern. Med.* **2015**, *2015*, 187575. [CrossRef] [PubMed]
22. Cuevas, A.; Saavedra, N.; Cavalcante, M.F.; Salazar, L.A.; Abdalla, D.S. Identification of microRNAs involved in the modulation of pro-angiogenic factors in atherosclerosis by a polyphenol-rich extract from propolis. *Arch. Biochem. Biophys.* **2014**, *557*, 28–35. [CrossRef] [PubMed]
23. Pandey, K.B.; Rizvi, S.I. Plant polyphenols as dietary antioxidants in human health and disease. *Oxid. Med. Cell. Longev.* **2009**, *2*, 270–278. [CrossRef] [PubMed]
24. Gu, H.F.; Mao, X.Y.; Du, M. Prevention of breast cancer by dietary polyphenols-role of cancer stem cells. *Crit. Rev. Food Sci. Nutr.* **2019**, 1–16. [CrossRef] [PubMed]
25. Potì, F.; Santi, D.; Spaggiari, G.; Zimetti, F.; Zanotti, I. Polyphenol Health Effects on Cardiovascular and Neurodegenerative Disorders: A Review and Meta-Analysis. *Int. J. Mol. Sci.* **2019**, *20*, 351. [CrossRef] [PubMed]
26. Gorinstein, S.; Leontowicz, H.; Leontowicz, M.; Jesion, I.; Namiesnik, J.; Drzewiecki, J.; Park, Y.S.; Ham, K.S.; Giordani, E.; Trakhtenberg, S. Influence of two cultivars of persimmon on atherosclerosis indices in rats fed cholesterol-containing diets: Investigation in vitro and in vivo. *Nutrition* **2011**, *27*, 838–846. [CrossRef]
27. Norata, G.D.; Marchesi, P.; Passamonti, S.; Pirillo, A.; Violi, F.; Catapano, A.L. Anti-inflammatory and anti-atherogenic effects of cathechin, caffeic acid and trans-resveratrol in apolipoprotein E deficient mice. *Atherosclerosis* **2007**, *191*, 265–271. [CrossRef]
28. D'Archivio, M.; Filesi, C.; Varì, R.; Scazzocchio, B.; Masella, R. Bioavailability of the polyphenols: Status and controversies. *Int. J. Mol. Sci.* **2010**, *11*, 1321–1342. [CrossRef]
29. Krook, M.A.; Hagerman, A.E. Stability of Polyphenols Epigallocatechin Gallate and Pentagalloyl Glucose in a Simulated Digestive System. *Food Res. Int.* **2012**, *49*, 112–116. [CrossRef]
30. Lesser, S.; Cermak, R.; Wolffram, S. Bioavailability of quercetin in pigs is influenced by the dietary fat content. *J. Nutr.* **2004**, *134*, 1508–1511. [CrossRef]
31. Milbury, P.E.; Vita, J.A.; Blumberg, J.B. Anthocyanins are bioavailable in humans following an acute dose of cranberry juice. *J. Nutr.* **2010**, *140*, 1099–1104. [CrossRef] [PubMed]
32. Manach, C.; Morand, C.; Gil-Izquierdo, A.; Bouteloup-Demange, C.; Rémésy, C. Bioavailability in humans of the flavanones hesperidin and narirutin after the ingestion of two doses of orange juice. *Eur. J. Clin. Nutr.* **2003**, *57*, 235–242. [CrossRef] [PubMed]
33. Simonetti, P.; Gardana, C.; Pietta, P. Caffeic acid as biomarker of red wine intake. *Methods Enzymol.* **2001**, *335*, 122–130. [PubMed]
34. Pandareesh, M.D.; Mythri, R.B.; Srinivas Bharath, M.M. Bioavailability of dietary polyphenols: Factors contributing to their clinical application in CNS diseases. *Neurochem. Int.* **2015**, *89*, 198–208. [CrossRef] [PubMed]
35. Canham, L.T. Bioactive silicon structure fabrication through nanoetching techniques. *Adv. Mater.* **1995**, *7*, 1033–1037. [CrossRef]
36. Martín-Palma, R.J.; Hernández-Montelongo, J.; Torres-Costa, V.; Manso-Silván, M.; Muñoz-Noval, Á. Nanostructured porous silicon-mediated drug delivery. *Expert Opin. Drug Deliv.* **2014**, *11*, 1273–1283. [CrossRef] [PubMed]
37. Hernández-Montelongo, J.; Torres-Costa, V.; Martín-Palma, R.J.; Muñoz-Noval, Á.; Manso-Silván, M. Silicon-Based Nanoparticles for Biosensing and Biomedical Applications. In *Encyclopedia of Inorganic and Bioinorganic Chemistry*; John Wiley & Sons, Ltd.: Hoboken, NJ, USA, 2015.
38. Maniya, N.H.; Patel, S.R.; Murthy, Z.V.P. Drug delivery with porous silicon films, microparticles, and nanoparticles. *Rev. Adv. Mater. Sci.* **2016**, *44*, 257–272.
39. Liu, D.; Zhang, H.; Herranz-Blanco, B.; Mäkilä, E.; Lehto, V.P.; Salonen, J.; Hirvonen, J.; Santos, H.A. Microfluidic assembly of monodisperse multistage pH-responsive polymer/porous silicon composites for precisely controlled multi-drug delivery. *Small* **2014**, *10*, 2029–2038. [CrossRef]
40. Gidwani, B.; Vyas, A. A Comprehensive Review on Cyclodextrin-Based Carriers for Delivery of Chemotherapeutic Cytotoxic Anticancer Drugs. *Biomed. Res. Int.* **2015**, *2015*, 198268. [CrossRef]

41. Hernández-Montelongo, J.; Oria, L.; Cárdenas, A.B.; Benito, N.; Romero-Sáez, M.; Recio-Sánchez, G. Nanoporous Silicon Composite as Potential System for Sustained Delivery of Florfenicol Drug. *Phys. Status Solidi Basic Res.* **2018**, *255*. [CrossRef]
42. Papadimitriou, S.; Bikiaris, D. Novel self-assembled core-shell nanoparticles based on crystalline amorphous moieties of aliphatic copolyesters for efficient controlled drug release. *J. Control. Release* **2009**, *138*, 177–184. [CrossRef] [PubMed]
43. Gibaldi, M.; Feldman, S. Establishment of sink conditions in dissolution rate determinations. Theoretical considerations and application to nondisintegrating dosage forms. *J. Pharm. Sci.* **1967**, *56*, 1238–1242. [CrossRef] [PubMed]
44. Higuchi, T. Mechanism of sustained-action medication. Theoretical analysis of rate of release of solid drugs dispersed in solid matrices. *J. Pharm. Sci.* **1963**, *52*, 1145–1149. [CrossRef] [PubMed]
45. Korsmeyer, R.W.; Gurny, R.; Doelker, E.; Buri, P.; Peppas, N.A. Mechanisms of solute release from porous hydrophilic polymers. *Int. J. Pharm.* **1983**, *15*, 25–35. [CrossRef]
46. Wu, S.H.; Lin, H.P. Synthesis of mesoporous silica nanoparticles. *Chem. Soc. Rev.* **2013**, *42*, 3862–3875. [CrossRef] [PubMed]
47. Meraz, K.A.S.; Vargas, S.M.P.; Maldonado, J.T.L.; Bravo, J.M.C.; Guzman, M.T.O.; Maldonado, E.A.L. Eco-friendly innovation for nejayote coagulation-flocculation process using chitosan: Evaluation through zeta potential measurements. *Chem. Eng. J.* **2016**, *284*, 536–542. [CrossRef]
48. Lian, T.; Peng, M.; Vermorken, A.J.M.; Jin, Y.; Luo, Z.; Van de Ven, W.J.M.; Wan, Y.; Hou, P.; Cui, Y. Synthesis and Characterization of Curcumin-Functionalized HP-beta-CD-Modified GoldMag Nanoparticles as Drug Delivery Agents. *J. Nanosci. Nanotechnol.* **2016**, *16*, 6258–6264. [CrossRef] [PubMed]
49. Hernandez-Montelongo, J.; Naveas, N.; Degoutin, S.; Tabary, N.; Chai, F.; Spampinato, V.; Ceccone, G.; Rossi, F.; Torres-Costa, V.; Manso-Silvan, M.; et al. Porous silicon-cyclodextrin based polymer composites for drug delivery applications. *Carbohydr. Polym.* **2014**, *110*, 238–252. [CrossRef] [PubMed]
50. McIntosh, I.M.; Nichols, A.R.L.; Tani, K.; Llewellin, E.W. Accounting for the species-dependence of the 3500 cm-1H2Otinfrared molar absorptivity coefficient: Implications for hydrated volcanic glasses. *Am. Mineral.* **2017**, *102*, 1677–1689. [CrossRef]
51. Singh, P.; Chauhan, K.; Priya, V.; Singhal, R.K. A greener approach for impressive removal of As(III)/As(v) from an ultra-low concentration using a highly efficient chitosan thiomer as a new adsorbent. *RSC Adv.* **2016**, *6*, 64946–64961. [CrossRef]
52. Alhosseini, S.N.; Moztarzadeh, F.; Mozafari, M.; Asgari, S.; Dodel, M.; Samadikuchaksaraei, A.; Kargozar, S.; Jalali, N. Synthesis and characterization of electrospun polyvinyl alcohol nanofibrous scaffolds modified by blending with chitosan for neural tissue engineering. *Int. J. Nanomed.* **2012**, *7*, 25–34.
53. Jafary, F.; Panjehpour, M.; Varshosaz, J.; Yaghmaei, P. Stability improvement of immobilized alkaline phosphatase using chitosan nanoparticles. *Braz. J. Chem. Eng.* **2016**, *33*. [CrossRef]
54. Salonen, J.; Laitinen, L.; Kaukonen, A.M.; Tuura, J.; Björkqvist, M.; Heikkilä, T.; Vähä-Heikkilä, K.; Hirvonen, J.; Lehto, V.P. Mesoporous silicon microparticles for oral drug delivery: Loading and release of five model drugs. *J. Control. Release* **2005**, *108*, 362–374. [CrossRef] [PubMed]
55. Perruchot, C.; Khan, M.A.; Kamitsi, A.; Armes, S.P.; Von Werne, T.; Patten, T.E. Synthesis of well-defined, polymer-grafted silica particles by aqueous ATRP. *Langmuir* **2001**, *17*, 4479–4481. [CrossRef]
56. Świsłocka, R. Spectroscopic (FT-IR, FT-Raman, UV absorption, 1H and 13C NMR) and theoretical (in B3LYP/6-311++G** level) studies on alkali metal salts of caffeic acid. *Spectrochim. Acta A Mol. Biomol. Spectros* **2013**, *100*, 21–30.
57. Skiba, M.; Lahiani-Skiba, M. Novel method for preparation of cyclodextrin polymers: Physico-chemical characterization and cytotoxicity. *J. Incl. Phenom. Macrocycl. Chem* **2013**, *74*, 341–349. [CrossRef]
58. Perrie, Y.; Rades, T. *FASTtrack Pharmaceutics: Drug Delivery and Targeting*; Pharmaceutical Press: Philadelphia, PA, USA, 2010; ISBN 9780857110596.
59. Carmeliet, P.; Jain, R.K. Angiogenesis in cancer and other diseases. *Nature* **2000**, *407*, 249–257. [CrossRef] [PubMed]
60. Chovatiya, R.; Medzhitov, R. Stress, inflammation, and defense of homeostasis. *Mol. Cell* **2014**, *54*, 281–288. [CrossRef] [PubMed]

© 2019 by the authors. Licensee MDPI, Basel, Switzerland. This article is an open access article distributed under the terms and conditions of the Creative Commons Attribution (CC BY) license (http://creativecommons.org/licenses/by/4.0/).

Article

# Retinol-Containing Graft Copolymers for Delivery of Skin-Curing Agents

Justyna Odrobińska [1], Katarzyna Niesyto [1], Karol Erfurt [2], Agnieszka Siewniak [2], Anna Mielańczyk [1] and Dorota Neugebauer [1,*]

[1] Department of Physical Chemistry and Technology of Polymers, Faculty of Chemistry, Silesian University of Technology, 44-100 Gliwice, Poland
[2] Department of Chemical Organic Technology and Petrochemistry, Faculty of Chemistry, Silesian University of Technology, 44-100 Gliwice, Poland
* Correspondence: dneugebauer@polsl.pl

Received: 28 June 2019; Accepted: 25 July 2019; Published: 2 August 2019

**Abstract:** The new polymeric systems for delivery in cosmetology applications were prepared using self-assembling amphiphilic graft copolymers. The synthesis based on "click" chemistry reaction included grafting of azide-functionalized polyethylene glycol (PEG-N$_3$) onto multifunctional polymethacrylates containing alkyne units. The latter ones were obtained via atom transfer radical polymerization (ATRP) of alkyne-functionalized monomers, e.g., ester of hexynoic acid and 2-hydroxyethyl methacrylate (AlHEMA) with methyl methacrylate (MMA), using bromoester-modified retinol (RETBr) as the initiator. Varying the content of alkyne moieties adjusted by initial monomer ratios of AlHEMA/MMA was advantageous for the achievement of a well-defined grafting degree. The designed amphiphilic graft copolymers P((HEMA-*graft*-PEG)-*co*-MMA), showing tendency to micellization in aqueous solution at room temperature, were encapsulated with arbutin (ARB) or vitamin C (VitC) with high efficiencies (>50%). In vitro experiments carried out in the phosphate-buffered saline solution (PBS) at pH 7.4 indicated the maximum release of ARB after at least 20 min and VitC within 10 min. The fast release of the selected antioxidants and skin-lightening agents by these micellar systems is satisfactory for applications in cosmetology, where they can be used as the components of masks, creams, and wraps.

**Keywords:** retinol; "click" chemistry; alkyne–azide reaction; ATRP; graft copolymers; amphiphilic copolymers; micellar carriers

---

## 1. Introduction

Innovative drug delivery systems (DDS) with polymeric carriers are designed to prolong and improve the action of biologically active substances, including pharmaceuticals, in the body, providing controlled and targeted therapies [1–4]. These polymers should be non-toxic, non-immunogenic, biocompatible with optional biodegradability, and chemically inert [5]. The well-fitted structures of well-defined polymers are synthesized to achieve the desired drug loading and release with efficient concentration at a proper time.

Biopolymers, such as chitosan [6,7], hyaluronic acid [8,9], collagen [10], or dextran [11], with the ability to aggregate, are often used as carriers in cosmetology. Among the synthesized amphiphilic polymers, the most common are those based on 2-hydroxyethyl methacrylate (HEMA) [12], *N*-isopropylacrylamide [13], 2-(diethylamino)ethyl methacrylate [14], methacrylic acid [15–17], and polyethylene glycol (PEG) [14,18–20]. These are mostly block copolymers [21], including star [22] and graft [23–25] topologies, obtained by the controlled polymerization methods for DDS applications. Specific graft copolymers [26] were achieved by combination of the backbone with side chains, which can be introduced by grafting from [27,28] and grafting through [29,30], or via a combination of both to attain block side chains [15], heterografted structures [31], or brush–block–brush [20].

The excellent biocompatibility, hydrophilicity, good blood compatibility, high water content, and permeability of the HEMA-based polymers [32] resulted in them finding numerous applications as biomaterials [33], for example, hydrogels [34], for manufacturing contact lenses with drug delivery [35], or artificial implants [36]. The presence of the hydroxyl group in HEMA is advantageous for modification using pre- or post-polymerization reactions. Pre-polymerization modification is usually performed to convert a hydroxyl group into another with particular properties, e.g., a bromoester group initiating an atom transfer radical polymerization (ATRP) reaction [37] or a trimethylsilyl protecting group [31,38]. Another opportunity is the preparation of azide-functionalized monomers [39] or amine-functionalized monomers via an alkyne–azide "click" chemistry reaction with the formation of a triazole ring [40]. However, the "click" approach is commonly applied after the polymerization, when the polymer is modified with "clickable" groups (post-polymerization modification). The introduction of alkyne groups into the polymer was helpful in further functionalization with specific groups, e.g., pentafluorobenzyl [41], or oligomers, e.g., polyhedral oligomeric silsesquioxanes [42], or to attach polymer chains, e.g., PEG-N$_3$ (grafting onto) [43].

In our previous work, HEMA-based polymers were successfully synthesized with the use of bromoester-functionalized retinol as a novel ATRP bioinitiator [44]. Retinol (vitamin A) is a well-known factor which stimulates collagen and glycosaminoglycan synthesis, supporting the reduction of wrinkles, acne, and hyperpigmentation. The aim of the current work was to prepare the alkyne-functionalized HEMA-based polymers for the "click" chemistry reaction. The alkyne group was introduced into the HEMA monomer (AlHEMA) via the esterification of the hydroxyl group with hexynoic acid (pre-polymerization modification), whereas the retinol-initiated alkyne-functionalized polymethacrylates resulting from ATRP were modified by grafting onto via an alkyne–azide "click" reaction to obtain the amphiphilic copolymers with PEG side chains. The used strategy is different from that reported in the literature because it provided polymers with adjustable amounts of alkyne groups (by proper ratios of AlHEMA to comonomer) as the guaranteed "click" sites. The grafted copolymers with self-assembling abilities were also examined for the encapsulation of active substances for skin treatment, such as arbutin or vitamin C, to show delivery activities of these potential systems for cosmetology.

## 2. Experimental

*2.1. Materials*

Methyl methacrylate (MMA, 99%, Alfa Aesar, Warsaw, Poland), 2-hydroxyethyl methacrylate (HEMA, 97%, Aldrich, Poznań, Poland), and anisole (99%, Alfa Aesar) were dried over molecular sieves and stored in a freezer under nitrogen. Copper (I) bromide (CuBr, 98%, Fluka, Steinheim, Germany) was purified by stirring in glacial acetic acid, followed by filtration and washing with ethanol and diethyl ether. After that, the solids were dried under vacuum. Additionally, 4,4-dinonyl-2,2-dipyridyl (dNdpy, 97%, Aldrich), N,N,N',N'',N''-pentamethyldiethylenetriamine (PMDTA, 98%, Aldrich), triethylamine (TEA, 99%, Aldrich), pyridine (99%, Aldrich), 2-bromoisobutyryl bromide (BriBuBr, 98%, Aldrich), ethyl α-bromoisobutyrate (EiB-Br, Aldrich, 98%), 5-hexynoic acid (HexA, 97%, Acros, Geel, Belgium) all-*trans*-retinol (RET, 95%, Acros), poly(ethylene glycol)methyl ether 2-bromoisobutyrate (PEG-Br, $M_n$ = 1200 g/mol, Aldrich), sodium azide (NaN$_3$, 99%, Acros), N,N'-dicyclohexylcarbodiimide (DCC, 99%, Acros), 4-dimethylaminopyridin (DMAP, 99%, Acros), N,N-dimethylformamide (DMF, 99%, Chempure, Piekary Śląskie, Poland), L(+)-ascorbic acid (VitC, 99%, Chempure), arbutin (ARB, 95%, Acros), and a 0.1 M sodium phosphate buffer solution (PBS; pH = 7.4, Aldrich) were used as received. All other chemicals were applied without purification.

## 2.2. Synthesis of Alkyne-Functionalized HEMA (2-(Prop-1-En-2-Carbonyloxy)Ethyl Hex-5-Ynate, AlHEMA)

HEMA (3.00 mL, 24.67 mmol) and DCC (5.67 g, 27.48 mmol) were dissolved into a 250-mL round-bottom flask with 50 mL of methylene chloride, yielding a colorless solution. Then, hexynoic acid (2.80 g, 24.97 mmol) was added dropwise to the solution. The reactor was cooled to 0 °C in an ice/water bath, and DMAP (0.1397 g, 1.14 mmol) in methylene chloride (2 mL) was added dropwise. The reaction mixture was stirred for 48 h at room temperature. After that, it was transferred into a separator with methylene chloride and extracted with $H_2O$ to neutral pH in the aqueous fraction. The organic phase was removed by rotary evaporation. The brown liquid product was dried under vacuum to constant mass. Yield: 61%. $^1$H-NMR (300 MHz, CDCl$_3$, ppm): 6.14 and 5.61 (2H, =CH$_2$), 4.35 (4H, –OCH$_2$CH$_2$O–), 2.52 (2H, –OC(=O)CH$_2$–), 2.28 (2H, –CH$_2$-C≡CH), 1.99 (1H, –C≡CH), 1.95 (3H, –CH$_3$), 1.81 (2H, –OC(=O)CH$_2$CH$_2$–). $^{13}$C-NMR (300 MHz, DMSO, ppm) (Supplementary Materials Figure S1): 172 (C7, –OC(=O)CH$_2$–), 166 (C4, –CC(=O)O), 136 (C2, CH$_2$=C–), 126 (C1, CH$_2$=C–), 83 (C11, –C≡CH), 72 (C12, –C≡CH), 63 (C5, –OCH$_2$CH$_2$O–), 62 (C6, –OCH$_2$CH$_2$O–), 32 (C8, –OC(=O)CH$_2$–), 27 (C9, –OC(=O)CH$_2$CH$_2$–), 18 (C10, –CH$_2$-C≡CH), 17 (C3, –CH$_3$). Electrospray ionization (ESI) MS (m/z): calculated for C$_{12}$H$_{16}$O$_4$, 224.0; found for [M + Na]$^+$, 247.1 (Supplementary Materials Figure S2).

## 2.3. Synthesis of 2-Bromoisobutyrate Derivative of Retinol (3,7-Dimethyl-9-(2,6,6-Trimethylcyclohex-1-En-1-Yl)Nona-2,4,6,8-Tetraen-1-yl 2-Bromo-2-Methylpropanoate, RET-Br)

The RET bioinitiator was prepared with a yield of 95% by esterification with BriBuBr (small excess in relation to molar amount of OH groups) in the presence of TEA according to a previously reported procedure [44]. $^1$H-NMR (600 MHz, DMSO, ppm): 6.50–6.60 (2H, 2* =CH–), 6.25–3.35 (1H, =CH–), 6.10–6.20 (2H, 2* =CH–), 5.60–5.65 (1H, =CH–), 4.09–4.12 (2H, –O–CH$_2$–), 1.98–2.02 (2H, –CH$_2$– ring), 2.00 (6H, –(CH$_3$)$_2$Br), 1.90–1.92 (3H, –CH$_3$ ring), 1.75–1.80 (3H, –CH$_3$ aliphat.), 1.65–1.70 (3H, –CH$_3$ aliphat.), 1.52–1.60 (2H, –CH$_2$– ring), 1.40–1.48 (2H, –CH$_2$– ring), 0.95–1.02 (6H, 2*-CH$_3$ ring). $^{13}$C-NMR (300 MHz, DMSO, ppm) (Figure S4, Supplementary Materials): 171 (C18, –OC=O), 137.7 (C10, =CH–), 137.6 (C11, –C(CH$_3$)–), 136.3 (C8, C14, –C(CH)= ring, =CH– ), 131 (C6, C12, =C(CH$_3$)– ring, =CH–), 126.8 (C9, –CH=), 126.4 (C15, =CH–), 124.5 (C13, –CH=), 80 (C16, –CH$_2$–O), 62 (C19, –C–Br), 46 (C3, –CH$_2$– ring), 34 (C2, –C(CH$_3$)$_2$– ring), 30 (C5, –CH$_2$– ring), 29 (2*C1, –CH$_3$ ring), 27 (C20, –(CH$_3$)$_2$Br), 21 (C4, –CH$_2$– ring), 19 (C7, –CH$_3$ ring), 9 (2*C17, –CH$_3$ aliphat.). ESI-MS (m/z): calculated for C$_{24}$H$_{35}$O$_2$Br, 434.9; found for [M + H]$^+$, 435.1 (Supplementary Materials Figure S5).

## 2.4. Synthesis of P(AlHEMA-co-MMA) with EiB-Br as Initiator (Example for III)

MMA (0.24 mL, 2.24 mmol), AlHEMA (1.50 mL, 6.70 mmol), anisole (0.5 mL, 30 vol.% of monomer), and PMDTA (4.66 µL, 0.022 mmol) were placed in a Schlenk flask and then degassed by two freeze–pump–thaw cycles. Then, EiB-Br (3.31 µL, 0.022 mmol) was added and degassed again. After that, CuCl (2.21 mg, 0.022 mmol) was added. The reaction flask was immersed in an oil bath at 60 °C. The polymerization was stopped by exposure to air. Then, the mixture was dissolved in acetone and passed through a neutral alumina column to remove the copper catalyst. The solution was concentrated by rotary evaporation. The polymer was precipitated by dropwise addition of a concentrated solution into diethyl ether. The product was isolated by decantation and dried under vacuum to constant mass.

## 2.5. Synthesis of P(AlHEMA-co-MMA) with RET-Br as Initiator (Example for VI)

RET-Br (14.17 mg, 0.033 mmol), dNdpy (26.70 mg, 0.065 mmol), MMA (0.35 mL, 3.27 mmol), AlHEMA (2.17 g, 9.69 mmol), and anisole (0.25 mL, 10 vol.% of monomer) were placed in a Schlenk flask and then degassed by three freeze–pump–thaw cycles. After that, CuBr (4.60 mg, 0.032 mmol) was added. The reaction flask was immersed in an oil bath at 60 °C. The next steps were performed according to above-described procedure for the synthesis of P(AlHEMA-co-MMA) with EiB-Br (Section 2.4).

## 2.6. Synthesis of P(HEMA-co-MMA) (VII–IX)

The series of HEMA-based copolymers with various compositions (HEMA/MMA = 75/25, 50/50, 25/75) were synthesized by ATRP (Table S1, Supplementary Materials) as reported earlier [44]. It was a similar procedure as for AlHEMA, where the copolymerization of HEMA and MMA was performed with the use of RET-Br in the ratio to monomer 1:400 and a CuBr/dNdpy 0.75/1.5 catalyst system in anisole at 60 °C. $^1$H-NMR (300 MHz, DMSO, ppm): 6.00–6.25 (2H$_{mon}$, –C$\underline{H}_2$–), 5.60–5.75 (2H$_{mon}$, –C$\underline{H}_2$–), 4.75–5.00 (1H, –O$\underline{H}$–), 4.00–4.20 (2H$_{mon}$, –C$\underline{H}_2$OH), 3.75–4.00 (2H$_{pol}$ –C$\underline{H}_2$OH), 3.75 (2H$_{mon}$, –COC$\underline{H}_2$CH$_2$OH), 3.65 (3H$_{mon}$, –OC$\underline{H}_3$), 3.50–3.60 (3H$_{pol}$, –OC$\underline{H}_3$; 2H$_{pol}$, –COC$\underline{H}_2$CH$_2$OH), 1.90–2.00 (3H$_{mon}$, –C$\underline{H}_3$; 2H$_{pol}$, –C$\underline{H}_2$– main chain), 0.50–1.50 (3H$_{pol}$, –C$\underline{H}_3$).

## 2.7. Synthesis of Poly(Ethylene Glycol)Methyl Ether 2-Azidoisobutyrate (PEG-N$_3$)

PEG-Br (1 g, 0.83 mmol) and NaN$_3$ (54.16 mg, 0.83 mmol) were dissolved in a 100-mL round-bottom flask with 20 mL of anhydrous DMF. The reaction mixture was stirred for 24 h at room temperature. After that, it was transferred into a separator with dichloromethane and extracted with NaHCO$_{3(aq)}$. The organic phase was removed by rotary evaporation. The brown liquid product was dried under vacuum to constant mass. Yield: 88%. $^1$H-NMR (300 MHz, DMSO, ppm) (Supplementary Materials Figure S8): 3.50 (n*4H, –[OCH$_2$CH$_2$]$_n$–), 3.24 (3H, –OCH$_3$), 1.88 (6H, –C(CH$_3$)$_2$N$_3$). $^{13}$C-NMR (300 MHz, DMSO, ppm) (Supplementary Materials Figure S9): 167 (C4, –OC(=O)–), 75 (C2 and C3, –OCH$_2$CH$_2$O–), 65 (C1, –OCH$_3$), 60 (C5, –C(CH$_3$)$_2$N$_3$), 41 (C6, –C(CH$_3$)$_2$N$_3$).

## 2.8. "Click" Chemistry Azide–Alkyne Reactions (Example for IVc)

Polymer IV (0.37 g, 0.03 mmol containing 0.696 mmol of AlHEMA units) was dissolved into a 100-mL round-bottom flask with 10 mL of DMF. Then, the equimolar amount of PEG-N$_3$ (0.90 g, 0.70 mmol) and 2.5-fold molar excess of PMDETA (0.36 mL, 1.74 mmol) were added. The reaction mixture was purged with an inert gas for 20 min. After that, CuBr (0.25 g, 1.74 mmol) was added and the reaction mixture was stirred for 48 h at room temperature without access to light. The mixture was purified from CuBr by means of cationite (Dowex) and concentrated by rotary evaporation. The product was precipitated in diethyl ether and dried under vacuum to constant mass. The reaction efficiency was calculated by integral area of the CH proton in the triazole ring (8.01 ppm, H$_M$) and the ≡CH proton from AlHEMA units that were not clicked (1.9–2.0 ppm, H$_J$) using the following equation:

$$E_{click} = \frac{H_M}{H_M + H_J} \times 100\%.$$

## 2.9. Incorporation of Active Substance into Polymeric Micelles

The amphiphilic copolymer and active substance were dissolved in methanol with the weight ratio of polymer to bioactive substance = 1:1; then, H$_2$O was added dropwise (200 vol.% of the solvent) under gentle stirring. The reaction was continued overnight. After that, the vial with sample was opened to evaporate the organic solvent. The sample was centrifuged to separate the unloaded active substance (4000 rpm for 10 min in room temperature), which was not dissolved. Next, the homogeneous aqueous fraction was collected and lyophilized by freezing. A solution of loaded micelles in MeOH (0.008 mg/mL) was prepared to determine the amount of entrapped substances by ultraviolet–visible light (UV–Vis) spectroscopy, measuring absorbance at λ = 282 nm for ARB and λ = 267 nm for VitC. Drug loading content (DLC) was calculated using the following equation:

$$DLC = \frac{\text{Weight of drug loaded into micelle}}{\text{Weight of total polymer and loaded drug}} \times 100\%.$$

*2.10. Active Substance Release Studies*

The loaded micelles were dissolved in PBS (pH = 7.4, 1.0 mg/mL). The solution was introduced into a dialysis cellulose membrane bag (molecular weight cut-off (MWCO) = 3.5 kDa), which was placed into glass vial with 50 mL of PBS and stirred at 37 °C in a water bath. The buffer solution sample (2.0 mL) was taken from the release medium, at appropriate time intervals, to determine the concentration of released drug by UV–Vis spectroscopy, measuring absorbance at λ = 282 nm for ARB and λ = 267 nm for VitC.

*2.11. Characterization*

$^1$H- and $^{13}$C-NMR spectra were recorded with a UNITY/INOVA (Varian) spectrometer operating at 300 MHz using dimethyl sulfoxide (DMSO) or CDCl$_3$ as a solvent and tetramethylsilane (TMS) as an internal standard. The monomer conversion was determined by gas chromatography (GC, Agilent Technologies 6850 Network GC System, Santa Clara, USA). The measurements were carried out in acetone as the solvent. The signals at different retention times corresponded to MMA (2.3 min), HEMA (8.5 min), AlHEMA (10.0 min), and anisole (4.9 min). Mass spectrometry (MS, Xevo G2 QTof, Waters Corporation, Milford, USA) was used to confirm the molecular masses of the modified retinol and functionalized HEMA. Molecular weights ($M_n$) and dispersity indices (Đ) were determined by gel permeation chromatography (GPC) equipped with an 1100 Agilent isocratic pump, autosampler, degasser, thermostatic box for columns, and differential refractometer MDS RI Detector. The measurements were carried out in tetrahydrofuran (THF) as the solvent at 30 °C with a flow rate of 0.8 mL/min. The GPC calculations were based on calibration with the use of linear polystyrene standards (580–300,000 g/mol). Fourier-transform infrared spectroscopy (FT-IR) was conducted with Perkin-Elmer Spectrum Two 1000 FT-IR Infrared Spectrometer using attenuated total reflection (ATR). The critical micelle concentration (CMC) was measured by fluorescence spectrophotometry (FL, Hitachi F-2500, Tokyo, Japan), using pyrene as a fluorescence probe. Excitation spectra of pyrene (λ = 390 nm) were recorded at a constant concentration of pyrene (3.0 × 10$^{-4}$ mol/L) and polymer concentrations in the range of 5 × 10$^{-4}$ to 1.0 mg/mL. The intensity ratio ($I_{336}/I_{332}$) from the pyrene excitation spectrum vs. logC (where C is the concentration in mg/mL) was plotted, where the cross-over point was estimated as the CMC value. The particle sizes and their distributions, that is, hydrodynamic diameter ($D_h$) and polydispersity index (PDI), were measured at 25 °C using dynamic light scattering (DLS, Zetasizer Nano-S90, Malvern Technologies). Each experiment was repeated three times to obtain the average value. The samples taken during the release process were analyzed by ultraviolet–visible light spectroscopy (UV–Vis, Thermo Fisher Scientific Evolution 300) to determine the DLC and the amount of released substance over time. The measurements were carried out in poly(methyl methacrylate) cells. DLC measurements for double-encapsulated systems were carried out using ultra high-performance liquid chromatography–mass spectrometry (UPLC–MS). Analysis was conducted on an ACQUITY UPLC system (Waters) equipped with an ACQUITY photodiode array (PDA) detector and a Waters ACQUITY UPLC®BEH C18 column (2.1 × 50 mm, 1.7 mm).

## 3. Results and Discussion

A strategy combining the controlled atom transfer radical polymerization (ATRP) and the Cu(I) catalyzed 1,3-dipolar azide-alkyne cycloaddition (CuAAC) was applied in the synthesis of amphiphilic graft copolymers with hydrophilic side chains, e.g., P((HEMA-*graft*-PEG)-*co*-MMA). A few-step procedure, which is presented in Figure 1, included (i) azidation of PEG, (ii) modification of HEMA to alkyne-functionalized monomer (AlHEMA), (iii) its copolymerization with MMA in the presence of different initiators (standard EiB-Br or RET-Br), (iv) the "click" reaction between P(AlHEMA-*co*-MMA) and PEG-N$_3$. Varying the content of alkyne groups (recognized as the "clicking" moieties) regulated by MMA units was advantageous to adjust the grafting degree of hydrophilic PEG, whereas a differential hydrophilic–hydrophobic balance influenced the behavior in aqueous solution. The self-assembly and delivery of the selected bioactive substances by the grafted copolymers were compared with the systems of linear amphiphilic copolymers of HEMA and MMA (combined with various proportions).

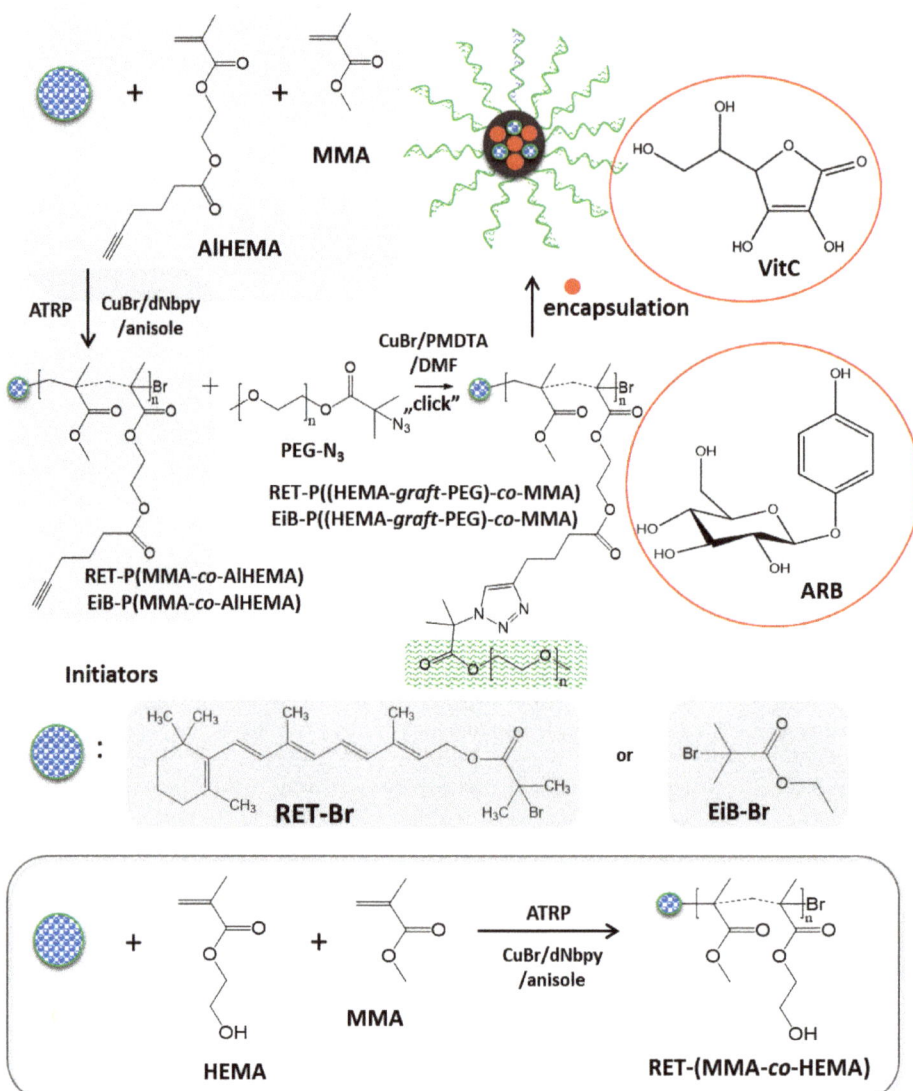

Figure 1. Synthesis of amphiphilic graft copolymers via grafting onto.

The alkyne derivative of HEMA (AlHEMA) was obtained by the coupling reaction via esterification between the OH group of HEMA and hexynoic acid. The structure of the resultant AlHEMA was confirmed by $^1$H-NMR, showing the shifted methylenoxy signals **C** and **D** at 3.61 and 4.11 ppm (Figure 2a) as the signal **F** at 4.35 ppm (Figure 2b) due to neighborhood changes following the formation of an ester group and introduction of the hexynoic moiety (Figure 2b: 2H, **H**: 1.8–1.9 ppm; 1H, **J**: 2.0 ppm; 2H, **I**: 2.2–2.4 ppm, and 2H, **G**: 2.4–2.6 ppm).

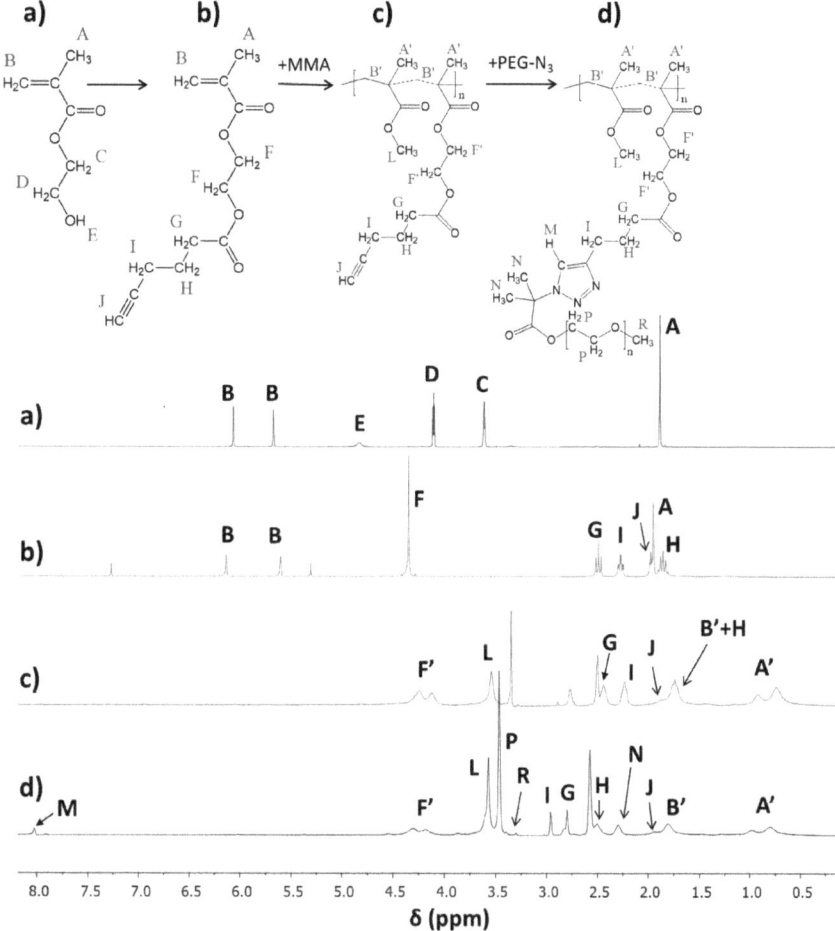

**Figure 2.** $^1$H-NMR spectra of (**a**) HEMA, and (**b**) after its modification AlHEMA, (**c**) P(MMA-*co*-AlHEMA) V, (**d**) P((AlHEMA-*graft*-PEG)-*co*-MMA) Vc.

Similarly, the retinol was modified by esterification to an ATRP bioinitiator with bromoester functionality. In the $^1$H-NMR spectrum, the signal of the hydroxyl group (Supplementary Materials Figure S3a; **p**: 4.7 ppm) disappeared after modification, whereas the signal of the methylene group in –CH$_2$OH was shifted (Supplementary Materials Figure S3a,b; **o**: from 4.1 to 4.8 ppm) due to the presence of the ester group. The successful modification was also confirmed by the $^{13}$C-NMR spectra (Supplementary Materials Figure S4) containing the signals, which corresponded to >C=O in the introduced ester group (**C18**: 170 ppm), carbon associated with the ester group (**C16**: 80 ppm), and tertiary carbon bonded to bromine (**C19**: 63 ppm). The lack of a broad band in the region of 3100–3600 cm$^{-1}$ corresponding to ν(O–H) stretching in the FT-IR spectra of esterified RET, the presence of the additional peak at 1260 cm$^{-1}$ from the stretching vibration of C–O, the strong peak at 1730 cm$^{-1}$ from the stretching vibration of C=O groups, and the strong peak at 800 cm$^{-1}$ from the stretching vibration of C–Br were evidence for the newly created bromoester group (Supplementary Materials Figure S6).

The prepared bromoester-functionalized retinol (RET-Br) and commercially available standard ATRP initiator, e.g., EiB-Br, were applied in the copolymerization of AlHEMA in the presence of a

CuBr/dNbpy or CuCl/PMDTA catalyst system in anisole at 60 °C (Table 1). The range of initiators was broadened by RET-Br to develop the biocompatibility of polymers, but it also motivated characterizing its influence on the reaction rate and properties of the copolymers, including the structural parameters (i.e., degree of polymerization (DP), dispersity index (Đ)) in comparison to that obtained with the use of EiB-Br. The introduction of alkyne groups into monomer prior to the polymerization reaction guaranteed that the alkyne functionality was contained in all HEMA units incorporated into the polymer chain, whereas, in the case of the post-polymerization modification of HEMA-based polymers, it was strongly dependent on the esterification efficiency. The used strategy of pre-polymerization modification was beneficial for the adjustment of a certain number of alkyne groups in the copolymer by the initial proportions of AlHEMA/MMA comonomers (25/75, 50/50, 75/25). The alkyne moieties in copolymers were observed on the FT-IR spectra (Figure 3b) as peaks at 550–700 cm$^{-1}$ and 3300 cm$^{-1}$ from the bending and stretching vibrations of ≡C–H, respectively. It was in contrast to the hydroxy-functionalized polymers (Figure 3a), which revealed a broad band in the region of higher values of wavelengths corresponding to $\nu$(O–H) stretching in HEMA units.

**Table 1.** Data for synthesis of AlHEMA/MMA copolymers by ATRP.

| | $M_1/M_2$ | Time (h) | Conversion (%) | | | | $DP_{n,GC}$ | $M_{n,GC}$ (g/mol) | $M_n$ [a] (g/mol) | Đ [a] |
| | | | NMR | | GC | | | | | |
| | | | $M_1$ | $M_2$ | $M_1$ | $M_2$ | | | | |
|---|---|---|---|---|---|---|---|---|---|---|
| I | 25/75 | 4.0 | 16 | 28 | 34 | 27 | 115 | 15,900 | 21,400 | 1.36 |
| II | 50/50 | 4.5 | 48 | 50 | 47 | 52 | 198 | 31,600 | 33,600 | 2.06 |
| III | 75/25 | 4.5 | 46 | 45 | 42 | 47 | 173 | 33,100 | 35,500 | 1.83 |
| IV | 25/75 | 24 | 33 | 29 | 23 | 22 | 89 | 12,200 | 25,400 | 1.68 |
| V | 50/50 | 24 | 27 | 15 | 18 | 17 | 71 | 12,000 | 17,800 | 1.72 |
| VI | 75/25 | 24 | 32 | 24 | 26 | 30 | 109 | 21,000 | 30,400 | 1.65 |

I: [AlHEMA+MMA]$_0$/[EiB-Br]$_0$/[CuBr]$_0$/[dNdpy]$_0$ = 400/1/0.75/1.5; II–III: [AlHEMA+MMA]$_0$/[EiB-Br]$_0$/[CuCl]$_0$/[PMDTA]$_0$ = 400/1/1/1; IV–VI: [AlHEMA+MMA]$_0$/[RET-Br]$_0$/[CuBr]$_0$/[dNdpy]$_0$ = 400/1/1/2; anisole 10 vol.% of mon., 60 °C; [a] determined by GPC in THF with polystyrene standards.

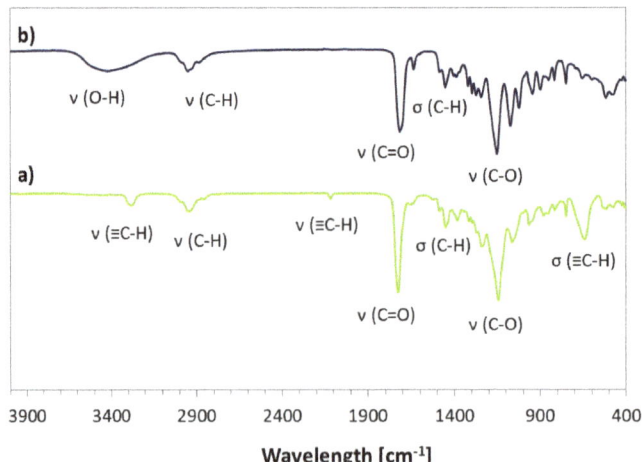

**Figure 3.** Fourier-transform infrared (FT-IR) spectra for copolymers (a) P(AlHEMA-co-MMA) III and (b) P(HEMA-co-MMA) IX.

The conversion of AlHEMA was calculated from the $^1$H-NMR spectrum for the reaction mixture (Supplementary Materials Figure S7) using protons in methylene groups via integration of signals corresponding to the monomer (CH$_2$=, **B**: 5.6 and 6.2 ppm) and polymer (–COO–CH$_2$–, **F'**: 4.1–4.2 ppm).

The determination of MMA conversion was based on protons in the methoxy group via integration of signals corresponding to the monomer (**L**: 3.75 ppm) and polymer (**L′**: 3.6 ppm). However, for further calculations of DP and molecular weight ($M_n$), the conversion by GC was selected due to the very good separation of signals, which was not always possible in the case of $^1$H-NMR analysis. Comparable values of monomer conversions (AlHEMA vs. MMA) allowed concluding the formation of statistical copolymers. The detailed data are summarized in Table 1, whereas the representative $^1$H-NMR spectrum of the purified copolymer P(AlHEMA-co-MMA) is presented in Figure 2c.

Dependent on the initiator (EiB-Br vs. RET-Br), the polymerization progress was characterized with different rates and total monomer conversions. The use of a standard EiB-Br initiator resulted in higher conversions within a shorter time compared to RET-Br (45% within 4.5 h vs. 25% within 24 h). However, the biological and non-toxic nature of the retinol starting unit can be beneficial for the improvement of skin treatment, whereas the resulting conversions are sufficient to obtain copolymers with the desired properties. In the case of HEMA polymerization initiated by RET-Br (Supplementary Materials Table S1), the reactions were significantly faster than the series of AlHEMA/RET-Br giving similar conversions in a shorter time (IV vs. VII = 23/22 vs. 18/18; VI vs. IX = 26/30 vs. 26/32 at 24 h vs. 0.5–4.5 h), which is rational due to more polar HEMA-based systems. The obtained copolymers were characterized by moderate dispersity indices usually exceeding 1.5, but their GPC traces were monomodal and symmetrical, showing some broadness and discrepancy with the increase in the content of alkyne or hydroxyl moieties (Figure 4).

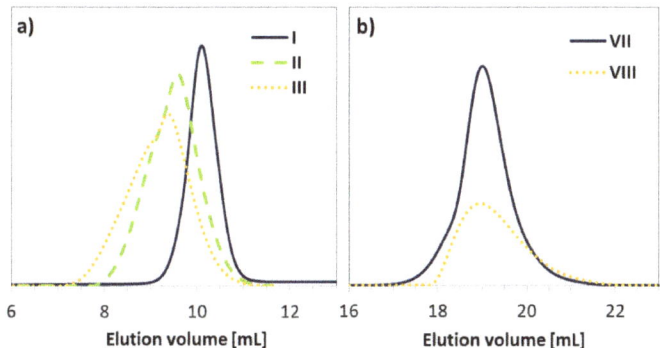

**Figure 4.** GPC traces for copolymers (**a**) P(AlHEMA-co-MMA) and (**b**) P(HEMA-co-MMA).

*Click Reactions*

A commercially available low-molecular-weight bromoester monofunctionalized PEG was converted by the substitution reaction of the bromine atom with an azide group. In the $^1$H-NMR spectrum, the signals coming from methyl groups adjacent to –Br or –$N_3$ were placed in the same range of chemical shifts giving multiplets at 1.82–1.92 ppm or a broad signal at 1.86–1.90 ppm, respectively (Supplementary Materials Figure S8). Because of the non-synonymous approval, the final identification of the end bromoester group in PEG was provided by $^{13}$C-NMR analysis, which confirmed the achievement of the azidation reaction. The signal derived from carbon of the carboxyl group was shifted toward lower chemical shifts (**C4**: 172 ppm to 168 ppm) due to the impact of a new group attached to the adjacent carbon (**C5**: 63 ppm to 60 ppm), whereas the signals of the methylene groups moved toward higher chemical shifts (**C6**: 37 ppm to 40 ppm) (Supplementary Materials Figure S9). The molecular weight of the PEG derivative determined by GPC was consistent with the weight calculated from $^1$H-NMR, giving similar values of $M_n$ for PEG-$N_3$ (1380 vs. 1300 g/mol by NMR vs. GPC, respectively) (Supplementary Materials Figure S10). Additionally, the impurities contained in the commercial PEG were removed by purification of the modified PEG, which resulted in the reduction of the dispersity index determined by GPC.

The hydrophilic PEG-N$_3$ chains were grafted onto the multifunctional P(MMA-co-AlHEMA) by a Huisgen "click" chemistry CuAAC reaction between azide and alkyne moieties catalyzed by CuBr/PMDTA in DMF with the formation of 1,4-substituted triazole rings (Figure 1). It was found that dispersity indices of the resulted amphiphilic graft copolymers were reduced in comparison to their backbones before the "click" reaction (Table 2, Figure 5). Generally, the discrepancy of $M_{n,GPC}$ from $M_{n,NMR}$ was more significant than for the linear copolymers because of the lower hydrodynamic volumes of nonlinear macromolecules, which were applied as standards for the calibration in GPC. Additionally, higher-molecular-weight distribution can influence such a deviation.

**Table 2.** Data for the synthesis of graft copolymers P((HEMA-*graft*-PEG)-*co*-MMA).

| | $DP_{AlHEMA}$ | $F_{AlHEMA}$ (%) | $E_{click}$ (%) | $n_{triazole}$ | DG (%) | $F_{hydrophilic}$ (wt%) | $M_{n,NMR}$ (g/mol) | $M_{n,GPC}$ [a] (g/mol) | Đ [a] | CMC (mg/mL) |
|---|---|---|---|---|---|---|---|---|---|---|
| Ic | 34 | 30 | 33 | 11 | 10 | 47 | 28,900 | 58,600 | 3.40 | 0.0159 |
| IIc | 94 | 47 | 31 | 29 | 15 | 54 | 69,300 | 24,100 | 1.34 | 0.0169 |
| IIIc | 126 | 73 | 27 | 34 | 20 | 57 | 77,300 | 36,500 | 1.50 | 0.0229 |
| IVc | 23 | 26 | 21 | 5 | 6 | 35 | 17,200 | 29,000 | 1.62 | 0.0836 |
| VIc | 79 | 72 | 64 | 51 | 47 | 76 | 75,000 | 35,900 | 1.42 | 0.4283 |

[a] Determined by GPC in THF with polystyrene standards; $DP_{AlHEMA}$—polymerization degree of AlHEMA; $F_{AlHEMA}$ and $F_{hydrophilic}$—content of AlHEMA and hydrophilic fraction in the copolymer, respectively; $E_{click}$—efficiency of "click" reaction; $n_{triazole}$—number of triazole moieties in the copolymer; DG—grafting degree.

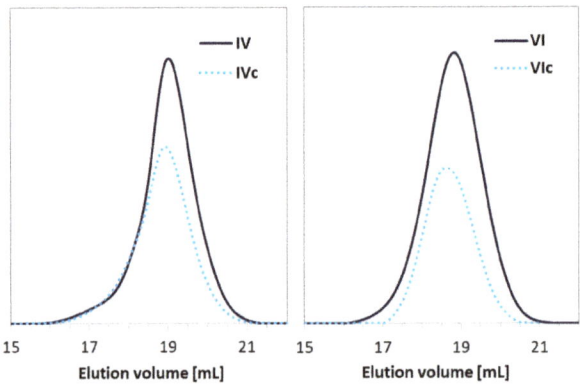

**Figure 5.** GPC traces for copolymers before (IV, VI) and after "click" reaction (IVc, VIc).

The presence of a triazole proton at 8.01 ppm ($H_M$) and three groups of signals from PEG (methoxy group at 3.29 ppm ($H_R$), methylene groups at 3.45 ppm ($H_P$), and methyl groups at 2.30 ppm ($H_N$)) in the $^1$H-NMR spectrum validated the success of the "click" reaction (Figure 2d). In the series Ic–IIIc, independently of the amount of alkyne groups, the "click" efficiency ($E_{click}$) reached the level of 30% (Table 2). It was related to grafting of 11–34 side PEG chains (synonymous with the number of triazole rings ($n_{triazole}$)), which was adjusted by the content of alkyne groups per backbone ($F_{AlHEMA}$). Various grafting degrees (DGs) also corresponded to an almost equimolar content of the introduced hydrophilic fraction (47–57 wt.%). Significant differences were observed for the RET series characterized with the lowest (IVc) and the highest (VIc) "click" efficiencies (21% vs 64%). The retinol-initiated copolymers IV and VI used for the "click" reaction were varied from the analogical I and III not only by the starting group (RET vs. EiB), but also with their structural parameters ($F_{AlHEMA}$ as the result of $DP_{AlHEMA}$ in relation to the chain length). A two-fold lower number of "clickable" alkyne moieties in the copolymer IV than in sample I ($DP_{AlHEMA}$ = 79 vs. 136) at a similar content of alkyne groups ($F_{AlHEMA}$ ~72%) appeared to be more effective in the PEG grafting.

Chemical structures of the graft copolymers were also confirmed using $^{13}$C-NMR and FT-IR spectroscopies. The presence of signals from carbons of the triazole ring ($C_1$, 130 ppm; $C_{10}$, 140 ppm) and the carbon signal of the carboxyl group at the PEG chain (176 ppm) indicated the success of the "click" reaction (Figure 6). In the FT-IR spectrum, the broad band in region of 3000–3600 cm$^{-1}$ corresponding to ν(N–H) stretching and the strong peak at 1650 cm$^{-1}$ from ν(N=N) in the triazole ring were observed (Figure 7).

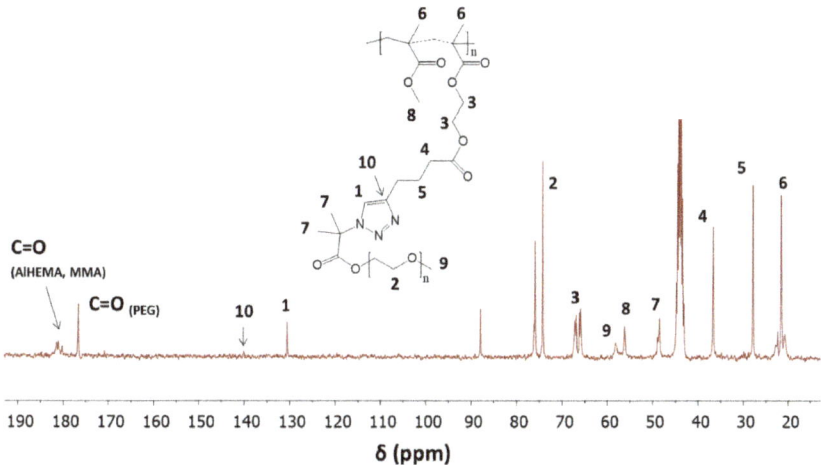

**Figure 6.** $^{13}$C-NMR of graft copolymer P((HEMA-*graft*-PEG)-*co*-MMA) by "click" reaction.

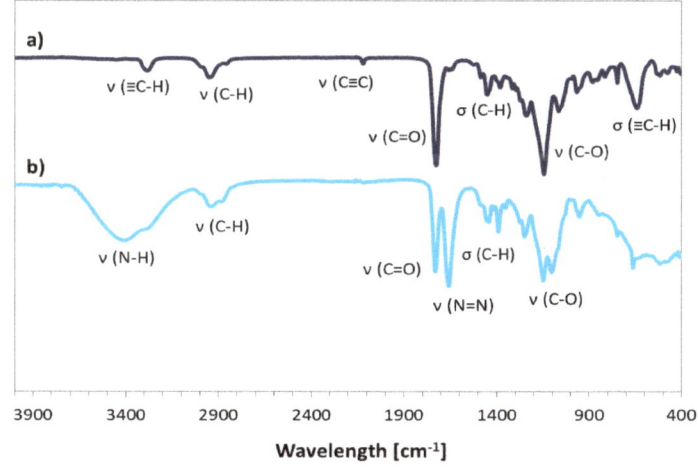

**Figure 7.** FT-IR spectra for P(AlHEMA-*co*-MMA) (a) and P((HEMA-*graft*-PEG)-*co*-MMA) (b).

The self-assembling behaviors of the graft copolymers were investigated by determination of the critical micelle concentration (CMC, Table 2). CMC values of copolymers were measured by a standard procedure using the emission spectra of pyrene to form the plot of $I_{336}/I_{332}$ vs. the logarithm of the copolymer concentration (Supplementary Materials Figure S11). It was noted that the CMC value decreased with the increase in hydrophobic fraction in the range of series (Ic–IIIc, and IVc vs. VIc), which is the general relationship in the self-assembly process. The highest CMC value was observed for the copolymer with a predominated content of hydrophilic fraction (VIc), which was generated by the highest degree of PEG grafting, demonstrating the lowest ability for micellization compared to the other systems. Copolymers Ic–IIIc with equimolar hydrophilic/hydrophobic fractions (47–57 wt.%) showed similar CMC values in the range of 0.01–0.02 mg/mL. Previously investigated linear hydroxyl-functionalized copolymers of P(HEMA-co-MMA) self-assembled at lower concentrations (0.002–0.04 mg/mL, Supplementary Materials Table S1) [44] compared to grafted copolymers with analogical backbones, which confirmed that the presence of PEG side chains improved the system solubility.

Satisfactory results obtained in the case of encapsulation of cosmetic substances (ferulic acid, VitC) into micelles of the linear HEMA-based copolymers [44] encouraged us to investigate the self-assembling graft polymers in the presence of ARB or VitC. These model bioactive substances are well-known components in cosmetics due to their antioxidant and skin-lightening activities. ARB prevents the formation of melanin-avoiding skin diseases, such as melanoma, and it is used as replacement of hydroquinone, whereas VitC inhibits the influence of free radicals, stimulates collagen synthesis, and is used with α-tocopherol for a synergistic effect. The efficiency of the single-drug encapsulation (performed in the ratio of polymer to drug 1:1) was verified by drug loading content (DLC, Table 3), which was determined by the use of UV–Vis spectra. ARB was encapsulated in larger amounts than VitC by grafted copolymers (~90% vs. ~15%), which was in contrast to the linear copolymers (~50–75% vs. ~80%). There was no effect of the grafting degree (Ic–IIIc, IVc vs. VIc) on the DLC, although the RET-based series was more efficient in ARB encapsulation than the EiB one (~90% vs. ~55%). The loading results allowed concluding that the structure of copolymer, including topology and localization of hydrophilic moieties in the polymer, and the nature of active substance are crucial factors when designing the encapsulated systems with an optimized amount of the cosmetic substance. The encapsulation of two active substances (VitC and ARB) at the same time was also attempted. The DLC calculations were performed by UHPLC–MS measurements because the bands of VitC and ARB overlapped in the UV–Vis spectrum. However, the results indicated that only ARB was encapsulated. This means that these systems are not sufficient for the dual delivery of these two specific active substances, whereas double encapsulation may be successful for other bioactive pairs.

Table 3. Characteristics of encapsulated particles and release of active substances.

| | $D_h$ [a] (nm) | | PDI | | DLC (%) | | Maximum Amount of Released Drug (%)/Time (h) | |
|---|---|---|---|---|---|---|---|---|
| | ARB | VitC | ARB | VitC | ARB | VitC | ARB | VitC |
| Ic | 427 | - | 0.508 ± 0.028 | - | 64 | - | 90/2.5 | - |
| IIc | 80 [b] | - | 0.717 ± 0.018 | - | 49 | - | 90/0.3 | - |
| IIIc | 7 [b] | - | 0.983 ± 0.016 | - | 56 | - | 99/0.5 | - |
| IVc | 420 | 369 | 0.241 ± 0.046 | 0.200 ± 0.063 | 99 | 16 | 94/1.5 | 88/0.17 |
| VIc | 310 [b] | 377 [b] | 0.634 ± 0.007 | 0.621 ± 0.085 | 87 | 13 | 65/4.5 | 49/0.17 |
| VII | 681 | 834 [c] | 0.093 ± 0.053 | 0.261 ± 0.081 | 55 | 87 [c] | 86/5.0 | 62/1.0 [c] |
| VIII | 316 [b] | 250 [c] | 0.904 ± 0.002 | 0.281 ± 0.060 | 48 | 78 [c] | 100/0.8 | 48/1.0 [c] |
| IX | 222 [b] | - | 0.691 ± 0.086 | - | 75 | - | 96/2.5 | - |

[a] Particle size distribution data based on intensity calculation method; [b] non-dominated fraction—the averaged value for the major fractions above 80%; [c] data presented in Reference [44]; Ic–VIc: graft copolymers; VII–IX: linear copolymers.

The resulting ARB or VitC encapsulated particles were analyzed by DLS in PBS solution (Table 3). Micelles containing ARB obtained from the grafted copolymers (Ic–VIc) were smaller compared to their linear counterparts (VII–IX). There was no significant effect for samples Ic and IVc differing with the initial unit (EiB vs. RET) and similar DG, which yielded one fraction of micelles; however, at a higher content of side chains, more fractions of the superstructures with different hydrodynamic diameters were observed (Figure 8). A few generations of loaded particles may indicate the presence of unimers, as well as the formation of micelles and aggregates (Figure 9, Supplementary Materials

Figure S12). Similar correlations were detected for the systems based on the linear amphiphilic analogs, although the size differences were more spectacular.

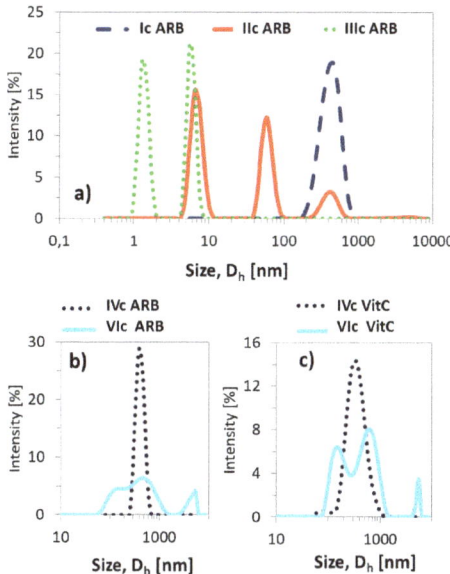

**Figure 8.** Size distribution plots by intensity for ARB- (**a,b**) or VitC (**c**) loaded polymer micelles in PBS at 25 °C.

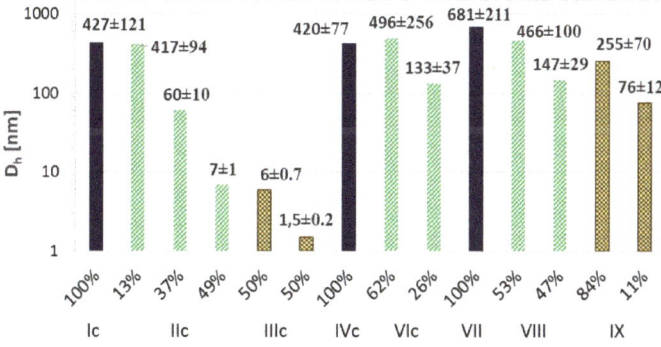

**Figure 9.** Particle size distribution data for ARB-loaded micellar systems based on intensity calculation method.

The release experiments were carried out in PBS at pH 7.4 for the ARB- or VitC-loaded systems, which demonstrated various release rates of bioactive substance dependent on the hydrophilic/hydrophobic balance and topology of carriers, including the length of the main chain/backbone (Table 3, Figure 10). The kinetic profiles showed a tendency of increasing rate of ARB release with the decrease in hydrophilic content for series of graft copolymers with backbone DPs ~100 (VIc: ($F_{hydrophilic}$ = 76%) ARB release 35%, Ic: (47%) 50%, and IVc: (35%) 60% within 30 min). However, this correlation was not valid for IIc and IIIc, showing significantly faster drug release than system Ic with a comparable amount of hydrophilic fraction (~100% vs. 50% within 30 min); however, contrary to the other graft copolymers, they contained two-fold longer backbones contributing to micelle core formation. For almost all systems based on the graft copolymers, the release of ARB was completed at ~90–100% with the exception of VIc with the largest grafting degree and the highest hydrophilic content. This discrepancy can be explained by the formation of micelles with a thicker outer layer of hydrophilic PEG, which decelerated the drug release. A similar effect was observed for VitC, although its release was faster than for ARB (for IVc 88% VitC vs. 32% ARB and for VIc 49% VitC vs. 14% ARB within 10 min, Figure 10b). In the case of linear hydroxyl-functionalized copolymers, the release of ARB ranged from 86–100% and it was strongly dependent on the chain length, showing the fastest release by the longest chain with an equimolar content of hydrophilic fraction (VIII DP = 136) and the slowest release by the shortest chains with a predominated hydrophobic fraction (VII DP = 73) (Figure 10c). The randomly distributed hydrophilic HEMA units along the polymeric chains were responsible for different correlations between release rate and hydrophilic content from that observed for the graft polymers containing side PEG segments. Surprisingly, both ARB and VitC were released faster from the graft copolymer systems than from their linear analogues. However, a short release time is beneficial from the point of view of applying the designed systems in cosmetic products due to the short time of application on the skin.

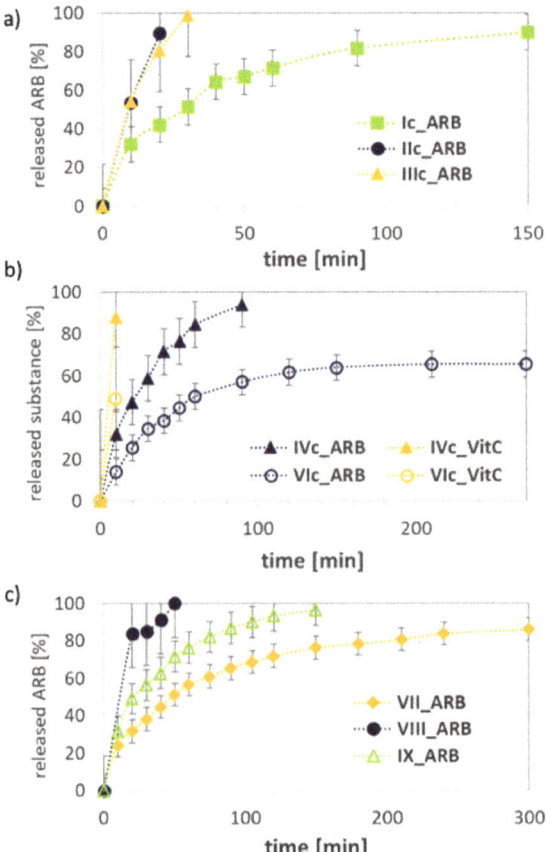

**Figure 10.** Release profiles for micellar systems formed by graft copolymers (**a,b**) and linear copolymers (**c**).

The release kinetics of ARB was also described by fitting to mathematical models, which were represented as semilogarithmic plots of remaining drug vs. time according to the first-order equation, and plots of the cumulative amount of released drug vs. square root of time according to the Higuchi model. Both types of plots (Supplementary Materials Figures S13 and S14) demonstrated good correlation coefficients, that is, 0.89–0.99 for Ic–VIc (graft copolymers), 0.93–0.99 for VII–IX (linear copolymers), 0.85–0.99 for Ic–VIc, and 0.87–0.96 for VII–IX. These results confirmed the concentration-dependent and diffusion-controlled mechanism.

## 4. Conclusions

Pre-polymerization modification was applied to obtain the alkyl-functionalized monomer originating from HEMA, which was used for copolymerization via ATRP with bromoester initiators, including the modified retinol. The pre-polymerization strategy provided much better control of the number of alkyne groups in the copolymer, which could be modified into the amphiphilic graft copolymer via a "click" reaction between the alkyne functionality in the HEMA-based copolymer and the azide-functionalized PEG. The self-assembling behavior in aqueous solution at room temperature was employed to encapsulate ARB or VitC into micelles with a relatively high efficiency for almost all systems (DLC > 50%). In vitro release was carried out indicating the maximum amount of released ARB after 20 min (up to 5 h) and VitC after 10 min. With respect to both encapsulation and release studies, the PEG graft copolymers seem to be good candidates for potential delivery applications. The micellar systems with a short release time (up to 30 min) can be effective in face masks, whereas the other ones delivering bioactive substances over a longer time are perfect for cream applications. All these systems need to be tested for toxicity and diffusion through artificial skin to verify their application in cosmetology (masks, under-eye patches, and wraps).

**Supplementary Materials:** The following are available online at http://www.mdpi.com/1999-4923/11/8/378/s1. Data for synthesis of P(HEMA-co-MMA) copolymers; $^1$H-NMR spectra of retinol before and after esterification, PEG-Br and PEG-N$_3$, reaction mixture of copolymerization AlHEMA/MMA; $^{13}$C-NMR spectra of AlHEMA, RET-Br, PEG-Br, and PEG-N$_3$; ESI-MS spectra of AlHEMA and RET-Br; FT-IR spectra and GPC traces for PEG-Br and PEG-N$_3$; CMC plots; particle size distribution for VitC-loaded systems; kinetic plots.

**Author Contributions:** Conceptualization, D.N.; data curation, J.O. and K.N.; formal analysis, J.O., K.N., K.E., A.S., and A.M.; funding acquisition, J.O.; investigation, J.O. and K.N.; methodology, D.N.; project administration, J.O. and D.N.; writing—original draft, J.O.; writing—review and editing, D.N.; supervision, D.N.

**Funding:** This work was financed from budget funds for science in the years 2017–2020 as a research project under the "Diamond Grant" program supported by the Ministry of Science and Higher Education (MNiSW, Poland), ID number: DI2016 008246.

**Conflicts of Interest:** The authors declare no conflict of interest.

## References

1. Xu, X.; Ho, W.; Zhang, X.; Bertrand, N.; Farokhzad, O. Cancer nanomedicine: From targeted delivery to combination therapy. *Trends Mol. Med.* **2015**, *21*, 223–232. [CrossRef] [PubMed]
2. Yu, X.; Trase, I.; Ren, M.; Duval, K.; Guo, X.; Chen, Z. Design of Nanoparticle-Based Carriers for Targeted Drug Delivery. *J. Nanomater.* **2016**, *2016*, 1087250. [CrossRef] [PubMed]
3. Habibi, N.; Kamaly, N.; Memic, A.; Shafiee, H. Self-assembled peptide-based nanostructures: Smart nanomaterials toward targeted drug delivery. *Nano Today* **2016**, *11*, 41–60. [CrossRef] [PubMed]
4. Rodzinski, A.; Guduru, R.; Liang, P.; Hadjikhani, A.; Stewart, T.; Stimphil, E.; Runowicz, C.; Cote, R.; Altman, N.; Datar, R.; et al. Targeted and controlled anticancer drug delivery and release with magnetoelectric nanoparticles. *Sci. Rep.* **2016**, *6*, 20867. [CrossRef] [PubMed]
5. Srivastava, A.; Yadav, T.; Sharma, S.; Nayak, A.; Kumari, A.; Mishra, N. Polymers in Drug Delivery. *J. Biosci. Med.* **2016**, *4*, 69–84. [CrossRef]
6. Mady, M.M.; Darwish, M.M.; Khalil, S.; Khalil, W.M. Biophysical studies on chitosan-coated liposomes. *Eur. Biophys. J.* **2009**, *38*, 1127–1133. [CrossRef]
7. Prado, A.G.S.; Santos, A.L.F.; Pedroso, C.P.; Carvalho, T.O.; Braga, L.R.; Evangelista, S.M. Vitamin A and vitamin E interaction behavior on chitozan microspheres. *J. Therm. Anal. Calorim.* **2011**, *16*, 415–420. [CrossRef]

8. Creuzet, C.; Kadi, S.; Rinaudo, M.; Auzély-Velty, R. New associative systems based on alkylated hyaluronic acid. Synthesis and aqueous solution properties. *Polymer* **2006**, *47*, 2706–2713. [CrossRef]
9. Olejnik, A.; Goscianska, J.; Zielinska, A.; Nowak, I. Stability determination of the formulations containing hyaluronic acid. *Int. J. Cosmet. Sci.* **2015**, *37*, 401–407. [CrossRef]
10. Avila Rodriguez, M.I.; Rodriguez Barroso, L.G.; Sanchez, M.L. Collagen: A review on its sources and potential cosmetic applications. *J. Cosmet. Dermatol.* **2018**, *17*, 20–26. [CrossRef]
11. Parisi, O.I.; Malivindi, R.; Amone, F.; Ruffo, M.; Malanchin, R.; Carlomagno, F.; Piangiolino, C.; Nobile, V.; Pezzi, V.; Scrivano, L.; et al. Safety and efficacy of dextran-rosmarinic acid conjugates as innovative polymeric antioxidants in skin whitening: What Is the Evidence? *Cosmetics* **2017**, *4*, 28. [CrossRef]
12. Li, Y.-Y.; Zhang, X.-Z.; Cheng, H.; Zhu, J.-J.; Cheng, S.-X.; Zhuo, R.-X. Self-Assembled, Thermosensitive PCL-g-P(NIPAAm-co-HEMA) Micelles for Drug Delivery. *Macromol. Rapid Commun.* **2006**, *27*, 1913–1919. [CrossRef]
13. Zhang, J.; Xu, X.D.; Wu, D.Q.; Zhang, X.Z.; Zhuo, R.X. Synthesis of thermosensitive P(NIPAAm-co-HEMA)/cellulose hydrogels via. "click" chemistry. *Carbohydr. Polym.* **2009**, *77*, 583–589. [CrossRef]
14. Yang, Y.Q.; Zhao, B.; Li, Z.D.; Lin, W.J.; Zhang, C.Y.; Guo, X.D.; Wang, J.F.; Zhang, L.J. PH-sensitive micelles self-assembled from multi-arm star triblock-co-polymers poly(e-caprolactone)-b-poly(2-(diethylamino) ethylmethacrylate)-b-poly(poly(ethylene glycol)methyl ether methacrylate) for controlled anticancer drug delivery. *Acta Biomater.* **2013**, *9*, 7679–7690. [CrossRef] [PubMed]
15. Cheng, G.L.; Boker, A.; Zhang, M.F.; Krausch, G.; Muller, A.H.E. Amphiphilic cylindrical core-shell brushes via. a "grafting from" process using ATRP. *Macromolecules* **2001**, *34*, 6883–6888. [CrossRef]
16. Neugebauer, D.; Odrobińska, J.; Bielas, R.; Mielańczyk, A. Design of systems based on 4-armed star-shaped polyacids for indomethacin delivery. *New J. Chem.* **2016**, *40*, 10002–10011. [CrossRef]
17. Chang, L.L.; Liu, J.J.; Zhang, J.H.; Deng, L.D.; Dong, A.J. PH-sensitive nanoparticles prepared from amphiphilic and biodegradable methoxy poly(ethylene glycol)-block-(polycaprolactone-graft-poly(methacrylic acid)) for oral drug delivery. *Polym. Chem.* **2013**, *4*, 1430–1438. [CrossRef]
18. Ruan, G.; Feng, S.S. Preparation and characterization of poly(lactic acid)–poly(ethylene glycol)–poly(lactic acid) (PLA–PEG–PLA) microspheres for controlled release of paclitaxel. *Biomaterials* **2003**, *24*, 5037–5044. [CrossRef]
19. Maksym-Bebenek, P.; Neugebauer, D. Study on Self-Assembled Well-Defined PEG Graft Copolymers as Efficient Drug-Loaded Nanoparticles for Anti-Inflammatory Therapy. *Macromol. Biosci.* **2015**, *15*, 1616–1624. [CrossRef]
20. Ishizu, K.; Satoh, J.; Sogabe, A. Architecture and solution properties of AB-type brush-block-brush amphiphilic copolymers via ATRP techniques. *J. Colloid Interface Sci.* **2004**, *274*, 472–479. [CrossRef]
21. Wolf, F.; Friedemann, N.; Frey, H. Poly(lactide)-block-Poly(HEMA) Block Copolymers: An Orthogonal One-Pot Combination of ROP and ATRP, Using a Bifunctional Initiator. *Macromolecules* **2009**, *42*, 5622–5628. [CrossRef]
22. Huang, L.-M.; Li, L.-D.; Shang, L.; Zhou, Q.-H.; Lin, J. Preparation of pH-sensitive micelles from miktoarm star block copolymers by ATRP and their application as drug nanocarriers. *React. Funct. Polym.* **2016**, *107*, 28–34. [CrossRef]
23. Cheng, R.; Wang, X.; Chen, W.; Meng, F.; Deng, C.; Liu, H.; Zhong, Z. Biodegradable poly(3-caprolactone)-g-poly(2-hydroxyethyl methacrylate) graft copolymer micelles as superior nano-carriers for "smart" doxorubicin release. *J. Mater. Chem.* **2012**, *22*, 11730–11738. [CrossRef]
24. Gromadzki, D.; Stepanek, P.; Makuska, R. Synthesis of densely grafted copolymers with tert-butyl methacrylate/2-(dimethylamino ethyl) methacrylate side chains as precursors for brush polyelectrolytes and polyampholytes. *Mater. Chem. Phys.* **2013**, *137*, 709–715. [CrossRef]
25. Bury, K.; Du Prez, F.; Neugebauer, D. Self-assembling linear and star shaped poly(ε-caprolactone)/poly[(meth)acrylic acid] block copolymers as carriers of indomethacin and quercetin. *Macromol. Biosci.* **2013**, *13*, 1520–1530. [CrossRef]
26. Neugebauer, D. Two decades of molecular brushes by ATRP. *Polymer* **2015**, *72*, 413–421. [CrossRef]
27. Neugebauer, D.; Bury, K.; Biela, T. Novel Hydroxyl-Functionalized Caprolactone Poly(meth)acrylates Decorated with tert-Butyl Groups. *Macromolecules* **2012**, *45*, 4989–4996. [CrossRef]
28. Maksym-Bębenek, P.; Biela, T.; Neugebauer, D. Water soluble well-defined acidic graft copolymers based on a poly(propylene glycol) macromonomer. *RSC Adv.* **2015**, *5*, 3627–3635. [CrossRef]

29. Cai, Y.L.; Hartenstein, M.; Muller, A.H.E. Synthesis of amphiphilic graft copolymers of n-butyl acrylate and acrylic acid by atom transfer radical copolymerization of macromonomers. *Macromolecules* **2004**, *37*, 7484–7490. [CrossRef]
30. Neugebauer, D.; Zhang, Y.; Pakula, T.; Sheiko, S.S.; Matyjaszewski, K. Densely-Grafted and Double-Grafted PEO Brushes via ATRP. A Route to Soft Elastomers. *Macromolecules* **2003**, *36*, 6746–6755. [CrossRef]
31. Neugebauer, D.; Zhang, Y.; Pakula, T.; Matyjaszewski, K. Heterografted PEO-PnBA brush copolymers. *Polymer* **2003**, *44*, 6863–6871. [CrossRef]
32. Neugebauer, D. Modifications of Hydroxyl-Functionalized HEA/HEMA and Their Polymers in the Synthesis of Functional and Graft Copolymers. *Curr. Org. Synth.* **2017**, *14*, 1–12. [CrossRef]
33. Montheard, J.P.; Chatzopoulos, M.; Chappard, D. 2-Hydroxyethyl Methacrylate (HEMA): Chemical Properties and Applications in Biomedical Fields. *J. Macromol. Sci. C* **1992**, *32*, 1–34. [CrossRef]
34. Passos, M.F.; Dias, D.R.C.; Bastos, G.N.T.; Jardini, A.L.; Benatti, A.C.B.; Dias, C.G.B.T.; Filho, R.M. PHEMA hydrogels. Synthesis, kinetics and in vitro tests. *J. Therm. Anal. Calorim.* **2016**, *125*, 361–368. [CrossRef]
35. Li, C.C.; Chauhan, A. Ocular transport model for ophthalmic delivery of timolol through p-HEMA contact lenses. *J. Drug Deliv. Sci. Technol.* **2007**, *17*, 69–79. [CrossRef]
36. Chirila, T.V.; Constable, I.J.; Crawford, G.J.; Vijayasekaran, S.; Thompson, D.E.; Chen, Y.-C.; Fletcher, W.A.; Griffin, B.J. Poly(2-hydroxyethyl methacrylate) sponges as implant materials: In Vivo and In Vitro evaluation of cellular invasion. *Biomaterials* **1993**, *14*, 26–38. [CrossRef]
37. Ling, J.; Zheng, Z.; Köhler, A.; Müller, A.H.E. Rod-Like Nano-Light Harvester. *Macromol. Rapid Commun.* **2014**, *35*, 52–55. [CrossRef]
38. Beers, K.L.; Gaynor, S.G.; Matyjaszewski, K.; Sheiko, S.S.; Moller, M. The Synthesis of Densely Grafted Copolymers by Atom Transfer Radical Polymerization. *Macromolecules* **1998**, *31*, 9413–9415. [CrossRef]
39. Sumerlin, B.S.; Tsarevsky, N.V.; Louche, G.; Lee, R.Y.; Matyjaszewski, K. Highly efficient "click" functionalization of poly(3-azidopropyl methacrylate) prepared by ATRP. *Macromolecules* **2005**, *38*, 7540–7545. [CrossRef]
40. Mishra, V.; Jung, S.-H.; Park, J.M.; Jeong, H.M.; Lee, H.-I. Triazole-Containing Hydrogels for Time-Dependent Sustained Drug Release. *Macromol. Rapid Commun.* **2014**, *35*, 442–446. [CrossRef]
41. Erol, F.E.; Sinirlioglu, D.; Cosgun, S.; Muftuoglu, A.E. Synthesis of Fluorinated Amphiphilic Block Copolymers Based on PEGMA, HEMA, and MMA via ATRP and CuAAC Click Chemistry. *Int. J. Polym. Sci.* **2014**, *2014*, 464806. [CrossRef]
42. Bach, L.G.; Islam, R.; Nga, T.T.; Binh, M.T.; Hong, S.S.; Gal, Y.S.; Taek, L.K. Chemical Modification of Polyhedral Oligomeric Silsesquioxanes by Functional Polymer via. Azide-Alkyne Click Reaction. *J. Nanosci. Nanotechnol.* **2013**, *13*, 1970–1973. [CrossRef] [PubMed]
43. Gao, H.; Matyjaszewski, K. Synthesis of Molecular Brushes by "Grafting onto" Method: Combination of ATRP and Click Reactions. *J. Am. Chem. Soc.* **2007**, *129*, 6633–6639. [CrossRef] [PubMed]
44. Odrobińska, J.; Neugebauer, D. Retinol derivative as bioinitiator in the synthesis of hydroxyl-functionalized polymethacrylates for micellar delivery systems. *Express Polym. Lett.* **2019**, *13*, 806–817. [CrossRef]

© 2019 by the authors. Licensee MDPI, Basel, Switzerland. This article is an open access article distributed under the terms and conditions of the Creative Commons Attribution (CC BY) license (http://creativecommons.org/licenses/by/4.0/).

*Article*

# Modulation of the Release of a Non-Interacting Low Solubility Drug from Chitosan Pellets Using Different Pellet Size, Composition and Numerical Optimization

Ioannis Partheniadis [1], Paraskevi Gkogkou [1], Nikolaos Kantiranis [2] and Ioannis Nikolakakis [1],*

[1] Department of Pharmaceutical Technology, School of Pharmacy, Faculty of Health Sciences, Aristotle University of Thessaloniki, 54124 Thessaloniki, Greece; ioanpart@pharm.auth.gr (I.P.); gkogkop@gmail.com (P.G.)
[2] Department of Mineralogy-Petrology-Economic Geology, School of Geology, Faculty of Sciences, Aristotle University of Thessaloniki, 54124 Thessaloniki, Greece; kantira@geo.auth.gr
* Correspondence: yannikos@pharm.auth.gr; Tel.: +30-2310997635

Received: 22 February 2019; Accepted: 8 April 2019; Published: 10 April 2019

**Abstract:** Two size classes of piroxicam (PXC) pellets (mini (380–550 μm) and conventional (700–1200 μm)) were prepared using extrusion/spheronization and medium viscosity chitosan (CHS). Mixture experimental design and numerical optimization were applied to distinguish formulations producing high sphericity pellets with fast or extended release. High CHS content required greater wetting liquid volume for pellet formation and the diameter decreased linearly with volume. Sphericity increased with CHS for low-to-medium drug content. Application of PXRD showed that the drug was a mixture of form II and I. Crystallinity decreased due to processing and was significant at 5% drug content. Raman spectroscopy showed no interactions. At pH 1.2, the dissolved CHS increased 'apparent' drug solubility up to 0.24 mg/mL while, at pH 5.6, the suspended CHS increased 'apparent' solubility to 0.16 mg/mL. Release at pH 1.2 was fast for formulations with intermediate CHS and drug levels. At pH 5.6, conventional pellets showed incomplete release while mini pellets with a CHS/drug ratio ≥2 and up to 21.25% drug, showed an extended release that was completed within 8 h. Numerical optimization provided optimal formulations for fast release at pH 1.2 with drug levels up to 40% as well as for extended release formulations with drug levels of 5% and 10%. The Weibull model described the release kinetics indicating complex or combined release (parameter '$b$' > 0.75) for release at pH 1.2, and normal diffusion for the mini pellets at pH 5.6 ('$b$' from 0.63 to 0.73). The above results were attributed mainly to the different pellet sizes and the extensive dissolution/erosion of the gel matrix was observed at pH 1.2 but not at pH 5.6.

**Keywords:** pellets 1; pellet diameter 2; crystallinity 3; sphericity 4; fast release 5; extended release 6

## 1. Introduction

Pharmaceutical pellets are multi-particulate drug delivery systems where the whole dose is divided into subunits with spherical shape and narrow particle size distribution within the range of 0.1 to 1.5 mm [1]. This multi-particulate presentation has several advantages over the tablet single-unit dosage form [2]. Pellets distribute uniformly in the gastrointestinal tract, which results in less variability in gastric emptying time, lower plasma level fluctuation, improved drug absorption, and a reduction of dose dumping. Additionally, their flowability enables processing of an automatic fast operation capsule and tableting machines, and their spherical shape makes them ideal for application of coatings [3–5].

For the preparation of pellets, extrusion/spheronization can be applied [6,7]. Product quality is controlled mainly by the composition while machine settings are less critical [8]. Microcrystalline cellulose (MCC) is usually part of the pellet base [9] due to its ability to retain large amounts of water in its structure, which provides elasto-plastic wet mass suitable for successful extrusion and good product quality [10–12]. However, in certain cases, incompatibilities of MCC have been reported because of its tendency to adsorb drugs on fibrils [13] and cause possible chemical interactions [14–16]. In addition, its use is obstructed by the prolonged and uncontrolled release of poorly soluble drugs [17]. An alternative is to replace part of the MCC with hydrophilic polymers, which aims to facilitate and control penetration of the aqueous dissolution medium into the pellet matrix by increasing hydrophilicity and swelling [18].

Chitosan (CHS) is a natural polysaccharide product of the deacetylation of chitin, which is a widely abundant polysaccharide. It dissolves in weakly acidic media by protonation of –NH2 groups, which forms a non-disintegrating gel matrix at high concentrations in water [19]. Its pharmaceutical importance as a functional excipient lies in its biocompatibility, biodegradability, and non-toxicity [20,21]. Additionally, it enhances the solubility of drugs, their permeation through the gastric mucosa, aids gastric protection due to its potent cytoprotective and healing action in gastric ulcers, and acts as a sustain-release agent [22–25]. Its effect on dissolution is influenced by the degree of deacetylation, the viscosity grade, its content in the formulations, and the drug solubility [19,23,26,27]. At low contents, it may act as a disintegrant, but, at high contents, it forms a hydrophilic gel, which delays drug release [28–30]. Due to its function as a 'molecular sponge', it can be used as an alternative to MCC [31]. Piroxicam (PXC) is a non-steroidal, anti-inflammatory, anti-rheumatoid, and analgesic drug [32,33] assigned to Biopharmaceutical Classification System (BCS) Class II due to its poor solubility in water [34]. Since CHS is positively charged in acidic media due to the protonation of amine [35,36] and PXC is cationic (pKa 5.3, Reference [35]) and is not ionized in deionized water, chemical interactions between these two are not expected and any effects on release should be due mainly to the contribution of the individual components.

The aim of this work was to prepare different CHS/MCC/PXC pellet formulations of two size classes (mini and conventional) by extruding through screens with small (0.5 mm) or large (1.0 mm) openings. Different CHS viscosity grades were initially compared and the one that prompted greater drug solubility was selected. It was expected that, due to gel formation and diffusional release, a reduction of pellet size will improve the release rate, and, thus, avoid the need to add hydrophilic excipients such as lactose [30]. However, extrusion through small orifice screens may adversely affect pellet shape and, for this reason, optimal pellet formulations with high sphericity, flowability, and instant or extended release were elucidated by applying a mixture experimental design followed by numerical optimization. The effect of processing on drug crystallinity was also examined. Since piroxicam has pH-dependent solubility, release was tested in both acidic and deionized water [37,38]. The release mechanisms were explained by analyzing the data using the Weibull model, which presents a relatively newer kinetic approach utilizing the entire drug release profile. This provides a more thorough description of the release mechanism. This model was described originally for extended release solid forms by Bonferoni et al. (1998) and interpreted by Papadopoulou et al. (2006) [29,39].

## 2. Materials and Methods

### 2.1. Materials

Microcrystalline cellulose (MCC, Avicel® PH-101, lot 6950C) was from FMC (Cork, Ireland) and chitosan (CHS) from Primex (Siglufjordur, Island). From the supplied CHS grades: TM 3493 (viscosity (η) 5 cps, deacetylation (DA) 90%), TM 3528 (η = 8 cps, DA = 96%), TM 3603 (η = 121 cps, DA = 90%), TM 3389 (η = 171 cps, DA = 95%), and TM 3425 (η = 463 cps, DA = 92%). The experimental CHS powders were prepared by mixing equal quantities of supplied TM 3493 with TM 3528 to give low viscosity experimental grade CHS1 (5–8 cps, DA = 93%), and TM 3603 with TM 3389 to give the medium viscosity experimental grade CHS2 (121–171 cps, DA = 92.5%). The high viscosity grade TM 3425 (463 cps, DA = 92%) was used as received and is denoted as the experimental grade CHS3.

Chemo Iberica S.A., Spain supplied Piroxicam (PXC). Polyvinylpyrrolidone (PVP) (K25, wt~21000) was gifted from BASF (Ludwigshafen, Germany).

*2.2. Solubility of PXC in pH 1.2 and 5.6 in the Presence of CHS*

To determine drug solubility in 0.1 N HCl, solutions with 0.01%, 0.05%, and 0.1% $w/v$ CHS were prepared in 50 mL 0.1 N HCl and excess drug (100 mg) was added to each. A saturated drug solution in 0.1 N HCl was also prepared for comparison. The solutions were kept at 37 °C for 24 h under agitation and, prior to analysis, they were centrifuged for 10 min at 4500 rpm (Labofuge 400R, Heraeus, Germany). The UV absorbance of the supernatant was measured at 334 nm (Pharma Spec UV-1700 Shimadzu, Kyoto, Japan) and converted to a concentration (mg/mL) from a reference curve [C = (Abs + 0.0085)/0.7437]. To measure solubility in deionized water (pH 5.6), 100 mg CHS (non-dissolving in pH 5.6) were suspended in 10 mL deionized water and an excess drug was added. The drug-saturated solutions were kept in closed containers at 37 °C for 24 h under agitation and centrifuged. The absorbance of supernatant was measured at 359 nm and converted to a concentration from a reference curve [C = (Abs + 0.0141)/0.4964].

*2.3. Preparation of Pellets*

Thirty-gram batches of CHS/MCC/PXC were blended in a Turbula® mixer (W.A. Bachofen, Muttenz, Switzerland) for 20 min and then transferred into a cylindrical vessel (0.8 L) fitted with a three-blade impeller. PVP 25 binder solution in water (7.5% $w/w$) was gradually added over 5 min to give 5% $w/w$ PVP concentration in the final dry pellets. PVP was added to the binder liquid to improve the consistency of wet mass [40]. Any further wetting liquid required was added as deionized water. The wet mass was immediately processed in a radial extruder (Model 20, Caleva Process Solutions, Dorset, UK) that was operated at 25 rpm and fitted with a 1-mm orifice screen for the production of conventional pellets or a 0.5-mm orifice screen for the mini pellets (both screens had 1.75-mm thickness). The extrudate was immediately processed for 5 min in a spheronizer (Model 120, Caleva Process Solutions) fitted with a 12-cm diameter cross-hatch friction plate (0.8-mm depth grooves and pyramidal protrusions), operated at 1250 rpm and corresponding to 7.85 m/s peripheral velocity. The pellets were dried (40 °C, 12 h) in a tray oven with air circulation (Hereaus, Germany).

*2.4. Characterization of Unprocessed Materials*

2.4.1. Particle Size

Particle size was determined using an image processing and an analysis system comprised of a microscope (Leitz Laborux S, Wetzlar, Germany), a video camera (VC-2512, Sanyo Electric, Osaka, Japan), and software (Quantimet 500, Cambridge, UK). Powder samples dispersed in liquid paraffin were examined at 40× total magnification. A mean particle diameter was expressed as an equivalent circle diameter (diameter of a sphere with the same projected area as the particle).

2.4.2. Pycnometric Density

Helium pycnometry was applied (Ultrapycnometer 1000, Quantachrome Instruments, Boynton Beach, Florida, FL, USA). The instrument was calibrated using a standard 7.0699 $cm^3$ steel ball. Samples were accurately weighed (3 decimals) and purged for 10 min before measurement. Sample volume (average of 10 runs) was measured from the displaced gas. Measurements were taken in triplicate and mean values and standard deviations were calculated.

2.4.3. Moisture Content

Samples of about 1 g were placed in an infrared radiation balance and heated at 105 °C (Halogen Moisture Analyzer HR73, Metler Toledo, OH, USA). Sample weight was automatically taken every 30 s and the process ended when the loss between two successive values was less than 0.01%. Moisture content (MC%) was expressed as the weight difference relative to the initial weight.

## 2.5. Characterization of Pellets

### 2.5.1. Size, Shape, and Density

Pellet size and shape were determined using an image processing and analysis system, as previously described [41]. Mean pellet diameter was expressed as an equivalent circle diameter and shape as the index $e_R$ (Equation (1)), which is sensitive to surface irregularity and pellet geometry, and its value increases with sphericity, which nearly reaches 0.75 for perfect spheres [42]. Examples of pellet shapes with corresponding $e_R$ values are shown in Figure 1.

$$e_R = (2 \times \pi \times \text{radius})/\text{perimeter} - \sqrt{(1 - (\text{width}/\text{length})^2)} \qquad (1)$$

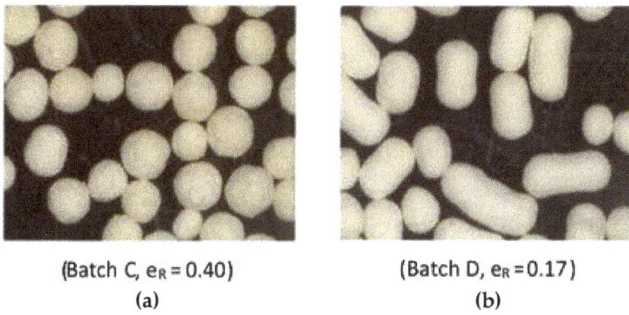

(Batch C, $e_R$ = 0.40)  (Batch D, $e_R$ = 0.17)
(a)  (b)

**Figure 1.** Images of pellet batches C (**a**) and D (**b**) showing different shapes.

The pycnometric density of the pellets was determined as described above for the unprocessed materials (Section 2.4.2).

### 2.5.2. Packing Ability, Flowability, and Porosity

Packing densities (bulk and tapped after 300 taps) were determined with a tester fitted with a 25 mL cylinder (14 mm drop, Erweka SVM 101, USP1, Heusenstamm, Germany). Carr's compressibility index (CC%), equal to the volumetric change relative to tapped volume, was calculated from the density values as the index of packing ability. Flowability was estimated with an apparatus constructed according to United States Pharmacopeia [43]. Samples were transferred into a cylinder of 1.5 cm internal diameter and a 5-mm orifice at the center of its base. Pellets flowing through the orifice were collected on the platform of a balance (Bel Engineering MARK330, Monza, Italy) located 5 cm underneath the cylinder. Weight data was recorded every 0.2 s and transferred to a computer via an RS-232 interface. Pellet porosity was expressed as $\varepsilon\% = [1 - (\text{pellet density}/\text{powder density})]$.

### 2.5.3. Physicochemical Characterization

Raman Spectroscopy

Raman spectra were recorded to detect possible interactions between the drug and excipients using a bench top Raman spectrometer (Agility, dual band 785/1064 nm model, BaySpec, San Jose, CA, USA) and supporting software (Agile 20/20). Unprocessed powders or pellet samples were placed in standard glass vials and scanned over the range of 100 to 2700 $cm^{-1}$ of Raman Shift using the laser excitation line 785 nm, exposure time of 1 s, and a power of incident laser beam of 150 mW. The recorded spectra were the average of 100 runs.

Powder X-ray Diffraction (PXRD)

Changes in the crystalline state and crystallographic characteristics of the drug were examined using PXRD (PHILIPS PW1710 diffractometer with CuKα, Ni-filtered 1.5418 Å radiation wavelength, Phillips, Eindhoven, The Netherlands). The samples were scanned over the 3°–43° 2θ range, at a speed of 1.2/min. Identification of the crystalline phases was made by comparing with published PXRD data based on the appearance and intensity of the reflections. Crystallinity was quantified as the crystallinity index (CI%) expressed by the ratio of the intensity of the strongest reflectance of the drug in the pellets at 26° 2θ, relative to that of the pure drug at the same 2θ. The crystallinity loss (LC%) due to processing was obtained as a percentage of the difference between the CI% of drug in physical mixtures (PM) and in pellets relative to that in PM (see Equation (2)).

$$LC\% = [(CI\% \text{ of drug in PM} - CI\% \text{ of drug in pellets})/CI\% \text{ of drug in PM}] \times 100 \quad (2)$$

### 2.6. In-Vitro Release

In-vitro release of PXC was tested using the USP II Apparatus at 100 rpm and 37 ± 0.5 °C. Pellet samples of the experimental batches corresponding to 20 mg of drugs were added into 900 mL dissolution fluid. Since PXC has pH-dependent solubility, tests were conducted in two media including HCL 0.1 N (pH 1.2) and deionized water (pH 5.6). In the last case, pH was measured at the beginning and the end of the test and no significant change was recorded. Pellets of MCC/PXC without chitosan were also tested for comparison. Aliquots were taken at timely intervals and analyzed by UV spectroscopy (Pharma Spec UV-1700 Shimadzu, Kyoto, Japan) at 334 nm for pH 1.2 and 359 nm for pH 5.6.

Kinetic Models

The drug release data were analyzed using the Weibull equation [44]

$$\ln[-\ln(1 - W/W_o)] = -\ln a + b\ln(t - t_o) \quad (3)$$

where $W$ is the drug released at time $t$, $W_o$ is the drug released at the end of the test, $t_o$ is the lag time before release as determined by trial and error for best line fitting, '$b$' is the constant characteristic of the shape of the release curve and the release mechanism [39], and a is a time-scale parameter defined as $a = (td)^b$ where td is the time required for 63.2% release.

### 2.7. Experimental Design and Optimization of Compositions for Instant or Extended Release

The influence of composition on the properties of pellets and drug release was studied separately for mini and conventional pellets. This doubled the number of experimental batches and, for this reason, the d-optimal mixture design including vertices, edge centers, centroid, and axial points was used as an efficient design applicable to constrained regions [45] (Table 1). The sum of components CHS (X1), MCC (X2), and PXC (X3) was 95% and the remaining 5% was PVP. Constraints were applied for CHS and MCC at 10% < X1, X2 < 80%, and for the drug at 5% < X3 < 70%, which provides a realistic design space. Consequently, the % weights were transformed into a 'real' scale (Equation (4)), where their sum is 1.0, and then into L-pseudo levels (Equation (5)) where their minimum value is 0 and the maximum is 1.

$$\text{Real} = \text{Actual/Total of Actuals} \quad (4)$$

$$\text{Pseudo} = (\text{Real} - Li)/(1 - L) \quad (5)$$

where Li is the lower constraint, and L is the sum of lower constraints.

Table 1. Experimental mixture design for pellet batches. The design was applied to mini and conventional pellet batches separately.

| Batch Code | Point in Design Space | Actual Values (%) | | | Real Values | | | L-Pseudo Values | | |
|---|---|---|---|---|---|---|---|---|---|---|
| | | CHS2 | MCC | Drug | CHS2 | MCC | Drug | CHS2 | MCC | Drug |
| A | Axial | 19.38 | 54.37 | 21.25 | 0.20 | 0.57 | 0.22 | 0.13 | 0.63 | 0.23 |
| B | Vertex | 80.00 | 10.00 | 5.00 | 0.84 | 0.11 | 0.05 | 1.00 | 0.00 | 0.00 |
| C | Center edge | 10.00 | 47.50 | 37.50 | 0.11 | 0.50 | 0.40 | 0.00 | 0.54 | 0.46 |
| D | Vertex | 15.00 | 10.00 | 70.00 | 0.16 | 0.11 | 0.74 | 0.07 | 0.00 | 0.93 |
| E | Centroid | 33.33 | 35.00 | 26.67 | 0.35 | 0.37 | 0.28 | 0.33 | 0.36 | 0.31 |
| F | Center edge | 45.00 | 45.00 | 5.00 | 0.47 | 0.47 | 0.05 | 0.50 | 0.50 | 0.00 |
| G [#] | Vertex | 10.00 | 80.00 | 5.00 | 0.11 | 0.84 | 0.05 | 0.00 | 1.00 | 0.00 |
| H | Axial | 54.37 | 19.38 | 21.25 | 0.57 | 0.20 | 0.22 | 0.63 | 0.13 | 0.23 |
| I | Center edge | 47.50 | 10.00 | 37.50 | 0.50 | 0.11 | 0.40 | 0.54 | 0.00 | 0.46 |

[#] This experimental batch was reproduced for testing lack of fit of the applied models.

Multiple linear regression analysis (backward elimination) based on Scheffé polynomial (Equation (6)) was used to derive model equations between the component levels and the following pellet properties: mean diameter, shape index, dissolution efficiency at pH 1.2, and drug release at pH 5.6 in 2 h and 8 h. The model includes linear and interaction terms.

$$Y = b_1 X_1 + b_2 X_2 + b_3 X_3 + b_{12} X_1 X_2 + b_{13} X_1 X_3 + b_{23} X_2 X_3 \tag{6}$$

Contour and trace plots were constructed from the regression equations. Contour plots show the effect of composition on a property. Each component line of the trace plots describes the effect of that component by moving along an imaginary straight path connecting the centroid of the experimental design (intersection of the three lines in the plots) to the vertex of that component in the triangle. This visualizes the effects of one single component while holding the ratio of the others constant. $p < 0.05$ was the statistically significant level, and $R^2$ and adjusted $R_{adj}^2$ indicate goodness of fit.

The derived regression equations were subsequently used to optimize formulations for good sphericity ($e_R$ in the range of 0.3–0.51), maximum flowability, and fast or extended release. The criteria for fast release were set as: maximum DE% and minimum td (time for 63.2% release, Weibull equation) for release at pH 1.2, and, for extended release, they were set as: minimum DE% for release at pH 1.2, which is less than 60% release in 2 h and a maximum after 8 h release at a pH of 5.6. Optimization was applied for preset drug levels. The respective equations for minimization and maximization of a response di were:

$$d_i = (Y_{max} - Y_i)/(Y_{max} - Y_{min}) \tag{7}$$

$$d_i = (Y_i - Y_{min})/(Y_{max} - Y_{min}) \tag{8}$$

where di is the desirability function of a response ranging from 0 to 1, Ymin is the lowest measured value, Ymax is the highest measured value, and Yi is any value. di = 0 if the response value is outside the desired range and di = 1 if the response value is within the desired range. For minimization or maximization, the di varies from 0 to 1. The overall desirability D (Equation (9)) denotes the geometric mean of the individual desirability of n responses.

$$D = (d_1 \times d_2 \times d_3 \times \ldots \times d_n)^{1/n} \tag{9}$$

Design Expert 8.0 (Stat-Ease, Minneapolis, MN, USA) was used to generate the experimental design, for statistical analysis, to draw the contour and trace plots, and for numerical optimization.

## 3. Results and Discussion

The effect of chitosan grade viscosity on the solubility of piroxicam in acidic pH 1.2 and deionized water was initially examined to select the grade that prompted greater 'apparent' solubility. The term 'apparent' is used to distinguish from the thermodynamic equilibrium solubility.

### 3.1. PXC Solubility and Influence of Chitosan

Figure 2 presents the 'apparent' solubility of piroxicam in water at both acidic pH 1.2 and deionized water (pH 5.6) measured in the presence of different CHS viscosity grades. In Figure 2a, the 'apparent' solubility of PXC at pH 1.2 (where CHS dissolves) was plotted against CHS concentration while the bars in Figure 2b show 'apparent' solubility at pH 5.6 in the presence of suspended CHS. Due to its primarily acidic character, PXC is more soluble in basic environments [35,36]. Its measured greater solubility in CHS-free water at pH 1.2 than at pH 5.6 (0.10, Figure 2a, compared with 0.023 mg/mL, Figure 2a,b) is attributed to its amphiphilic character that allows some ionization at pH 1.2 (protonation of -NH2) but not at pH 5.6. From Figure 2a, it can be seen that, at pH 1.2, the 'apparent' solubility increased remarkably from 0.10 mg/mL in CHS-free water to 0.165 mg/mL in the 0.01 $w/w$ CHS solution. Thereafter, the increase differed for each grade. For the low viscosity CHS1, it was small and reached 0.194 mg/mL at the highest 0.1% $w/w$ CHS while, for the medium CHS2 and high CHS3 viscosity grades, it increased exponentially. It reached 0.238 and 0.222 mg/mL. Similarly, Figure 2b shows that the presence of CHS as suspended polymer in deionized water also increased drug solubility, but, to a lesser extent than pH 1.2, and the differences between the drug solubilities in the three CHS suspensions were small (0.138, 0.157, and 0.150 mg/mL for CHS1, CHS2, and CHS3, respectively).

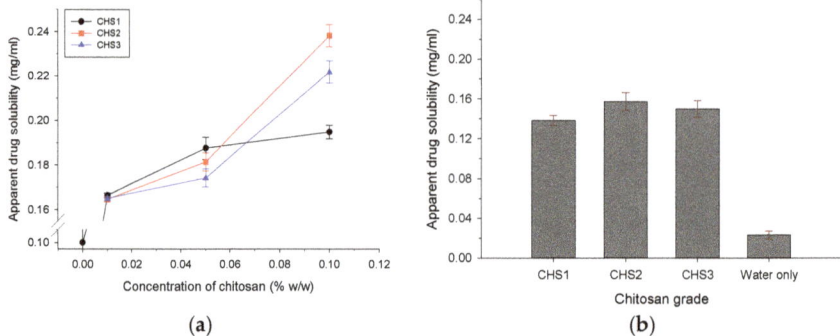

**Figure 2.** (a) Apparent solubility of piroxicam at pH 1.2 in the presence of increasing concentrations of low (CHS1), medium (CHS2), and high viscosity (CHS3) chitosan grades (error bars are standard deviations, $n = 3$). (b) Apparent solubility of piroxicam in deionized water (pH 5.6) in the presence of 100 mg chitosan suspended in 10 mL water.

The greater effect of CHS on 'apparent' drug solubility at pH 1.2 compared to pH 5.6 is attributed to the solubility of CHS at pH 1.2 but not in deionized water (pH 5.6). At pH 5.6, polymer fibers sorb water but do not dissolve, and, hence, only those drug molecules adsorbed onto fibers benefit from wetting, which results in limited increased solubility. On the other hand, in the acidic environment, the dissolved CHS chains form water-soluble units with the drug molecules attached to the chains, which considerably increases 'apparent' drug solubility [46]. A possible explanation for the formation of chitosan/drug water soluble units could be weak hydrogen bonding between hydroxyl groups present in the chitosan with the hydroxyl of the protonated in the acidic pH zwitterionic tautomer of piroxicam [47]. The greater drug solubility improvement shown by the high viscosity CHS grades may be due to network formation of the CHS chains where the CHS-PXC water-soluble units are stabilized.

The chains of low viscosity CHS are unable to form networks due to their mobility. Therefore, the CHS2 grade that prompted higher solubility was chosen to develop CHS/MCC/PXC pellets.

## 3.2. Technological Characteristics

### 3.2.1. Wetting Liquid Volume, Pellet Size, and Shape

The measured properties of the experimental materials are given in Table 2. From Table 2, it can be seen that the experimental powders had similar median (d50) particle diameters for CHS2 21 µm, for PXC 11 µm, and MCC 31 µm, and similar pycnometric densities for CHS2 1.64 g/cc, MCC 1.69 g/cc, and PXC 1.58 g/cc. The similarity in particle diameters and densities favors good mixing and homogeneous paste formation [48,49]. The moisture contents of the unprocessed materials fell within the expected ranges.

Table 2. Properties of the unprocessed experimental materials.

| Material | Moisture Content (%) | Particle Density (g/cc) [#] | Particle Size Distribution Parameters [#] (µm) | | |
|---|---|---|---|---|---|
| | | | d10 | d50 | d90 |
| CHS2 | 8.78 ± 0.71 | 1.64 ± 0.01 | 10.3 | 21.0 | 83.2 |
| MCC | 5.38 ± 0.22 | 1.69 ± 0.01 | 10.7 | 31.0 | 100.6 |
| PXC | 0.32± 0.01 | 1.58 ± 0.01 | 6.4 | 11.0 | 28.6 |

[#] d10, d50, and d90 diameters correspond to 10%, 50%, and 90% of the particle size distribution, respectively.

Table 3 presents the volumes of added liquid and the paste consistencies that, after extrusion/spheronization, gave the highest sphericity for mini and conventional pellets with a mean pellet diameter and shape indexes. Batches B, F, and H with high CHS2 (80%, 45%, and 54.37%) showed high consumption. Although both CHS and MCC consist of fibrils in a sponge-like structure, chitosan-rich pellets, e.g., batch B with 80% CHS2, consumed larger amounts of liquid than MCC-rich pellets, e.g., batch G with 80% MCC (58.0 mL compared to 42.0 mL). However, the wet mass of batch B had low plasto-elasticity, that separated under light pressure, and gave low sphericity pellets after spheronization ($e_R$ 0.27 and 0.32 for the mini and conventional pellets, respectively, Table 3) with a large proportion of small-sized pellets, as reflected by the low mean diameters (367 and 712 µm, Table 3). This indicates that only part of the consumed liquid was used to form a mass suitable for extrusion, with the remainder residing loosely in the interior [31]. Batch D with the highest drug content (70%) gave a creamy paste despite a low consumption value (21 mL), which shows an inability to take up liquid and form an extrudable paste. This was also indicated by the low pellet sphericity ($e_R$ 0.17 and 0.15 for mini and conventional pellets).

From Table 3, it can be seen that the mean diameters of the mini pellets were about half the value of the conventional pellets (367 to 585 µm compared to 712 to 1206 µm), as expected. Batches B, F, and H with high CHS2 content and greater liquid consumption gave smaller pellets than batches A, C, and G with high MCC (mean diameters of 367, 411, and 384 µm when compared to 532, 516, and 489 µm for the mini pellets, and 712, 984, 967 µm when compared to 1010, 1013, and 1021 µm for the conventional pellets, respectively). The relationship between liquid consumption and pellet diameter (not shown graphically) was linear and inversely proportional ($R^2$ 0.689 for the mini and 0.820 for the conventional pellets), which generally agrees with previous reports [26]. Mini pellets were less spherical ($e_R$ range 0.15–0.44 compared to 0.16–0.51, Table 3), and this can be attributed to the more efficient absorption of centrifugal and frictional forces exerted during spheronization by the conventional pellets due to their larger size and mass.

**Table 3.** Wetting liquid consumption, paste consistency, pellet size, and pellet shape of the experimental mini and conventional pellet batches.

| Batch Code | Wetting Liquid (mL) | Paste Consistency | Mini Pellets | | Conventional Pellets | |
|---|---|---|---|---|---|---|
| | | | Mean Pellet Diameter (µm) | Shape Index $e_R$ | Mean Pellet Diameter (µm) | Shape Index $e_R$ |
| A | 40.0 | Good | 532 | 0.44 | 1010 | 0.39 |
| B | 58.0 | Fragile | 367 | 0.27 | 712 | 0.31 |
| C | 29.0 | Good | 516 | 0.40 | 1013 | 0.41 |
| D | 21.0 | Creamy | 585 | 0.17 | 1206 | 0.15 |
| E | 41.0 | Good | 449 | 0.27 | 1001 | 0.51 |
| F | 52.0 | Good | 411 | 0.32 | 984 | 0.43 |
| G | 42.0 | Plastic | 429 | 0.34 | 1021 | 0.42 |
| H | 50.0 | Good | 384 | 0.28 | 967 | 0.40 |
| I | 42.0 | Good | 499 | 0.15 | 1002 | 0.16 |

### 3.2.2. Packing, Flowability, and Porosity

Table 4 presents pycnometric (Ps) and packing densities (bulk Pb and tap Pt), porosity ($\varepsilon$%), Carr's compressibility index (CC%), and flowability. Tap density expresses the extent while CC% expresses the easiness of packing. The properties of the mini and conventional pellets were correlated linearly ($R^2$ for: Ps = 0.856, Pb = 0.818, Pt = 0.815, $\varepsilon$% = 0.939, and flowability of 0.619). Mini pellets showed greater Pt values (from 0.67 to 0.87) compared to the conventional pellets (0.57 to 0.79), which is attributed to their smaller size, and, therefore, smaller inter-particle voids less occupied volume. Batches B and D with irregular pellet shape gave the lowest $p_t$ values. The parameter CC% presented low values (<15%) for both mini and conventional pellets, which indicates good packing ability.

**Table 4.** Pycnometric densities, packing densities, packing index, porosity, and flowability of the experimental mini and conventional pellets.

| Batch Code | Mini Pellets | | | | | | Conventional Pellets | | | | | |
|---|---|---|---|---|---|---|---|---|---|---|---|---|
| | Ps (g/cc) | Pb (g/cc) | Pt (g/cc) | CC% | $\varepsilon$% | Flowability (g/s) | Ps (g/cc) | Pb (g/cc) | Pt (g/cc) | CC% | $\varepsilon$% | Flowability (g/s) |
| A | 1.469 | 0.76 | 0.87 | 12.50 | 10.14 | 2.65 | 1.436 | 0.68 | 0.76 | 10.00 | 12.13 | 1.90 |
| B | 1.470 | 0.58 | 0.65 | 10.81 | 6.53 | 2.26 | 1.489 | 0.51 | 0.58 | 11.48 | 5.35 | 1.63 |
| C | 1.556 | 0.72 | 0.80 | 9.52 | 4.29 | 2.37 | 1.594 | 0.65 | 0.70 | 6.94 | 1.97 | 1.74 |
| D | 1.572 | 0.60 | 0.68 | 12.07 | 0.31 | 1.93 | 1.574 | 0.53 | 0.61 | 13.43 | 0.20 | 1.41 |
| E | 1.476 | 0.71 | 0.82 | 13.16 | 8.19 | 2.32 | 1.516 | 0.66 | 0.75 | 11.59 | 5.75 | 1.95 |
| F | 1.434 | 0.69 | 0.80 | 13.04 | 11.48 | 2.58 | 1.421 | 0.73 | 0.77 | 4.84 | 12.30 | 2.24 |
| G | 1.460 | 0.79 | 0.86 | 8.47 | 12.45 | 2.81 | 1.433 | 0.72 | 0.79 | 9.72 | 14.10 | 1.99 |
| H | 1.452 | 0.69 | 0.74 | 6.78 | 8.47 | 2.14 | 1.455 | 0.64 | 0.75 | 14.66 | 8.27 | 1.81 |
| I | 1.514 | 0.61 | 0.67 | 8.20 | 3.89 | 1.78 | 1.526 | 0.51 | 0.57 | 11.94 | 3.08 | 1.40 |

Ps: pycnometric density. Pb: bulk density. Pt: tap density. CC: Carr's compressibility index. $\varepsilon$: porosity. Standard deviations for Ps < 0.015. Pb < 0.01. Pt < 0.01. CC% < 0.03 and for $\varepsilon$% < 0.83 (n = 3).

The porosity ranges of the mini and conventional pellets were similar (0.31–12.45% and 0.20–14.10%, respectively). Batches B, E, and H with high CHS2 and up to 21.25% PXC showed low $\varepsilon$% (6.53%, 8.19% and 8.47% for the mini and 5.35%, 5.75 and 8.27% for the conventional pellets), which implies a denser structure that is attributed to the binder action of CHS [40]. However, while batches A, F, and G with high MCC showed higher $\varepsilon$% values (10.14%, 11.48%, and 12.45% for mini pellets and 12.13%, 12.30%, and 14.10% for conventional pellets), which is similar to those previously reported for MCC pellets [50].

Batches C, D, and I with high PXC showed unusually low porosities (4.29%, 0.31%, and 3.89% for the mini and 1.97%, 0.20%, and 3.08% for the conventional pellets).

Flowability depends on pellet size, shape, and orifice diameter. The 6-mm orifice was more than six times larger than the diameter of the pellets (means of 452 µm and 992 µm for mini and conventional, respectively, Table 3) and was selected to avoid blocking [51]. From Table 4, it can be seen that the flowability of the mini pellets was considerably greater than conventional pellet flowability (from 1.76 to 2.82 g/min when compared to the 1.35 to 1.99 g/min) despite their lower overall sphericity, which is due to their smaller diameter and easier movement through the orifice.

### 3.2.3. Analysis of Mixture Design for the Effect of Composition on Pellet Size, Shape, and Flowability

Statistically significant model equations with fitting indices derived from the analysis of the experimental design that describe the effects of composition on the pellet diameter, the shape, and flowability for the mini and conventional pellets are presented in Table 5. In all cases there was no lack of fit. Respective contour and trace plots are presented in Figures 3–5. The plots in Figure 3 show that increasing CHS2 decreased the mean pellet diameter for both mini and conventional pellets while PXC caused a diameter increase and MCC had a minimal effect. The model equations were linear ($R^2$ 0.837 and 0.742 for mini and conventional pellets respectively). The decrease observed at high CHS2 content is associated with high wetting liquid consumption followed by shrinkage during drying. The increase of the pellet diameter at high PXC is due to the irregular, elongated pellet shape (Figure 1b).

**Table 5.** Results of regression analysis of the experimental mixture design based on Scheffé quadratic models.

| Response | Significance of Terms ($p$-Values) | | | | Model Equation in Terms of Actual Components | $p$-sign. | $R^2$ | $R_{adj}^2$ |
|---|---|---|---|---|---|---|---|---|
| | Linear Mixture | $X_1X_2$ | $X_1X_3$ | $X_2X_3$ | | | | |
| $D_{50}$/mini | 0.002 | - | - | - | $+3.45X_1 + 4.67X_2 + 7.10X_3$ | 0.002 | 0.837 | 0.742 |
| $D_{50}$/conv. | <0.001 | 0.004 | 0.058 | 0.006 | $+7.71X_1 + 10.80X_2 + 13.64X_3$ | 0.003 | 0.802 | 0.746 |
| $e_R$/mini | 0.021 | - | - | 0.010 | $+2.79 \times 10^{-3}X_1 + 3.12 \times 10^{-3}X_2 - 4.82 \times 10^{-3}X_3 + 1.44 \times 10^{-4}X_2X_3$ | 0.010 | 0.832 | 0.748 |
| $e_R$/conv. | 0.036 | - | - | 0.120 | $+3.86 \times 10^{-3}X_1 + 4.56 \times 10^{-3}X_2 - 7.37 \times 10^{-4}X_3 + 1.21 \times 10^{-4}X_2X_3$ | 0.042 | 0.721 | 0.581 |
| Flowability/mini | 0.015 | 0.273 | 0.163 | - | $+0.023X_1 + 0.027X_2 + 0.023X_3 + 1.826 \times 10^{-4}X_1X_2 - 2.815 \times 10^{-4}X_1X_3$ | 0.033 | 0.838 | 0.708 |
| Flowability/conv. | <0.001 | 0.001 | - | - | $+0.013X_1 + 0.019X_2 + 0.013X_3 + 3.518 \times 10^{-4}X_1X_2$ | <0.001 | 0.956 | 0.934 |
| DE%/pH1.2/mini | 0.298 | - | 0.012 | - | $+0.22X_1 + 0.72X_2 + 0.26X_3 + 0.03X_1X_3$ | 0.042 | 0.722 | 0.582 |
| DE%/pH1.2/conv. | 0.004 | 0.172 | 0.001 | - | $+0.07X_1 + 0.72X_2 + 0.44X_3 + 5.61 \times 10^{-3}X_1X_2 + 0.02X_1X_3$ | 0.002 | 0.945 | 0.902 |
| DE%/pH5.6/mini | <0.001 | 0.084 | 0.015 | 0.001 | $+0.70X_1 + 0.47X_2 + 0.57X_3 + 5.70 \times 10^{-3}X_1X_2 - 0.01X_1X_3 - 0.02X_2X_3$ | 0.001 | 0.983 | 0.962 |
| DE%/pH5.6/conv. | 0.042 | - | - | 0.029 | $+0.16X_1 + 0.25X_2 + 0.71X_3 - 0.01X_2X_3$ | 0.026 | 0.764 | 0.646 |
| td/pH1.2/mini | 0.543 | 0.072 | 0.009 | - | $1.44X_1 + 0.702X_2 + 1.188X_3 - 0.025X_1X_2 - 0.054X_1X_3$ | 0.039 | 0.825 | 0.685 |
| td/pH1.2/conv. | 0.069 | - | 0.042 | - | $1.079X_1 + 0.439X_2 + 0.676X_3 - 0.030X_1X_3$ | 0.044 | 0.717 | 0.576 |
| td/pH5.6/mini | 0.027 | - | 0.054 | 0.103 | $0.234X_1 + 1.108X_2 + 0.478X_3 - 0.036X_1X_3 + 0.024X_2X_3$ | 0.035 | 0.833 | 0.699 |
| td/pH5.6/conv. | <0.001 | - | <0.001 | <0.001 | $0.084X_1 + 0.6118X_2 - 0.171X_3 + 0.058X_1X_3 + 0.044X_2X_3$ | <0.001 | 0.979 | 0.963 |
| %Released 2h/pH5.6 | <0.001 | - | 0.016 | 0.002 | $0.817X_1 + 0.513X_2 + 0.570X_3 - 0.01X_1X_2 - 0.02X_2X_3$ | <0.001 | 0.964 | 0.935 |
| %Released 8h/pH5.6 | 0.002 | 0.029 | - | 0.002 | $0.737X_1 + 0.790X_2 + 0.747X_3 + 0.013X_1X_2 - 0.025X_2X_3$ | 0.002 | 0.953 | 0.916 |

$X_1$: Chitosan. $X_2$: Microcrystalline cellulose. $X_3$: Drug. $D_{50}$: median pellet diameter. $e_R$: shape coefficient. DE%: dissolution efficiency. td: parameter in the Weibull equation.

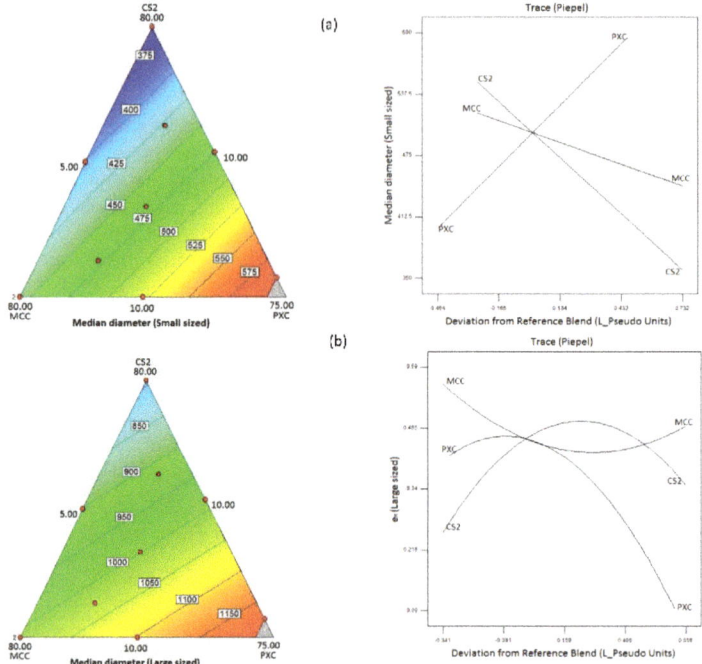

**Figure 3.** Contour and trace plots of mean pellet diameters for (**a**) mini and (**b**) conventional pellets.

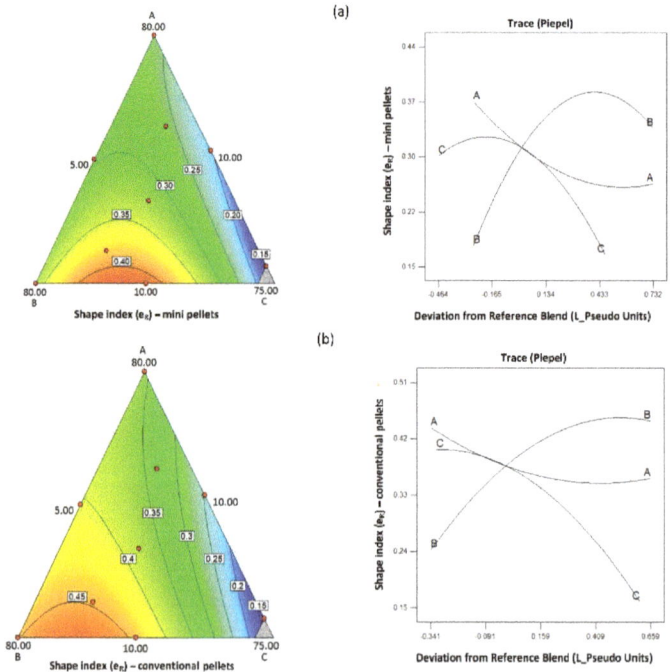

**Figure 4.** Contour and trace plots of the shape index $e_R$ for (**a**) mini and (**b**) conventional pellets.

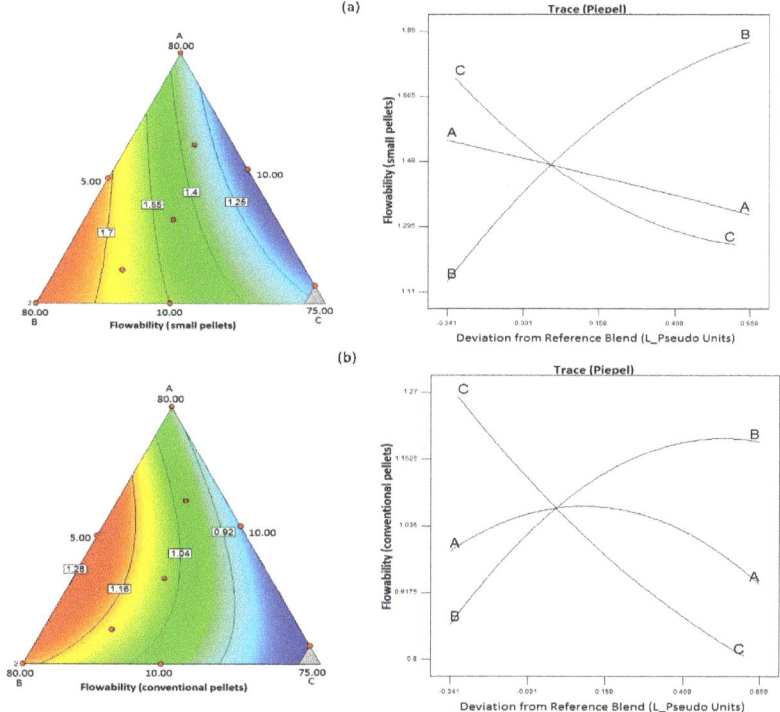

**Figure 5.** Contour and trace plots of flowability for (**a**) mini and (**b**) conventional pellets.

Considering the pellet shape, from the trace plots in Figure 4, it can be seen that sphericity was adversely affected by PXC for both mini and conventional pellets. It was positively affected by MCC while CHS2 had a small effect. The increased irregularity at high drug content is due to the lack of sufficient excipients resulting in a creamy paste even at low volumes of added liquid (batch D, Table 3). The regression model for both mini and conventional pellets was reduced quadratic ($R^2$ 0.885 and 0.901) with an interaction of the effects of MCC and PXC, i.e., at low MCC levels, the effect of drug content was small but strongly negative at high MCC (Figure 4b). Furthermore, from Figure 5, it can be seen that MCC had a positive effect, whereas the drug had a negative effect on flowability, which can be explained by their respective effects on sphericity (Figure 4). The model describing the effects of composition on flowability was reduced quadratic for both mini and conventional pellets ($R^2$ 0.838 and 0.956), which was indicated by curvilinear plots.

### 3.3. Physicochemical Evaluation

#### 3.3.1. Raman Spectroscopy

Raman spectra provide information regarding interactions of drugs with excipients. In Figure 6, the spectra of chitosan (CHS2), microcrystalline cellulose (MCC), piroxicam (PXC), and batches C, l with 37.5% drug and low or high CHS/MCC ratio (10/47.5 and 47.5/10), respectively, are presented. The CHS2 spectrum showed two small peaks at 119 $cm^{-1}$ and 583 $cm^{-1}$. The MCC spectrum showed one small double peak at 1196/1120 $cm^{-1}$. The PXC spectrum showed a large fingerprint region in the range of 1090 to 1661 $cm^{-1}$ (enclosed by vertical dotted lines), which is due to the complex patterns of C–C, C–N, and aromatic ring vibrations [52]. The drug peaks at 1543 and 1570 $cm^{-1}$ are due to ring stretching and C=C symmetric stitching vibrations, respectively (indicated by arrows), are present in both C, I spectra and are characteristic of form II [53]. The spectra of both C and I showed the same peaks in the fingerprint region as the drug spectrum, which indicates no polymorphic transformation or interaction due to processing. Furthermore, peaks at 1007 and 1401 cm$^{-1}$ that are characteristic of the PXC monohydrate [35,53] are not seen in the spectra of drug or pellet batches, which indicates an absence of the monohydrate drug.

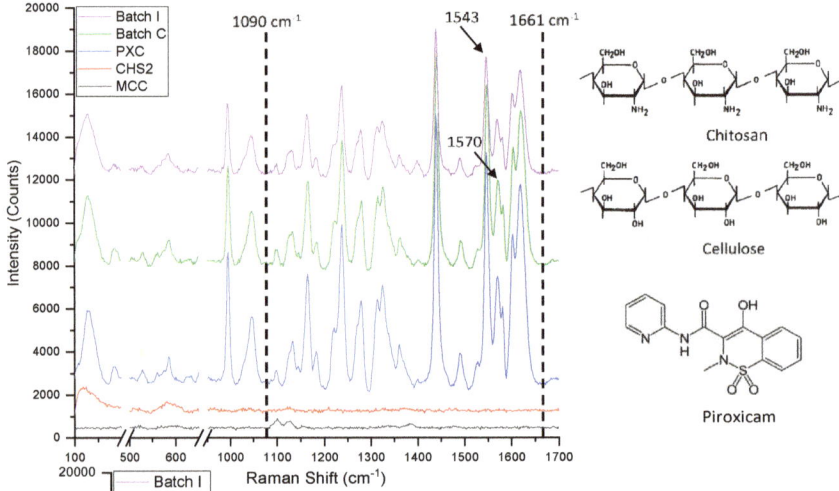

**Figure 6.** Chemical structures and Raman spectra of microcrystalline cellulose (MCC), chitosan (CHS2), unprocessed drug (PXC), and experimental batch I (CHS2 47.5%, MCCI 10%, PXC 37.5%).

3.3.2. Powder X-ray Diffraction (PXRD)

PXRD patterns provide detailed information about the crystallinity of the drug in the pellet batches. Figure 7 presents the patterns of CHS2, MCC, and PXC powders, pellet batches B and G with low 5% drug and high (80%) or low (10%) CHS2, and batches H and A with higher 21.25% drug and high (54.37%) or low CHS2 (19.38%). The CHS2 pattern showed one small reflection at 10.9° 2θ and one strong reflection at 20.0° 2θ, whereas the MCC pattern showed three reflections with one broad between 15° and 16.5° 2θ, one strong at 22.3° 2θ, and one small at 34.2° 2θ.

The PXC pattern showed sharp reflections at 9.1° and 10.2° 2θ, four consecutive reflections between 15° and 17° 2-θ, and strong, sharp, reflections at 15.9°, 26.1°, and 27.1° 2θ. The strong reflection at 9.1° and the consecutive reflections at 26.1°/27.1° 2θ confirm the presence of the anhydrous form, which is in agreement with the Raman spectroscopy results. Additionally, from the PXRD pattern of PXC and using the database of the International Center of Diffraction Data (ICDD 2003) [54], it can be inferred that the form-II (44-1839 ICDD card) of the piroxicam structure is the major phase, while the form-I (40-1982 ICDD card) is identified as the minor phase. More specifically, using the mass absorption coefficient, the density, and the specific reflections of each form, it is estimated that form-II constitutes 69% $w/w$ of the drug, while form-I is 31% $w/w$. From the patterns of the pellets shown in Figure 7, it appears that the drug reflections at 9.1°/10.2° and 26.1°/27.1° 2θ are still discernible in the PXRDs of batches G and B (though small due to the low drug content), and are clearly seen in the PXRDs of batches A and H with higher drug content, which indicates the predominantly crystalline state of the drug.

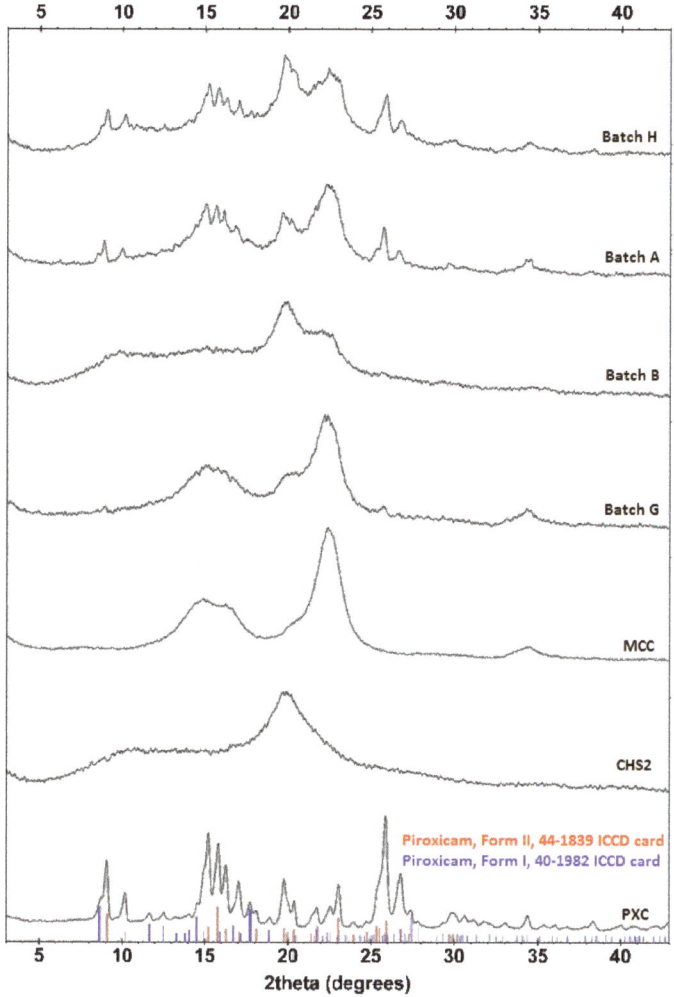

**Figure 7.** PXRDs for unprocessed materials and batches B and G with 5% drug content, and batches A and H with 21.25% drug content.

Crystallinity indices of the drug for the experimental pellet batches with 5% (B, G), 21.25% (A, H), and 37.5% drug (C, I), and their corresponding physical mixtures are presented in Table 6. In all cases, there is a loss of percentage in crystallinity due to processing, which is greater for the low drug batches B and G (21.14–22.57% compared to 2.51–3.87% for A and H, and 2.54–3.98% for C and I). It is documented that CHS2 is able to amorphize drugs processed by wet granulation [25]. (PVP is also an amorphizer but, because it is included with a low content of 5%, its contribution is negligible [55]). This effect should result from the dissolution of some drug content during wetting/extrusion, which is subsequently converted to an amorphized form after drying. Since the effect of CHS on solubility at a pH of 5.6 is independent of CHS content (Figure 1b), the amount of amorphized crystalline drug should not differ between batches, which explains the greater percentage of crystallinity loss observed in batches B and G with low drug content. Comparing batches with the same PXC content, it can be seen that the differences in crystallinity loss are small, regardless of the CHS/MCC ratio.

**Table 6.** Crystallinity index (CI%) of the drug in the experimental pellet batches and the corresponding physical mixtures (PM). (Intensity of the drug at 25.8 2θ degrees 2132.55 a.u.).

| Batch Code/Drug% | CI% of Drug in PM | CI% of Drug in Pellet | Crystallinity Loss (%) |
|---|---|---|---|
| B/5 | 24.73 ± 0.31 | 19.15 ± 0.05 | 22.56 ± 0.09 |
| G/5 | 21.62 ± 0.02 | 17.05 ± 0.18 | 21.14 ± 0.17 |
| A/21.25 | 35.96 ± 0.21 | 34.57 ± 0.15 | 3.87 ± 0.42 |
| H/21.25 | 33.84 ± 0.03 | 32.99 ± 0.04 | 2.51 ± 0.48 |
| C/37.5 | 55.29 ± 0.18 | 53.09 ± 0.16 | 3.98 ± 0.26 |
| I/37.5 | 58.21 ± 0.19 | 56.73 ± 0.21 | 2.54 ± 0.34 |

*3.4. In-Vitro Release*

In Figure 8, release profiles of mini and conventional pellets at two pH media (pH 1.2 and 5.6) are presented together with data from MCC/PXC pellets without chitosan for comparison purposes. Although the stay of the drug in the stomach delivered by conventional formulations is usually less than 120 min., this time may increase considerably for the present pellet formulations due to the muco-adhesive properties of chitosan [56]. For this reason, release studies of up to 500 min were conducted since the release exceeded this period for some pellet batches. Representative images of pellets before and after dissolution at pH 1.2 are shown in Figure 9 for batches A and C. Figure 10 presents images of all experimental batches before and after dissolution in deionized water (pH 5.6). The dissolution efficiency (DE%) and similarity factors (f2) comparing drug release from mini and conventional sized pellets at pH 1.2 and pH 5.6 are given in Table 7.

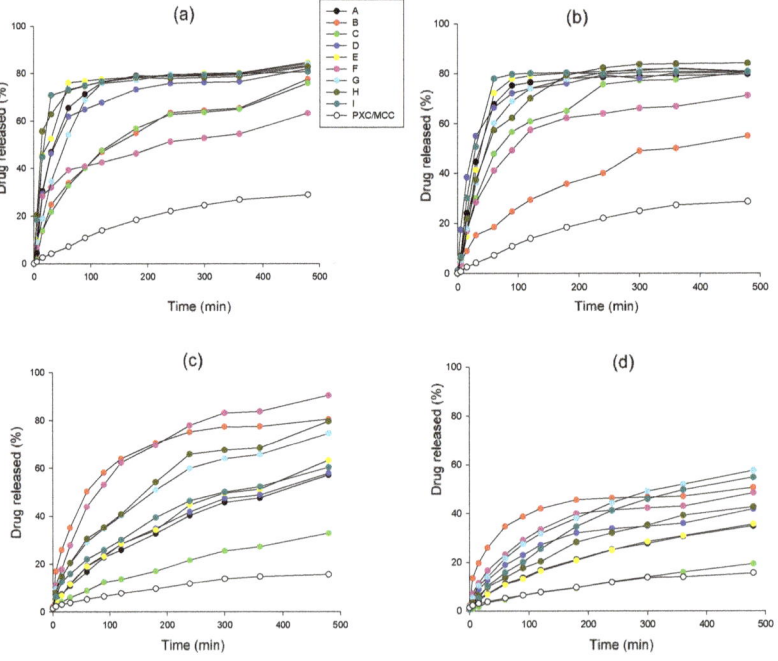

**Figure 8.** Release profiles of: (**a**) mini pellets at pH 1.2, (**b**) conventional pellets at pH 1.2, (**c**) mini pellets at 5.6, and (**d**) conventional pellets at pH 5.6. Release of drug from pellets prepared without chitosan (PXC/MCC) is also shown for comparison purposes.

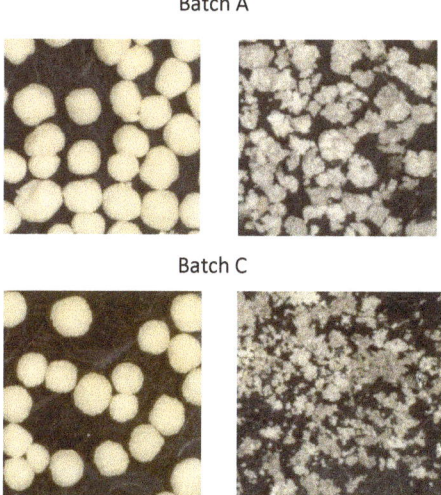

**Figure 9.** Images of pellets from batches A and C before (left) and after (right) the dissolution test at pH 1.2.

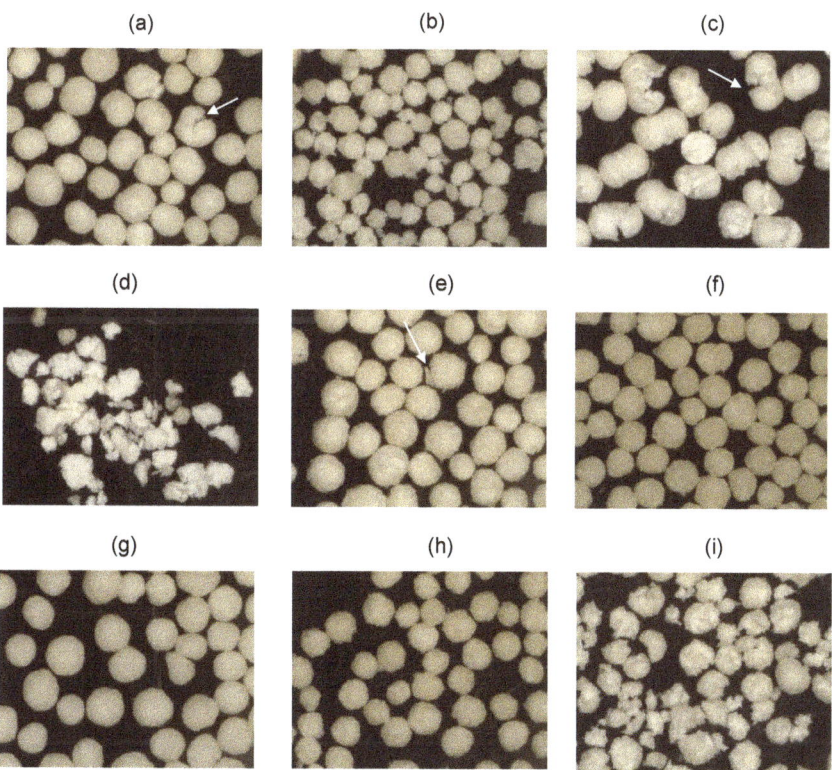

**Figure 10.** Images of pellets from all experimental batches (**a–i**) (Table 1) collected and dried after dissolution testing at pH 5.6 (Arrows indicate areas of disruption).

**Table 7.** Dissolution efficiency (DE%) (mean ±SD, $n$ = 3) and similarity factors (f2) comparing drug release from mini and conventional pellets at pH 1.2 and 5.6.

| Batch Code | DE% for Mini Pellets at: | | DE% for Conventional Pellets at: | | Similarity Factor (f2) Comparing Mini and Conventional Pellets at: | |
|---|---|---|---|---|---|---|
| | pH 1.2 | pH 5.6 | pH 1.2 | pH 5.6 | pH 1.2 | pH 5.6 |
| A | 67.56 ± 0.74 | 28.26 ± 0.17 | 69.48 ± 1.12 | 19.73 ± 0.39 | 77.90 | 51.93 |
| B | 38.85 ± 1.86 | 61.43 ± 1.52 | 30.44 ± 2.56 | 17.75 ± 1.89 | 44.44 | 37.00 |
| C | 44.87 ± 1.42 | 15.09 ± 1.05 | 58.53 ± 2.74 | 9.40 ± 1.14 | 50.83 | 61.14 |
| D | 64.46 ± 0.23 | 29.85 ± 0.68 | 70.16 ± 1.54 | 43.24 ± 0.27 | 62.07 | 53.94 |
| E | 70.70 ± 0.41 | 30.73 ± 0.37 | 71.01 ± 1.32 | 19.73 ± 0.84 | 49.30 | 47.75 |
| F | 43.78 ± 1.36 | 60.23 ± 1.89 | 52.44 ± 1.95 | 13.45 ± 1.07 | 52.48 | 31.74 |
| G | 60.64 ± 0.65 | 40.10 ± 0.41 | 66.85 ± 1.10 | 10.76 ± 0.31 | 69.82 | 45.43 |
| H | 70.41 ± 0.59 | 45.46 ± 1.24 | 67.03 ± 0.89 | 25.26 ± 0.27 | 41.26 | 35.77 |
| I | 71.84 ± 0.61 | 33.25 ± 1.18 | 73.05 ± 1.27 | 31.49 ± 0.02 | 53.02 | 65.12 |

Comparing Figure 8a with Figure 8c–d, it can be seen that release at pH 1.2 is generally faster than at pH 5.6. This can be explained partly by the greater 'apparent' solubility of the drug at pH 1.2 than at pH 5.6 in the presence of CHS2 (0.238 compared to 0.157 mg/mL, Figure 2), but mainly due to the different state of the pellet matrix in the two pH media. Since CHS dissolves in pH 1.2 but not in pH 5.6, the pellet matrix dissolves and erodes during dissolution testing, in parallel to diffusion (Figure 9). However, in deionized water, erosion occurs to a much lower extent (Figure 10). Additionally, from Figure 8a–d, it can be seen that the pellet batch prepared with MCC/PXC only (black line open circles) showed much lower release, which underlines the importance of chitosan as already reported for other drugs [57].

From Figure 8a,b, it can be seen that, in six of nine cases, drug release at pH 1.2 completes within 3 h. Faster release was obtained from batches H and I with high CHS/medium or high drug (Table 1) (60–70% within 30 min from mini and about 50% from conventional pellets. Together with batch E, these batches also gave higher DE% values for both mini and conventional pellets (Table 7). Batches B and F with high CHS/low PXC and batch C with low CHS/high PXC, respectively, showed the lowest DE% (Table 7) and extended release. However, since erosion was involved (Figure 9), the extended release observed was highly dependent on composition and difficult to control. For example, the curves of batches B and C are located closer together than those of batches A and C (Figure 8a,b), despite the greater similarity in compositions of the latter two (Table 1).

The plateaus in the curves in the 8 h dissolution-testing period corresponded to less than 100% release, which may be partly ascribed to changes in solubility due to transformation into a hydrate form of lower solubility during dissolution [35], but also to the continued slow release after 8 h. In fact, measurements over a longer dissolution time showed that drug released (%) was still increasing and reaching above 85% after 24 h for both mini and conventional pellets, with the exception of conventional pellet batch B (70% release). For the purpose of this study, the end of the test period was considered to be 8 h. The DE% values for the mini and conventional pellets were linearly correlated ($R^2$ = 0.880) and differences between their release profiles were small. Only batches B and H had similarity factors f2 < 50. These contained high CHS (80% and 54.37%, Table 1). Therefore, gel formation, diffusional release, and pellet size had higher significance than for other batches [30].

From Figure 8c,d, it can be seen that, at pH 5.6, the profiles of the mini pellets differed greatly from the conventional pellet profiles with the latter displaying much slower release (<50% after 8 h), which precludes practical use. From the values of the similarity factors f2 in Table 7, it appears that only batches A, C, D, and I showed similarity between mini and conventional pellets (f2 > 50). These batches all had high PXC content (37.5% or 70%) and exhibited erosion after dissolution testing (Figure 10), which alters the diffusional release. Figure 8c shows that the release curves of the mini pellets fall into three groups where: one group has a faster release (over 80% in 8 h) and comprises batches B and F, with high CSH2/PXC ratios of 16.0 and 9.0, respectively; a second group comprises batches G and H with CSH2/PXC ratios of 2.0 and 2.5 and releases 70–80% after 8 h; a the third group comprises batches A, D, E, and I with CSH2/PXC ratios of 0.91, 0.21, 1.25, and 1.27, respectively, showing low release between 50% and 60%. Batch C that had a CSH2/PXC ratio of 0.27, showed poor release. The latter five batches with reduced or poor release exhibited erosion and loss of matrix integrity after testing (Figure 10), which altered the beneficial release effect. On the other hand, batches B, F, G, and H of mini pellets, which did not disintegrate, showed good extended release over 8 h. Of these, batches B, F, and G also gave extended release at pH 1.2 and, therefore, appear to be promising extended-release formulations.

From the above, it appears that, when the CSH2/PXC ratio was ≥2, the pellet matrix retains the characteristics of the chitosan network, which results in extended and complete release after 8 h.

Table 5 shows statistically significant model equations derived from regression analysis of the experimental design describing the effect of composition on the dissolution efficiency (DE%) at pH 1.2 for mini and conventional pellets. The respective contour and trace plots are presented in Figure 11. From the contour plots, it appears that, at pH 1.2, greater DE% is obtained for intermediate levels of the three components, and the trace plots show clearly that CHS exerts the greatest effect. Statistically significant model equations for the effect of composition on DE% at pH 5.6 for mini and conventional pellets are also given in Table 5, with their respective contour and trace plots in Figure 12. It can be seen that, contrary to the effect of CHS2 at pH 1.2, at pH 5.6, greater DE% is obtained for the mini pellets at high CHS and low drug content and that these two components had the greatest effect (Figure 12). The conventional pellets showed very low DE% (values 9.40% to 31.49%, Table 7), which indicates poor release. The trace plots in Figure 12 show that for the conventional pellets at pH 5.6 the drug component had the greater effect on the release.

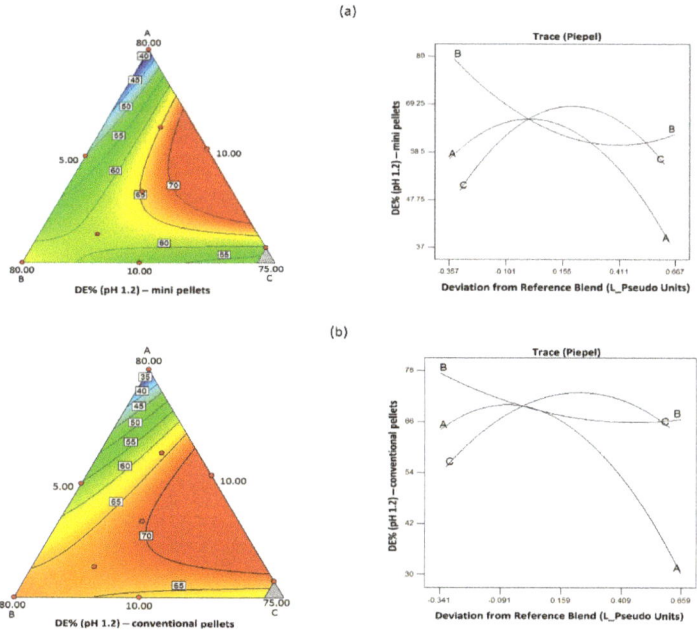

**Figure 11.** Contour and trace plots of dissolution efficiency for (**a**) mini and (**b**) conventional pellets at pH 1.2.

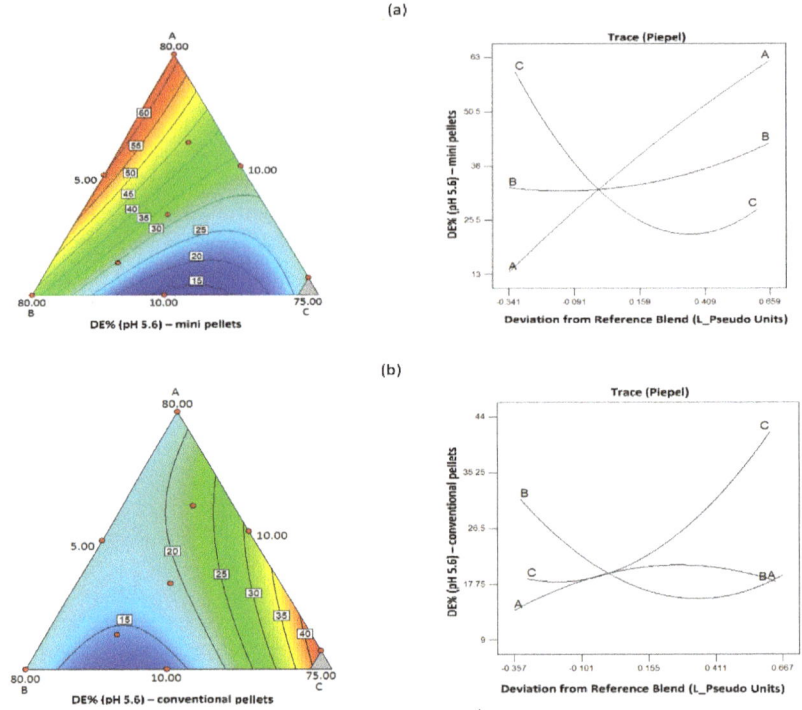

**Figure 12.** Contour and trace plots of dissolution efficiency for (**a**) mini and (**b**) conventional pellets at pH 5.6.

In-Vitro Release Mechanisms

Further elucidation of the release mechanisms is provided by examining the values of the '*b*' parameter from the Weibull model, as presented in Table 8 for release at pH 1.2 and pH 5.6 for mini and conventional pellets. The generally high $R^2$ values of the non-linear regressions indicate good model fitting. Considering the release times recorded at pH 1.2, it can be seen from Table 8 that, for the mini pellet batches A, B, C and G and the conventional pellet batches A, E, F, G, H, and I, the parameter '*b*' is >1.0. This indicates a sigmoidal curve, i.e., a small initial increase up to the infection point (not visible in the curves since it occurs within in the first 15 min) and followed thereafter by an increase asymptotically to maximum.

Concerning release from the mini pellet batches at pH 5.6, the data in Table 8 show that, with the exception of batches G and D, '*b*' parameter values ranged between 0.69 and 0.75, which indicates normal (Fickian) diffusion. Batch G presented $b = 0.63$, which indicates release by diffusion in a disordered matrix structure. This may be attributed to the high MCC or low CHS2 (10%) resulting in a non-homogenous structure and possible existence of 'dry' regions inside the gel during dissolution. Batch D showed $b = 0.76$, which indicates Fickian diffusion enhanced by a further release mechanism. This may be attributed to the high drug content and erosion during dissolution (Figure 10d). Regarding release at pH 5.6 for the conventional pellets, it can be seen from Table 8 that the '*b*' values varied between 0.69 and 0.96, which implies operation of normal or combined diffusion. However, because less than 50% of the drug was released in the 8-h test period, a detailed analysis of the mechanisms involved would be unreliable and of limited value.

**Table 8.** Weibull equation parameters (mean ±SD, n = 3) for the dissolution of mini and conventional pellets at pH 1.2 and 5.6.

| Batch Code | Parameters at pH 1.2 | | | | | | Parameters at pH 5.6 | | | | | |
|---|---|---|---|---|---|---|---|---|---|---|---|---|
| | Mini Pellets | | | Conventional Pellets | | | Mini Pellets | | | Conventional Pellets | | |
| | $b$ [#] | td | $R^2$ | $b$ | td | $R^2$ | $b$ | td | $R^2$ | $b$ | td | $R^2$ |
| A | 1.16 | 40.3 ± 1.7 | 0.996 | 1.14 | 36.8 ± 1.5 | 0.993 | 0.73 | 116.2 ± 1.9 | 0.969 | 0.71 | 107.1 ± 1.2 | 0.990 |
| B | 1.05 | 81.3 ± 1.1 | 0.988 | 0.78 | 83.4 ± 2.6 | 0.982 | 0.69 | 42.3 ± 1.2 | 0.879 | 0.67 | 32.2 ± 0.7 | 0.968 |
| C | 1.15 | 79.2 ± 0.6 | 0.992 | 0.96 | 69.7 ± 1.9 | 0.985 | 0.70 | 121.1 ± 2.6 | 0.980 | 0.85 | 120.6 ± 1.9 | 0.992 |
| D | 0.74 | 26.9 ± 2.3 | 0.912 | 0.74 | 26.9 ± 2.9 | 0.912 | 0.76 | 105.9 ± 1.8 | 0.938 | 0.79 | 88.2 ± 2.1 | 0.994 |
| E | 0.96 | 26.9 ± 1.4 | 0.997 | 1.36 | 42.7 ± 3.5 | 0.981 | 0.75 | 127.9 ± 1.7 | 0.931 | 0.70 | 114.4 ± 1.7 | 0.963 |
| F | 0.69 | 41.9 ± 2.5 | 0.973 | 1.10 | 55.8 ± 2.7 | 0.987 | 0.74 | 73.8 ± 0.9 | 0.915 | 0.76 | 59.2 ± 1.1 | 0.956 |
| G | 1.12 | 56.1 ± 3.2 | 0.987 | 1.10 | 51.0 ± 2.3 | 0.989 | 0.63 | 111.2 ± 1.4 | 0.838 | 0.79 | 64.6 ± 0.3 | 0.972 |
| H | 0.70 | 18.7 ± 1.3 | 0.921 | 0.99 | 55.6 ± 2.3 | 0.991 | 0.69 | 112.2 ± 1.8 | 0.935 | 0.78 | 104.3 ± 1.3 | 0.970 |
| I | 0.81 | 19.5 ± 2.1 | 0.981 | 1.18 | 33.1 ± 1.8 | 0.981 | 0.69 | 96.1 ± 1.4 | 0.962 | 0.96 | 118.2 ± 1.8 | 0.991 |

[#] For parameter $b$, SD < 0.01.

### 3.5. Optimization of Formulations

The numerical optimization was based on the modulation of the parameters $e_R$ (>0.30) and flowability (maximize) for optimal technological properties and release as follows. For fast release in pH 1.2 solution, the DE% was maximized and the td was minimized. For extended release, DE% in pH 1.2 was minimized. Release within 2 h in pH 5.6 was set to 60%, and release within 8 h at pH 5.6 was maximized. As seen from the regression analysis results in Table 5, the above parameters were described by statistically significant models with high values of fitting indices. Therefore, reliable prediction and optimization of the formulations with optimal technological performance and fast or extended release can be computed. The results of the numerical optimization are presented in Table 9 as desirability values for fast (A) and extended release (B). Due to the poor release from the conventional size pellets at pH 5.6, optimization is presented only for release at pH 1.2.

**Table 9.** Desirability values as computed by numerical optimization for fast (A) and extended release (B) pellets (Criteria for $e_R$, Flowability, DE%, and td are described in the text. Drug content was set to target levels shown in the fourth column).

| Pellet Size | CHS2 | MCC | Drug | $e_R$ | Flowability | DE% | td (pH 1.2) | Desirability |
|---|---|---|---|---|---|---|---|---|
| A. Numerical Optimization Solutions for Instant Release ||||||||||
| Mini Pellets | 10.55 | 79.45 | 5.0 | 0.33 | 2.70 | 63.97 | 47.94 | 0.925 |
| | 25.84 | 59.16 | 10.0 | 0.34 | 2.53 | 58.51 | 36.77 | 0.914 |
| | 35.04 | 44.96 | 15.0 | 0.33 | 2.40 | 64.30 | 32.39 | 0.925 |
| | 37.71 | 37.29 | 20.0 | 0.32 | 2.32 | 70.44 | 30.06 | 0.934 |
| | 36.34 | 33.66 | 25.0 | 0.32 | 2.27 | 74.57 | 28.91 | 0.932 |
| | 36.35 | 28.65 | 30.0 | 0.30 | 2.21 | 77.36 | 26.25 | 0.926 |
| | 32.14 | 27.86 | 35.0 | 0.30 | 2.18 | 77.60 | 28.12 | 0.919 |
| | 27.74 | 27.26 | 40.0 | 0.30 | 2.15 | 75.72 | 32.11 | 0.909 |
| Conventional | 46.81 | 43.19 | 5.0 | 0.40 | 1.95 | 54.33 | 52.51 | 0.889 |
| | 43.41 | 41.59 | 10.0 | 0.40 | 1.92 | 59.17 | 50.24 | 0.902 |
| | 39.24 | 40.76 | 15.0 | 0.40 | 1.88 | 63.37 | 48.28 | 0.910 |
| | 34.75 | 40.25 | 20.0 | 0.40 | 1.85 | 66.55 | 47.12 | 0.914 |
| | 30.10 | 39.90 | 25.0 | 0.40 | 1.81 | 68.57 | 46.94 | 0.912 |
| | 31.80 | 33.20 | 30.0 | 0.37 | 1.75 | 70.54 | 44.38 | 0.909 |
| | 32.91 | 27.09 | 35.0 | 0.34 | 1.68 | 72.69 | 41.08 | 0.904 |
| | 28.45 | 26.55 | 40.0 | 0.33 | 1.65 | 73.05 | 40.73 | 0.897 |

B. Numerical Optimization Solutions for Extended Release
Criteria: Shape index > 0.30. Flowability maximize. DE% at pH 1.2 maximize. Release in pH 5.6 in less than 2 h. Release in 5.6 at 8 h more than 70%. Drug content set to different target levels (column 4).

| | CHS2 | MCC | Drug | $e_R$ | Flowability | DE% | Release (2 h) | Release (8 h) | Desirability |
|---|---|---|---|---|---|---|---|---|---|
| Mini Pellets | 46.02 | 43.98 | 5.0 | 0.30 | 1.71 | 49.9 | 60.0 | 93.3 | 0.857 |
| | 47.32 | 37.68 | 10.0 | 0.33 | 1.61 | 54.0 | 53.1 | 85.4 | 0.726 |

Considering optimization at pH 1.2 for different preset drug levels, it can be seen that high values of desirability function above 0.919 (in a scale 0 to 1) were obtained for drug levels up to 40%. These represented pellets with acceptable shape ($e_R > 0.30$) and flowability (>2.15 g/min) that released drugs with dissolution efficiency between 63.97% and 77.60% for mini pellets and 54.33% to 73.05% for conventional pellets within a time-scale of 28.12 to 47.94 min and 40.73 to 52.51 min, respectively. Considering optimization at pH 5.6, it can be seen that good values of desirability function of 0.857 and 0.726 were obtained for pellet formulations with 5% or 10% drug content, with CHS2%/MCC%/PXC% compositions of 46.0/43.98/5.0 and 47.3/37.7/10.0, respectively (the analysis did not give meaningful results for drug levels exceeding 10%). These showed DE% values of less than 49.9% and 54% at pH 1.2 and pH 5.6, respectively, and released less than 60% and 53.1% within 2 h, and 93.3% and 85.4% after 8 h. Therefore, these formulations could be considered suitable for the further development of extended-release chitosan pellets of a poorly soluble non-interacting drug.

## 4. Conclusions

Medium viscosity chitosan grade prompted greater 'apparent' solubility improvement of piroxicam in both pH 1.2 and deionized water. However, due to the coherent gel consistency, drug release from conventional-sized pellets obtained with a 1-mm extrusion screen was extremely slow and excluded any possibility for practical use. On the contrary, use of mini pellets obtained with a 0.5 mm extrusion screen gave complete and extended release in deionized water following Fickian diffusion, as indicated by analysis of 8-h release data, according to the Weibull kinetic model. Batches of mini pellets prepared with a chitosan/microcrystalline cellulose ratio ≥2 and drug content up to 21.25% showed the best extended release that was nearly complete after 8 h and could be potentially used for further development. Optimization of the results of technological properties and drug release pointed out optimal formulations for fast-release mini pellets with CHS between 10.55% and 37.71%, MCC between 27.26% and 79.45%, and drugs up to 40%. For extended release, the optimal formulations were mini pellets with 5% drug, CHS2 = 46.02% and MCC% = 43.98%, and mini pellets with 10% drug, CHS2 = 47.32%, and MCC = 37.68%.

**Author Contributions:** Conceptualization, I.N. and P.G. Methodology, I.N. Software, I.N. and I.P. Experimentation and data processing, P.G., I.P. and N.K. Writing—original draft preparation, I.P., P.G. and I.N.

**Funding:** The purchase of Raman spectrometer was funded by the Research Program for the Development of Industrial Research in Greece and Technology Growth (NSRF2013) granted to VIANEX SA, Athens, Greece to whom we are grateful.

**Conflicts of Interest:** The authors declare no conflict of interest.

## References

1. Ghebre-Selassie, I.; Martin, C. *Pharmaceutical Extrusion Technology*; Taylor & Francis: Abingdon, UK, 2003.
2. Nikolakakis, I.; Partheniadis, I. Self-Emulsifying Granules and Pellets: Composition and Formation Mechanisms for Instant or Controlled Release. *Pharmaceutics* **2017**, *9*, 50. [CrossRef] [PubMed]
3. Akhgari, A.; Sadeghi, F.; Garekani, H.A. Combination of time-dependent and pH-dependent polymethacrylates as a single coating formulation for colonic delivery of indomethacin pellets. *Int. J. Pharm.* **2006**, *320*, 137–142. [CrossRef] [PubMed]
4. Prabhakaran, L.; Prushothaman, M.; Sriganesan, P. Pharmaceutical micropellets: An overview. *Pharm. Rev.* **2009**, *7*, 727–731.
5. Yoburn, B.C.; Chen, J.; Huang, T.; Inturrisi, C.E. Pharmacokinetics and pharmacodynamics of subcutaneous morphine pellets in the rat. *J. Pharmacol. Exp. Ther.* **1985**, *235*, 282–286. [PubMed]
6. Muley, S.; Nandgude, T.; Poddar, S. Extrusion–spheronization a promising pelletization technique: In-depth review. *Asian J. Pharm. Sci.* **2016**, *11*, 684–699. [CrossRef]
7. Reddy, P.N.S.; De, A.; Nagasamy Venkatesh, D. Pelletization process and techniques. *Pharma Times* **2015**, *47*, 22–27.
8. Pałkowski, Ł.; Karolak, M.; Kubiak, B.; Błaszczyński, J.; Słowiński, R.; Thommes, M.; Kleinebudde, P.; Krysiński, J. Optimization of pellets manufacturing process using rough set theory. *Eur. J. Pharm. Sciences* **2018**, *124*, 295–303. [CrossRef] [PubMed]
9. Sarkar, S.; Liew, C.V.; Soh, J.L.P.; Heng, P.W.S.; Wong, T.W. Microcrystalline cellulose: An overview. In *Functional Polymeric Composites: Macro Nanoscales*; Apple Academic Press: New York, NY, USA, 2017. [CrossRef]
10. Newton, J.M. Extrusion and extruders. In *Encyclopedia of Pharmaceutical Technology*; Swarbrick, J., Boylan, J.C., Eds.; Marcel Dekker Inc.: New York, NY, USA; Basel, Switzerland, 2002; pp. 1220–1236.
11. Shah, R.D.; Kabadi, M.; Pope, D.G.; Augsburger, L.L. Physico-mechanical characterization of the extrusion-spheronization process. Part II: Rheological determinants for successful extrusion and spheronization. *Pharm. Res.* **1995**, *12*, 496–507. [CrossRef] [PubMed]
12. Sonaglio, D.; Bataille, B.; Ortigosa, C.; Jacob, M. Factorial design in the feasibility of producing Microcel MC 101 pellets by extrusion/spheronization. *Int. J. Pharm.* **1995**, *115*, 53–60. [CrossRef]
13. Okada, S.; Nakahara, H.; Isaka, H. Adsorption of drugs on microcrystalline cellulose suspended in aqueous solutions. *Chem. Pharm. Bull.* **1987**, *35*, 761–768. [CrossRef]
14. Awa, K.; Shinzawa, H.; Ozaki, Y. The effect of microcrystalline cellulose crystallinity on the hydrophilic property of tablets and the hydrolysis of acetylsalicylic acid as active pharmaceutical ingredient inside tablets. *AAPS PharmSciTech* **2015**, *16*, 865–870. [CrossRef]
15. Basit, A.W.; Newton, J.M.; Lacey, L.F. Formulation of ranitidine pellets by extrusion-spheronization with little or no microcrystalline cellulose. *Pharm. Dev. Technol.* **1999**, *4*, 499–505. [CrossRef]
16. Brandl, M.; Magill, A.; Rudraraju, V.; Gordon, M.S. Approaches for improving the stability of ketorolac in powder blends. *J. Pharm. Sci.* **1995**, *84*, 1151–1153. [CrossRef] [PubMed]

17. O'Connor, R.E.; Schwartz, J.B. Spheronization II: Drug Release from Drug-Diluent Mixtures. *Drug Dev. Ind. Pharm.* **1985**, *11*, 1837–1857. [CrossRef]
18. Otero-Espinar, F.J.; Luzardo-Alvarez, A.; Blanco-Méndez, J. Non-MCC materials as extrusion-spheronization aids in pellets production. *J. Drug Deliv. Sci. Technol.* **2010**, *20*, 303–318. [CrossRef]
19. Goskonda, S.R.; Hileman, G.A.; Upadrashta, S.M. Controlled release pellets by extrusion-spheronization. *Int. J. Pharm.* **1994**, *111*, 89–97. [CrossRef]
20. Cheung, R.C.; Ng, T.B.; Wong, J.H.; Chan, W.Y. Chitosan: An Update on Potential Biomedical and Pharmaceutical Applications. *Mar. Drugs* **2015**, *13*, 5156–5186. [CrossRef]
21. Muzzarelli, R.A.A. Chitins and chitosans as immunoadjuvants and non-allergenic drug carriers. *Mar. Drugs* **2010**, *8*, 292–312. [CrossRef]
22. Castellsague, J.; Riera-Guardia, N.; Calingaert, B.; Varas-Lorenzo, C.; Fourrier-Reglat, A.; Nicotra, F.; Sturkenboom, M.; Perez-Gutthann, S. Individual NSAIDs and upper gastrointestinal complications: A systematic review and meta-analysis of observational studies (the SOS project). *Drug Saf.* **2012**, *35*, 1127–1146. [CrossRef]
23. Imai, T.; Shiraishi, S.; Saitô, H.; Otagiri, M. Interaction of indomethacin with low molecular weight chitosan, and improvements of some pharmaceutical properties of indomethacin by low molecular weight chitosans. *Int. J. Pharm.* **1991**, *67*, 11–20. [CrossRef]
24. Ito, M.; Ban, A.; Ishihara, M. Anti-ulcer effects of chitin and chitosan, healthy foods, in rats. *Jpn. J. Pharmacol.* **2000**, *82*, 218–225. [CrossRef]
25. Zerrouk, N.; Mennini, N.; Maestrelli, F.; Chemtob, C.; Mura, P. Comparison of the effect of chitosan and polyvinylpyrrolidone on dissolution properties and analgesic effect of naproxen. *Eur. J. Pharm. Biopharm.* **2004**, *57*, 93–99. [CrossRef]
26. Charoenthai, N.; Kleinebudde, P.; Puttipipatkhachorn, S. Use of chitosan-alginate as alternative pelletization aid to microcrystalline cellulose in extrusion/spheronization. *J. Pharm. Sci.* **2007**, *96*, 2469–2484. [CrossRef] [PubMed]
27. Sakkinen, M.; Seppala, U.; Heinanen, P.; Marvola, M. In vitro evaluation of microcrystalline chitosan (MCCh) as gel-forming excipient in matrix granules. *Eur. J. Pharm. Biopharm.* **2002**, *54*, 33–40. [CrossRef]
28. Bernkop-Schnurch, A.; Dunnhaupt, S. Chitosan-based drug delivery systems. *Eur. J. Pharm. Biopharm.* **2012**, *81*, 463–469. [CrossRef] [PubMed]
29. Bonferoni, M.C.; Rossi, S.; Ferrari, F.; Bertoni, M.; Bolhuis, G.K.; Caramella, C. On the employment of λ carrageenan in a matrix system. III. Optimization of a λ carrageenan–HPMC hydrophilic matrix. *J. Controll. Release* **1998**, *51*, 231–239. [CrossRef]
30. Tapia, C.; Buckton, G.; Newton, J.M. Factors influencing the mechanism of release from sustained release matrix pellets, produced by extrusion/spheronisation. *Int. J. Pharm.* **1993**, *92*, 211–218. [CrossRef]
31. Agrawal, A.M.; Manek, R.V.; Kolling, W.M.; Neau, S.H. Water distribution studies within microcrystalline cellulose and chitosan using differential scanning calorimetry and dynamic vapor sorption analysis. *J. Pharm. Sci.* **2004**, *93*, 1766–1779. [CrossRef]
32. Izdes, S.; Orhun, S.; Turanli, S.; Erkilic, E.; Kanbak, O. The effects of preoperative inflammation on the analgesic efficacy of intraarticular piroxicam for outpatient knee arthroscopy. *Anesth. Analg.* **2003**, *97*, 1016–1019. [CrossRef] [PubMed]
33. Mohammadi-Samani, S.; Zojaji, S.; Entezar-Almahdi, E. Piroxicam loaded solid lipid nanoparticles for topical delivery: Preparation, characterization and in vitro permeation assessment. *J. Drug Deliv. Sci. Technol.* **2018**, *47*, 427–433. [CrossRef]
34. Shohin, I.E.; Kulinich, J.I.; Ramenskaya, G.V.; Abrahamsson, B.; Kopp, S.; Langguth, P.; Polli, J.E.; Shah, V.P.; Groot, D.W.; Barends, D.M.; et al. Biowaiver monographs for immediate release solid oral dosage forms: Piroxicam. *J. Pharm. Sci.* **2014**, *103*, 367–377. [CrossRef] [PubMed]
35. Paaver, U.; Lust, A.; Mirza, S.; Rantanen, J.; Veski, P.; Heinamaki, J.; Kogermann, K. Insight into the solubility and dissolution behavior of piroxicam anhydrate and monohydrate forms. *Int. J. Pharm.* **2012**, *431*, 111–119. [CrossRef] [PubMed]
36. Sanka, K.; Bandari, S.; Jukanti, R.; Veerareddy, P. Colon-Specific Microparticles of Piroxicam: Formulation and Optimization Using 3 2 Factorial Design. *J. Disper. Sci. Technol.* **2011**, *32*, 1396–1403. [CrossRef]
37. Abdulkarim, M.F.; Abdullah, G.Z.; Chitneni, M.; Salman, I.M.; Ameer, O.Z.; Yam, M.F.; Mahdi, E.S.; Sattar, M.A.; Basri, M.; Noor, A.M. Topical piroxicam in vitro release and in vivo anti-inflammatory and analgesic effects from palm oil esters-based nanocream. *Int. J. Nanomed.* **2010**, *5*, 915–924. [CrossRef] [PubMed]

38. Mohammed, M.A.; Syeda, J.T.M.; Wasan, K.M.; Wasan, E.K. An Overview of Chitosan Nanoparticles and Its Application in Non-Parenteral Drug Delivery. *Pharmaceutics* **2017**, *9*, 53. [CrossRef]
39. Papadopoulou, V.; Kosmidis, K.; Vlachou, M.; Macheras, P. On the use of the Weibull function for the discernment of drug release mechanisms. *Int. J. Pharm.* **2006**, *309*, 44–50. [CrossRef] [PubMed]
40. Santos, H.; Veiga, F.; Pina, M.; Podczeck, F.; Sousa, J. Physical properties of chitosan pellets produced by extrusion-spheronisation: Influence of formulation variables. *Int. J. Pharm.* **2002**, *246*, 153–169. [CrossRef]
41. Matsaridou, I.; Barmpalexis, P.; Salis, A.; Nikolakakis, I. The influence of surfactant HLB and oil/surfactant ratio on the formation and properties of self-emulsifying pellets and microemulsion reconstitution. *AAPS PharmSciTech* **2012**, *13*, 1319–1330. [CrossRef]
42. Podczeck, F.; Newton, J.M. A Shape Factor to Characterize the Quality of Spheroids. *J. Pharm. Pharmacol.* **1994**, *46*, 82–85. [CrossRef]
43. General Chapters: <1174> Powder Flow. Available online: http://www.pharmacopeia.cn/v29240/usp29nf24s0_c1174.html (accessed on 2 November 2018).
44. Langenbucher, F. Linearization of dissolution rate curves by the Weibull distribution. *J. Pharm. Pharmacol.* **1972**, *24*, 979–981. [CrossRef]
45. Myers, R.H.; Montgomery, D.C. *Process and Product Optimization Using Designed Experiments*; Wiley Series in Probability and Statistics; John Wiley and Sons: New York, NY, USA, 1995; pp. 590–591.
46. Drebushchak, V.A.; Shakhtshneider, T.P.; Apenina, S.A.; Medvedeva, A.S.; Safronova, L.P.; Boldyrev, V.V. Thermoanalyticalinvestigation of drug-excipient interaction. *J. Therm. Anal. Calorim.* **2006**, *86*, 303–309. [CrossRef]
47. Ivanova, D.; Deneva, V.; Nedeltcheva, D.; Kamounah, F.S.; Gergov, G.; Hansen, P.E.; Kawauchid, S.; Antonova, L. Tautomeric transformations of piroxicam in solution: A combined experimental and theoretical study. *RSC Adv.* **2015**, *5*, 31852–31860. [CrossRef]
48. Oka, S.; Emady, H.; Kašpar, O.; Tokárová, V.; Muzzio, F.; Štěpánek, F.; Ramachandran, R. The effects of improper mixing and preferential wetting of active and excipient ingredients on content uniformity in high shear wet granulation. *Powder Technol.* **2015**, *278*, 266–277. [CrossRef]
49. Tang, P.; Puri, V.M. Methods for Minimizing Segregation: A Review. *Part. Sci. Technol.* **2004**, *22*, 321–337. [CrossRef]
50. Balaxi, M.; Nikolakakis, I.; Kachrimanis, K.; Malamataris, S. Combined effects of wetting, drying, and microcrystalline cellulose type on the mechanical strength and disintegration of pellets. *J. Pharm. Sci.* **2009**, *98*, 676–689. [CrossRef]
51. Harwood, C.F.; Pilpel, N. The flow of granular solids through circular orifices. *J. Pharm. Pharmacol.* **1969**, *21*, 721–730. [CrossRef]
52. Kogermann, K.; Aaltonen, J.; Strachan, C.J.; Pollanen, K.; Veski, P.; Heinamaki, J.; Yliruusi, J.; Rantanen, J. Qualitative in situ analysis of multiple solid-state forms using spectroscopy and partial least squares discriminant modeling. *J. Pharm. Sci.* **2007**, *96*, 1802–1820. [CrossRef]
53. Liu, G.; Hansen, T.B.; Qu, H.; Yang, M.; Pajander, J.P.; Rantanen, J.; Christensen, L.P. Crystallization of Piroxicam Solid Forms and the Effects of Additives. *Chem. Eng. Technol.* **2014**, *37*, 1297–1304. [CrossRef]
54. International Center for Diffraction Data. *Standard Cards of Piroxicam, PDF Numbers 40-1982 and 44-1839*; ICDD: Newtown Square, PA, USA, 2003.
55. Albertini, B.; Cavallari, C.; Passerini, N. Evaluation of b-lactose, PVP K12 and PVP K90 as excipients to prepare piroxicam granules using two wet granulation techniques. *Eur. J. Pharm. Biopharm.* **2003**, *56*, 479–487. [CrossRef]
56. Khattab, A.; Zaki, N. Optimization and Evaluation of Gastroretentive Ranitidine HCl Microspheres by Using Factorial Design with Improved Bioavailability and Mucosal Integrity in Ulcer Model. *AAPS PharmSciTech* **2017**, *18*, 957–973. [CrossRef] [PubMed]
57. Nejati, L.; Kalantari, F.; Bavarsad, N.; Saremnejad, F.; Moghaddam, P.T.; Akhgari, A. Investigation of using pectin and chitosan as natural excipients in pellet formulation. *Int. J. Biol. Macromol.* **2018**, *120*, 1208–1215. [CrossRef] [PubMed]

© 2019 by the authors. Licensee MDPI, Basel, Switzerland. This article is an open access article distributed under the terms and conditions of the Creative Commons Attribution (CC BY) license (http://creativecommons.org/licenses/by/4.0/).

*Article*

# Thermally-Responsive Loading and Release of Elastin-Like Polypeptides from Contact Lenses

**Wan Wang** [1,†], **Changrim Lee** [1,†], **Martha Pastuszka** [1], **Gordon W. Laurie** [2] **and J. Andrew MacKay** [1,3,4,*]

1. Department of Pharmacology and Pharmaceutical Sciences, School of Pharmacy, University of Southern California, Los Angeles, CA 90033, USA; leilawangwan@gmail.com (W.W.); changril@usc.edu (C.L.); martha.past@gmail.com (M.P.)
2. Department of Cell Biology, School of Medicine, University of Virginia, Charlottesville, VA 22908, USA; gwl6s@virginia.edu
3. Department of Biomedical Engineering, Viterbi School of Engineering, University of Southern California, Los Angeles, CA 90089, USA
4. Department of Ophthalmology, USC Roski Eye Institute, Keck School of Medicine, University of Southern California, Los Angeles, CA 90089, USA
* Correspondence: jamackay@usc.edu; Tel.: +1-323-442-4119
† Authors contributed equally to this research.

Received: 25 March 2019; Accepted: 24 April 2019; Published: 7 May 2019

**Abstract:** Contact lenses are widely prescribed for vision correction, and as such they are an attractive platform for drug delivery to the anterior segment of the eye. This manuscript explores a novel strategy to drive the reversible adsorption of peptide-based therapeutics using commercially available contact lenses. To accomplish this, thermo-sensitive elastin-like polypeptides (ELPs) alone or tagged with a candidate ocular therapeutic were characterized. For the first time, this manuscript demonstrates that Proclear Compatibles™ contact lenses are a suitable platform for ELP adsorption. Two rhodamine-labelled ELPs, V96 (thermo-sensitive) and S96 (thermo-insensitive), were employed to test temperature-dependent association to the contact lenses. During long-term release into solution, ELP coacervation significantly modulated the release profile whereby more than 80% of loaded V96 retained with a terminal half-life of ~4 months, which was only 1–4 days under solubilizing conditions. A selected ocular therapeutic candidate lacritin-V96 fusion (LV96), either free or lens-bound LV96, was successfully transferred to HCE-T cells. These data suggest that ELPs may be useful to control loading or release from certain formulations of contact lenses and present a potential for this platform to deliver a biologically active peptide to the ocular surface via contact lenses.

**Keywords:** elastin-like polypeptide (ELPs); contact lens; lacritin; protein therapeutics; drug delivery

## 1. Introduction

As growth factors and peptides derived from the tear proteome are explored as novel therapies for the anterior segment [1], it may be worthwhile to explore new drug delivery platforms that can be integrated with contact lenses [2]. New platforms may benefit from being biocompatible, biodegradable, and compatible with existing medical devices [3]. One such platform explored by our group and others are the elastin-like polypeptides (ELPs) [4]. ELPs are composed of repeated pentameric peptides, (Val-Pro-Gly-Xaa-Gly)$_n$. They reversibly phase separate from aqueous solution above a transition temperature ($T_t$) which can be tuned by adjusting the identity of a guest amino acid (Xaa) and the length ($n$) [5]. Like parent ELPs, ELP fusion proteins 'coacervate' above $T_t$; furthermore, this assembly process can functionalize pharmacologically drug carriers [6] or imaging probes [7]. Our group previously demonstrated the encapsulation [8] and fusion [9] ability of thermo-responsive elastin-like polypeptides (ELPs) carrying either small molecules or protein treatments to ocular tissue. Now, this manuscript reports the surprising discovery that ELPs significantly attach and dramatically extend release from a commercially available contact lens, Proclear Compatibles™. Using this discovery, two hypotheses were tested: 1) coacervation enhances attachment and slows the detachment of ELPs to and from contact lens; 2) ELP fusions with a biologically active peptide can transfer proteins from the lens to a cell-culture model of the corneal epithelium. By involving two types of ELPs, V96 (thermo-sensitive) and S96 (thermo-insensitive), the data show that the attachment and release of ELPs to contact lenses is both ELP and incubation temperature dependent. As a proof of the concept that ELPs can deliver a fusion protein, we modified the lens with a prosecretory mitogenic fusion called LV96 and demonstrated that the proximity between the LV96 on the contact lens enhances transfer to cultured human corneal epithelial cells.

## 2. Materials and Methods

### 2.1. Synthesis, Expression and Purification of ELPs

cDNAs encoding either ELPs V96, S96, or LV96 were cloned into the pET-25b(+) vector that was originally purchased from Novagen (#69753, Madison, WI, USA) and further modified for ELP or ELP fusion cloning [10]. The cloned constructs were sequenced, transformed into and expressed in BLR(DE3) competent *Escherichia* Coli (*E.* Coli) (#69053, Novagen, Madison, WI, USA). For V96 and S96, both were fermented in terrific broth media for 16–18 h at 37 °C without isopropyl β-D-1-thiogalactopyranoside (IPTG). For LV96, it was fermented in terrific broth media for 4 h at 37 °C followed by 0.5 mM IPTG induction. The temperature was immediately decreased to 25–30 °C and it was fermented for another 5–6 h. For all ELPs, the supernatant was subjected to ELP-mediated phase separation in 2 M sodium chloride at 37 °C after bacterial cell lysis and clarification of cell debris by centrifugation. Coacervates were immediately pelleted after the phase separation was observed (hot-spin). After centrifugation, soluble impurities (supernatant) were removed and coacervates (pellet) were resolubilized in clean ice-cold phosphate buffered saline (PBS). Thoroughly resolubilized ELPs were centrifuged to remove any insoluble impurities (cold-spin). After the cold-spin, the supernatant was transferred to a clean tube. Cycles of hot-spin followed by cold-spin were repeated 3 times to achieve the necessary purity. LV96 was further subjected to size exclusion chromatography to remove the cleaved byproduct.

### 2.2. ELPs Inverse Phase Transition Characterization

The $T_t$-concentration phase diagrams for rhodamine-labeled ELPs or ELP fusion proteins were characterized by optical density observation at 350 nm (OD 350nm) using a DU800 UV–Vis spectrophotometer (Beckman Coulter, CA, USA) as a function of solution temperature. Different concentrations of ELPs (5, 10, 25, 50, and 100 µM) were heated at 1 °C/min from 10 to 85 °C and OD 350 nm was recorded every 0.3 °C. $T_t$ was defined at the point of the maximum first derivative. The $T_t$ from each concentration was used to plot the phase diagram and fit with Equation 1.

$$T_t = b - m \, \log_{10}[C_{ELP}] \tag{1}$$

### 2.3. Rhodamine Labeling of V96, S96 and LV96, and Decoration of Proclear Compatibles™ Contact Lenses

ELPs were covalently modified with N-hydroxysuccinimide (NHS)-Rhodamine (Thermo Fisher Scientific Inc, Rockford, IL, USA). The conjugation was performed in 100 mM borate buffer (pH 8.5) overnight at 4 °C to covalently conjugate amine reactive NHS-esters to the primary amine at the ELP amino terminus. Excess fluorophore was removed using a desalting PD-10 column (GE Healthcare, Piscataway, NJ, USA) and overnight dialysis against PBS at 4 °C. For the initial screening study, contact lenses were either incubated with 50 µM labeled ELPs overnight at 37 °C in a 24-well plate or spot-decorated with concentrated, labeled ELPs using a 20-µL pipette at 37 °C. Proclear Compatibles™ contact lenses (CooperVision, Inc., Lake Forest, CA, USA) were incubated in 100 µM rhodamine-labeled

V96 or S96 for 48 h at 4 or 37 °C. After a gentle rinse with ddH$_2$O at 4 or 37 °C, contact lenses were immediately imaged using Zeiss 510 confocal laser scanning microscopy (Carl Zeiss AG, Oberkochen, Germany), respectively at 37 or at 4 °C, and quantified using ImageJ. Decoration of LV96 onto contact lenses in a ring shape was achieved by overnight incubation of 50 µM rhodamine-labeled LV96 with Proclear Compatibles™ contact lens at 37 °C followed by washing off LV96 attached at the center of the lens using ice-cold PBS and pipetting out.

### 2.4. Characterization of Release Kinetics of ELPs from Proclear Compatibles™ Contact Lenses

Contact lenses were incubated in 100 µM rhodamine-labeled V96 or S96 for 24 h at 4 °C or 37 °C. After one gentle rinse with PBS at 4 or 37 °C, contact lenses were immediately placed into 4 mL of PBS at 4 or 37 °C for 1 week. Small aliquots of the solution (100 µL) were withdrawn at predetermined intervals (5, 15, 30 min, 1, 2, 4, 8, 24, 48, 72, 96, 120, 168 h) and kept at −20 °C. After one week, lenses were thoroughly washed in PBS at 4 °C for 24 h to detach ELPs. Fluorescence intensity of collected samples was measured spectrophotometrically (Ex: 525 nm, Em: 575 nm) using a Synergy™ H1m Monochromator-Based Multi-Mode Microplate Reader (BioTek Instruments, Inc., Winooski, VT, USA) and analyzed using built-in Gen5 2.01 Data Analysis Software (BioTek). Total fluorescence on the lens was calculated using Equation 2. Since the measurement of the contact lens-bound fraction at each time point was distorted due to the convex shape of the contact lenses, the percent of retention on the lens was defined at each time point using Equation 3. Using GraphPad Prism (Prism Software, Irvine, CA, USA), these retention data failed to fit one-phase dissociation model; however, a two-phase dissociation model (Equation 4) fit well to the observed profiles. Goodness of fit and predicted values are reported.

$$Total\ I_{rhodamine} = I_{release\_Total} + I_{wash\_Total} \tag{2}$$

$$Retention(t) = \frac{Total\ I_{rhodamine} - \sum_{t=0}^{t} I_{release\_t}}{Total\ I_{rhodamine}} \times 100\% \tag{3}$$

$$Retention(t) = Percent_{fast}\ e^{-k_{fast}t} + \left(100 - Percent_{fast}\right)e^{-k_{slow}t} \tag{4}$$

$$AUC_{0-Infinity} = AUC_{0-168h} + \%_{last}/k_{slow} \tag{5}$$

### 2.5. Human Corneal Epithelial Cells-Transformed with SV40 (HCE-T) Uptake Study

HCE-T cellular uptake was conducted on 35-mm glass coverslip-bottomed dishes. Briefly, HCE-T cells were grown to 70–80% confluence and gently rinsed with warm fresh medium before changing to fresh media containing either rhodamine-labeled lacritin (10 µM, protein concentration), LV96 (10 µM) or contact lenses loaded with rhodamine-labeled LV96. After incubation at 37 °C for 1 h, cells were rinsed with fresh media, incubated with 4′,6-diamidino-2-phenylindole (DAPI) for 15 min to stain nuclei, and then imaged using a Zeiss 510 confocal microscope system (Carl Zeiss AG, Oberkochen, Germany) with quantification by ImageJ (National Institutes of Health, Bethesda, MD, USA). To evaluate the transfer of contact lens-bound LV96 to the monolayer of HCE-T cells, images from different zones were directly obtained at the edge of the lens where the highest likelihood of direct contact between the lens and the monolayer occurred.

### 2.6. Statistical Analysis

All experiments were repeated at least three times. Statistical analysis was performed by Student's t-test or one-way ANOVA followed by Tukey's post-hoc test using statistical software IBM SPSS Statistics v21 (IBM Corp., Armonk, NY, USA). A $p$ value of less than 0.05 was considered statistically significant.

## 3. Results

### 3.1. Expression and Purification of ELPs

All ELPs involved in this study, V96, S96, and LV96, were heterologously expressed from a seamlessly cloned synthetic gene in *E. coli* (Table 1). Purification was done via inverse transition cycling [10], which is a non-chromatographic purification method that utilizes ELP-mediated phase separation from clarified bacterial lysates supplemented with 1~2 M NaCl to induce phase separation [11]. The final material after purification yielded ~90 mg/L of V96, ~40 mg/L of S96, and ~10 mg/L of LV96 with > 98% purity, as verified by SDS-PAGE (Figure 1A). The precise determination of molecular weight by MALDI-TOF for V96, S96, and LV96 was reported previously [10,12]. To determine the $T_t$ of ELPs, optical density at 350 nm over a range of temperatures was measured (Figure 1B). All ELPs tested showed a negative correlation between the $T_t$ and the ELP concentration [13], and the phase diagram was fit by Equation 1 (Table 1).

Table 1. Summary of the elastin-like polypeptides (ELPs) involved in this study.

| Label | Amino Acid Composition | *MW (kDa) | $T_t$ (°C) at 25 μM | Phase Diagram | |
|---|---|---|---|---|---|
| | | | | Slope, $m$ [°C/$\log_{10}$(μM)] | y-intercept, $b$ [°C] |
| S96 | G(VPGSG)$_{96}$Y | 38.4 | 57.6 | −1.669 | 59.31 |
| V96 | G(VPGVG)$_{96}$Y | 39.5 | 31.6 | −3.252 | 36.06 |
| LV96 | **Lacritin-G(VPGVG)$_{96}$Y | 52.3 | 26.8 | −1.192 | 28.56 |

*MW determined by MALDI-TOF analysis. **Lacritin (12.7 kDa) amino acid sequence: EDASSDSTGADPAQEAGTSKPNEEISGPAEPASPPETTTTAQETSAAAVQGTAKVTSSRQELNPLKSIVEKSILLTE QALAKAGKGMHGGVPGGKQFIENGSEFAQKLLKKFSLLKPWA.

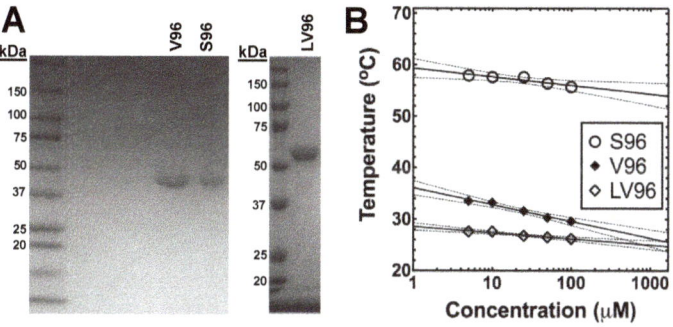

**Figure 1.** The purity, size, and temperature-dependent phase behavior of V96, S96, and LV96 evaluated for this study. (**A**) Identity and purity of V96, S96 and LV96 were analyzed by SDS-PAGE using Coomassie blue staining. (**B**) The phase transition temperature was plotted vs. concentration as a phase diagram, below which ELPs remains soluble, and fit with Equation 1. Solid line: Fit; Dashed line: 95% confidence interval of mean.

### 3.2. ELPs Display Differential Attachment to Commercially Available Contact Lenses

Discovery of ELPs' attachment to contact lenses came from a quick screen of four commonly marketed contact lenses, including Acuvue Oasys®, Acuvue Advance Plus®, Dailies AquaComfort Plus™ and Proclear Compatibles™ (Table 2). Surprisingly, rhodamine-labeled V96 selectively attached to Proclear Compatibles™ contact lenses at 37 °C after overnight incubation in PBS solution. This attachment remained stable at 37 °C in PBS solution for more than 24 h (Figure 2A). Motivated by the rationale that the delivery system itself should not scatter light within the central visual field, we investigated whether it was possible to arrange the ELP only around the periphery of the contact lens using a cold wash. Interestingly, by controlling the location of cold washing and warm spotting, the final deposition pattern on the lens could be controlled (Figure 2B).

**Table 2.** Summary of the contact lenses involved in this study.

| Brand Name | Manufacturer | Polymer | Monomer | ELP Attachment |
|---|---|---|---|---|
| Proclear Compatibles™ | CooperVision | Omafilcon A | pHEMA/PC | + |
| Dailies AquaComfort Plus™ | CIBA Vision | Nelfilcon A | HPMC/PEG/PVA | − |
| Acuvue Oasys® | Johnson & Johnson | Senofilcon A | pHEMA + DMA + mPDMS + siloxane macromer + TEGDMA + PVP | − |
| Acuvue Advanced Plus® | Johnson & Johnson | Galyfilcon A | pHEMA + DMA + mPDMS + siloxane macromer + EGDMA + PVP | − |

pHEMA: poly(hydroxyethyl methacrylate); HPMC: Hydroxypropyl methylcellulose; PC: phosphorylcholine; mPDMS: monofunctional poly(dimethylsiloxane); DMA: N,N-dimethylacrylamide; EGDMA: ethyleneglycol dimethacrylate; TEGDMA: tetraethyleneglycol dimethacrylate; PVP: poly(vinyl pyrrolidone); PVA: poly(vinyl alcohol); PEG: poly(ethylene glycol).

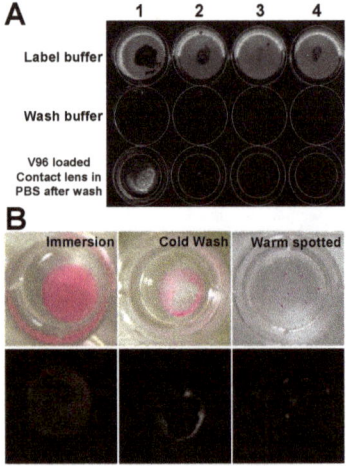

**Figure 2.** ELP selectively phase separate onto Proclear compatibles™ contact lens. (**A**) Among four types of contact lenses tested, rhodamine-labeled V96 preferentially phase separated onto Proclear compatibles™ contact lens. 1: Proclear compatibles™; 2: Dailies AquaComfort Plus™; 3: Acuvue OASYS®; 4: Acuvue Advance Plus®. Label buffer: 50 µM rhodamine-labeled V96 in PBS; Wash buffer: ddH$_2$O used for gentle wash after contact lens incubation with label buffer. White circles: each well in 12-well plate; yellow circles: contact lens in the well. (**B**) Different spatial deposition patterns for rhodamine-labelled V96 on Proclear compatibles™ contact lens were evaluated. The entire lens can be labeled during complete immersion in a warm solution (immersion), the central field can be depleted by a cold PBS wash (cold wash), or individual positions can be labeled by warm pipet spotting. Upper: white light; lower: fluorescence.

## 3.3. ELP-Mediated Phase Separation Enhances Attachment to Proclear Compatibles[TM]

To test whether V96 attachment is due to coacervation of V96 at 37 °C, lenses were visualized following overnight V96 incubation at 37 °C (above $T_t$) or 4 °C (below $T_t$). There was a striking and significant difference in V96 deposition in response to coacervation (Figure 3A). Contact lenses were then incubated with V96 at 37 °C overnight and cut into halves. The first half was incubated at 4 °C and the second half was incubated at 37 °C. Incubation at 4 °C resulted in rapid dissociation of V96, whereas V96 was retained at 37 °C (Figure 3B). To test the effects of $T_t$ and incubation temperature on ELPs' affinity to contact lenses, V96 ($T_t$ = 29.6 °C, 100 µM, Equation 1) was compared to a heat-insensitive control S96 ($T_t$ = 60.0 °C, 100 µM, Equation 1). After 24 h incubation, total attachment of V96 at 37 °C was about 5-fold higher than that of S96 at 37 °C; and 59-fold higher than that of V96 at 4 °C and 8-fold higher than that of S96 at 4 °C (Figure 3C,D). The contact lens association with S96 at 37 °C, V96 at 4 °C, and S96 at 4 °C did not differ significantly from each other ($p > 0.50$). The difference in contact lens association between V96 and S96 at 37 °C was confirmed using confocal microscopy (Figure 3E). Heat-insensitive S96 washed away immediately prior to imaging. However, V96 coacervates decorated the lens uniformly, even after 3 days of incubation at 37 °C (Figure 3E). Although the specific biophysical interactions between ELPs and contact lenses remains to be explored, the ProClear lens composition (Table 2) clearly demonstrated both a non-specific association with S96 and a coacervate-dependent association with V96 when incubated above its transition temperature.

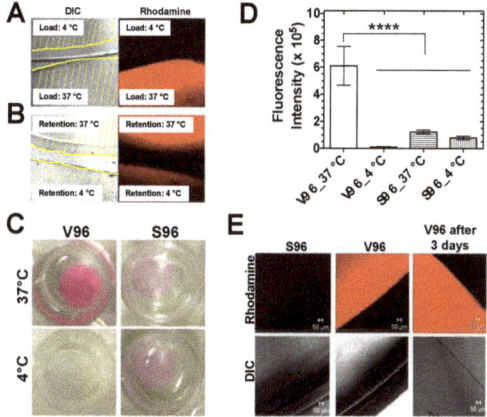

**Figure 3.** Coacervation of a temperature-responsive ELP enhances loading onto Proclear Compatibles[TM] contact lenses. (**A**) Contact lenses were loaded overnight at 4 or 37 °C with rhodamine-labeled V96 (100 µM) and imaged side by side using confocal microscopy. Each lens' location is depicted by yellow lines upon the differential interference contrast (DIC) channel. (**B**) A contact lens loaded with V96 (100 µM) at 37 °C was cut into halves and incubated at 4 or 37 °C in ddH$_2$O overnight. Side-by-side confocal microscopy shows that the half incubated at 37 °C retains most of the V96 label. (**C**) Shown are representative pictures of lenses loaded with rhodamine-labeled V96 or S96 at 37 or 4 °C for 24 h and washed. (**D**) Total fluorescence intensity associated with lenses loaded overnight with V96 or S96 at different incubation temperature. Mean ± SD, N = 3, ****$p < 0.0001$. Significant differences between: V96_37 °C vs. V96_4 °C ($p = 0.00004$); V96_37 °C vs. S96_37 °C ($p = 0.0002$); V96_37 °C vs. S96_4 °C ($p = 0.00009$). (**E**) Confocal microscopy was used to observe lenses incubated overnight with rhodamine-labeled S96 and V96 at 37 °C and gently washed. Even after 3 days at 37 °C, the V96 remained associated with the lens. Scale bar: 50 µm.

## 3.4. Coacervation Prolongs the Retention of ELPs on Proclear Compatibles[TM] Contact Lenses

Having demonstrated that ELP phase separation enhances loading of V96, the retention of ELPs was explored following washing. Five groups were evaluated for the retention of rhodamine-labeled

ELPs and lenses: group 1) load V96 at 37 °C and retention at 37 °C (V96_37 °C → 37 °C); group 2) load V96 at 37 °C and retention at 4 °C (V96_37 °C → 4 °C); group 3) load V96 at 4 °C and retention at 4 °C (V96_4 °C → 4°C); group 4) load S96 at 37 °C and retention at 37 °C (S96_37 °C → 37 °C); group 5) load S96 at 4 °C and retention at 4 °C (S96_4 °C → 4 °C). After one week of lens retention testing in PBS, group 1 (V96_37 °C → 37 °C) retained ~ 80% of the initial fluorescence, that was mostly lost from all others (Figure 4A,B). Groups 3, 4, and 5 showed similar retention profiles, while group 2 lost ~ 75% of initial signal during the first 24 h. When both incubation and retention temperatures were below $T_t$, little difference was observed in either total fluorescence loaded (Figure 3D) or retention (groups 3, 4, and 5). The relationship between ELP retention and coacervation is most evident by comparison of groups 1 and 2. To understand retention kinetics, we attempted to fit each dataset first by a one-phase and then by a two-phase decay model. The two-phase disassociation model was best ($p < 0.0001$) and was applied to the estimation of the terminal half-life and percentage of material lost to washing through fast release (Table 3). Most notably, the area under the curve (AUC) of group 1 during a one-week period (AUC$_{0-120}$) was about 4-fold higher than group 2, 2-fold higher than group 3, 3-fold higher than group 4, and 2-fold higher than group 5. The extrapolated total AUC (AUC$_{0-Inf}$) for group 1 was about 119-fold higher than group 2, 45-fold higher than group 3, 55-fold higher than group 4, and 44-fold higher for group 5.

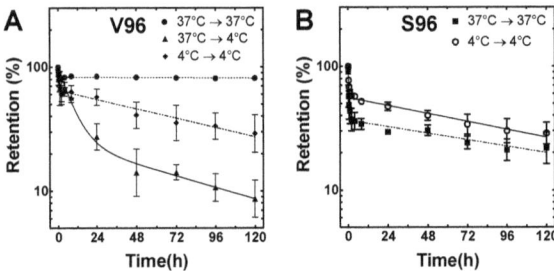

**Figure 4.** ELP retention on Proclear Compatibles™ lenses depends on ELP $T_t$ and incubation temperature. (**A**) Release profiles of group 1 (V96_37 °C → 37 °C), group 2 (V96_37 °C → 4 °C), and group 3 (V96_4 °C → 4 °C) were shown. (**B**) Release profiles of group 4 (S96_37 °C → 37 °C) and group 5 (V96_4 °C → 4 °C) were shown. Small aliquots of the incubation solution were sampled over time and the fluorescence intensity of these samples were measured to estimate lens retention (Equation 3). Lines joining data points represent a best-fit to a biexponential decay model (Equation 4). Mean ± SD, $N = 3$.

**Table 3.** Release kinetics of ELPs from Proclear Compatibles™ contact lenses.

| Parameters | Group 1 | Group 2 | Group 3 | Group 4 | Group 5 |
| --- | --- | --- | --- | --- | --- |
| ELP | V96 | V96 | V96 | S96 | S96 |
| Label Temp (°C) | 37 | 37 | 4 | 37 | 4 |
| Release Temp (°C) | 37 | 4 | 4 | 37 | 4 |
| Percent Fast (%) | 16.8 (15.6~18.0) | 75.0 (63.7~86.2) | 35.1 (27.9~42.2) | 63.3 (59.3~67.3) | 55.9 (52.1~59.6) |
| $k_{fast}$ (h$^{-1}$) | 2.9 (2.0~3.9) | 0.1 (0.06~0.2) | 3.3 (0.0~6.7) | 3.4 (2.0~4.7) | 2.4 (1.4~3.7) |
| $t_{1/2,fast}$ (h) | 0.2 (0.18~0.35) | 5.8 (4.0~10.9) | 0.2 (0.1~inf.) | 0.2 (0.1~0.3) | 0.3 (0.2~0.5) |
| $k_{slow}$ (h$^{-1}$) | 0.0002 (0.0~0.0004) | 0.009 (0.004~0.01) | 0.007 (0.005~0.009) | 0.005 (0.003~0.007) | 0.006 (0.005~0.007) |
| $t_{1/2,slow}$ (h) | 4615 (1815~inf.) | 78.3 (49.8~183.7) | 96.2 (76.9~128.4) | 137.1 (101.6~210.7) | 112.5 (95.9~136.0) |
| AUC$_{0-120h}$ | 9938 | 2565 | 5156 | 3245 | 4660 |
| AUC$_{0-Inf}$ | 418564 | 3525 | 9302 | 7608 | 9436 |
| $R^2$ | 0.89 | 0.88 | 0.81 | 0.82 | 0.93 |

*Fast* and *slow* represent the fast and slow exponential decay phase in the two-phase dissociation (decay) model, respectively. Values indicate the mean (95% CI).

## 3.5. Co-Incubation of LV96 with Proclear Compatibles™ Enables Transfer to Cultured HCE-T Cells

To explore cellular delivery from ELP-loaded contact lenses, lacritin, an abundant protein from normal human tears [14], was selected. Topical lacritin, including lacritin-ELP [12,15], promotes basal tearing and corneal wound repair in rabbit and mouse models [15] which makes it a potential treatment for dry eye disease and cornea wound healing. We first added rhodamine-labeled lacrtin-V96 (LV96) or recombinant lacritin (Lacrt) to HCE-T cells (Figure 5A). After 60 min, high levels of rhodamine-labeled lacritin had become internalized, whereas LV96 remained associated with the cell surface in lower relative amounts (Figure 5B). Accordingly, the average nuclei to closest rhodamine pixel distance was significantly greater for LV96 versus recombinant lacritin after both 10 and 60 min (Figure 5C), possibly due to steric hindrance with lacritin ligand syndecan-1 [16] on the cell surface and/or with endocytic machinery. These observations are in accordance with our previous report of comparably low cellular targeting and delay on cellular uptake of LV96 compared to Lacrt, mainly due to the fusion of V96 and its ability to coacervate at 37 °C [12]. Nonetheless, evidence of LV96 cell targeting was clearly apparent. We next tested delivery from LV96-decorated contact lenses in which rhodamine-labeled LV96 was restricted to a peripheral ring. HCE-T cells growing directly under (zone 1), adjacent (zone 2) or outside (zone 3) the ring were scrutinized after one hour (Figure 6A). Most zone 1 cells were covered with LV96, versus progressively less coverage of zones 2 and 3 cells (Figure 6B) with zone 3 showing negligible targeting and uptake (Figure 6C).

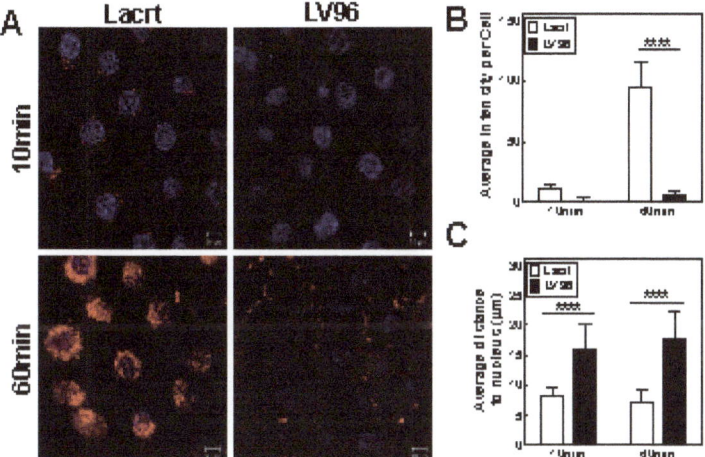

**Figure 5.** HCE-T cells associate with both recombinant Lacritin (Lacrt) and a Lacritin-V96 fusion (LV96). (**A**) Representative pictures showing live-HCE-T cell targeting and uptake of 10 μM rhodamine-labeled lacritin (Lacrt) or LV96 over 1 h at 37 °C in complete media. Red: rhodamine-labeled Lacrt or LV96; Blue: DAPI-stained nuclei. Bar = 10 μm. (**B**,**C**) Image analysis was used to quantify (**B**) integrated intensity per cell and (**C**) average distance to the nucleus of LV96 vs. Lacrt. Mean ± SD, $N$ = 9 measurements, ****$p$ < 0.0001.

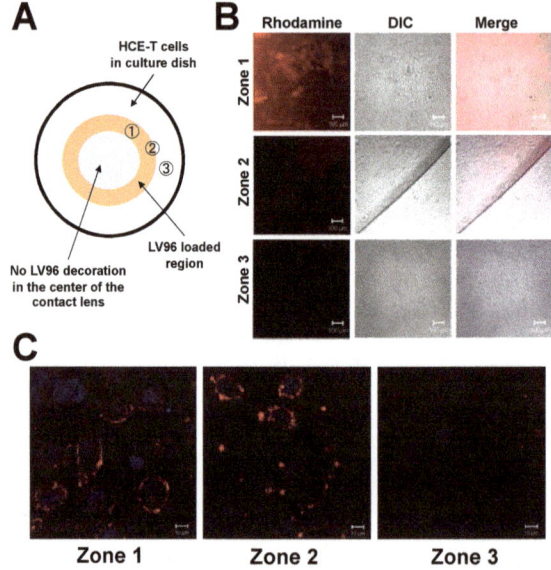

**Figure 6.** Spatial proximity is required for the efficient transfer of LV96 from Proclear Compatibles™ contact lenses to cultured HCE-T cells. (**A**) Cartoon showing contact lens loaded with a ring pattern of LV96 with three zones indicated: 1) under the lens; 2) at the edge of the lens, and 3) distal to the lens. (**B**) Confocal imaging was performed to confirm the location of the rhodamine-labeled LV96 bound to a lens above cultured HCE-T cells, which were incubated for 1 h at 37 °C in complete media. Bar = 100 µm. (**C**) High magnification images show efficient association of rhodamine-labeled LV96 with cultured HCE-T cells in zones 1 and 2. In zone 3, less labeling was apparent. Red: rhodamine-labeled LV96; Blue: DAPI-stained nuclei. Bar = 10 µm. Images shown are representative from at least three independent experiments.

## 4. Discussion

Drugs delivered by drops on the eye can suffer from an inefficient pharmacokinetic profile beginning with an initial transient overdose, followed by a prolonged period of drug insufficiency [17], further diminished by blinking, reflex tearing, and nasolacrimal system drainage. Only 1~7% by volume generally targets the eye [18]. Emerging drug delivery systems include: ophthalmic ointments, viscous polymer vehicles, nanoparticles, in situ gel-forming systems, iontophoresis, and modified punctal plugs [19–21]. Problems include the lack of optically transparency, instability, difficulty inserting and discomfort [22]. Druggable contact lenses offer an attractive alternative [23] as they are conceptually simple, and should not impair vision [24]. Strategies include simple immersion in drug [25,26], inclusion of drug-loaded colloidal nanoparticles [27,28] and molecular imprinting [29] with a focus on small molecule therapeutics, including cyclosporine A [30], timolol [31,32] and Latanoprost [33]. Also, adipose-derived stem-cells loaded contact lens were tested for the treatment of acute alkaline burns [34]. In general, none have succeeded in exerting full spatiotemporal control over drug delivery towards eliminating a drug bolus on application, and consequential side effects. Further, none have successfully developed a method to slowly deliver protein therapeutics [35,36], which is more challenging due to steric hindrance and complex template design. ELP fusion proteins offer a solution, and custom contact lenses are not required. We report here the selective adsorption of thermo-responsive ELPs to a commercially available contact lens, elucidate the $T_t$ dependence for attachment and detachment, and we further showed proof of the concept by the spatiotemporal delivery of model ocular protein drug lacritin via contact lens to HCE-T cells. Although ELP-contact lens delivery systems have not met the desirable zero-order release, 80% retention (group 1: V96_37 °C → 37 °C) after initial release may provide a way to maintain therapeutic dose for an extended time on the ocular surface that can overcome short half-lives, low tear bioavailability, and prolonged sub-therapeutic concentrations accompanying eye drop instillation. In addition, by changing the ELP composition and raising its transition temperature closer to that at the ocular surface, future studies may show that it is possible to titrate the rate of release necessary for optimal therapy.

The core questions that have to be answered would be the mechanism behind ELP-contact lens adsorption that is uniquely observed with Proclear Compatibles™ and further stabilization of ELP adsorption upon coacervation. As can be seen from the S96 association with the lenses, even below their transition temperature ($T_t$), ELPs nonspecifically adsorb to ProClear contact lenses. Upon an increase in temperature, we hypothesize that adsorbed ELPs nucleate coacervation, which recruits additional ELPs from solution. The exact interaction that leads to the adsorption of ELPs to ProClear remains unknown. Electrostatic interactions are ruled out because neither the ELP nor the ProClear formulation contain excess charged groups. Van-der Waals interactions are possible between the hydrophobic moiety on the abundant Valine residues on the ELP and hydrophobic groups on the ProClear formulation. A third possibility is that hydrogen bonding may play an important role for this adsorption. Two representative biopolymer modalities used to study temperature-dependent phase transition behavior are ELPs and Poly(N-isopropylacrylamide) (PNIPAM) [37]. Both polymers undergo coacervation above their transition temperature ($T_t$). During coacervation, highly unordered PNIPAM remains unordered with negligible amount of additional hydrogen bond formation [38]; however, highly unordered ELPs are thought to form more ordered structure (type II β-turns, β-spirals, or distorted β-sheets) that nucleate the growth of coacervate phases [39]. This process involves the formation of a hydrogen bond called '1–4 hydrogen bond' (C=O of the first residue (valine) and the NH of the fourth residue (guest residue)) within the ELP pentapeptide [40]. We propose that these abundant hydrogen bonds may participate in hydrogen bonding with organic phosphates on the phosphorylcholine (PC)-coated contact lens [41]. Given that every ELP pentamer forms one additional hydrogen bond upon coacervation, and each ELP contains 96 pentameric units, even small contributions from these hydrogen bonds to phosphate-mediated hydrogen bonding between ELP coacervates and contact lens may promote ELP adsorption and coacervation. It should be emphasized that the adsorption of V96 coacervates to the contact lens does not indicate this particular brand is sub-optimal in preventing protein adsorption because ELPs and their coacervates do not have the amino acid compositions, biophysical properties, or affinity for typical proteins [42]. Given that only Proclear Compatibles™ contains phosphorylcholine on the contact lens surface among those tested, further investigations comparing the adsorption of ELP coacervates to pHEMA vs. pHEMA+PC may give more insights towards the molecular bonding that links ELPs to this formulation of lenses.

Several aspects have to be considered when using ELP coacervates on the lens for future therapeutic purposes. First, the presence of the ELP layer may alter $O_2$ permeability which may lead to insufficient oxygen supply to the cornea surface. Second, ELP coacervates may affect visual acuity. These possible limitations can be solved by loading ELP coacervates in a defined region, e.g., on the edge of the ring (Figures 2 and 6). This may guarantee sufficient oxygen supply to the cornea surface with clear vision, while ELP drugs diffuse into the cornea surface.

The delivery of therapeutic ELP fusion proteins to the ocular system and its compatibility, biodistribution, or therapeutic efficacy were studied by us and other groups. Examples include the delivery of intravitreal αB crystallin ELP fusion to a mouse model of age-related macular degeneration [43], topical lacritin-ELP as an eyedrop for healing of mouse corneal wound model [44], intravitreal injection of poly(VPAVG) particles to a normal rabbit model [45], and cell-penetrating peptide ELP fusions to a normal rabbit model [46]. As these modalities have shown ELPs as a highly promising ocular drug delivery platform, it is necessary to have a unique set of evidence that the delivery of ELPs or ELP fusions can be rerouted to the ocular system via contact lenses in a more hassle-free fashion.

Different from other reported contact lens-mediated drug delivery systems, our study represents a 'drug refillable system,' which would enable refill of a drug at home, by the patient [47,48]. Given the panel of ELP fusion modalities developed in our laboratory, the data presented in this study indicate that this system's application catalogue can be broadened to anti-inflammation agents, antibiotics, polypeptides and diverse protein/antibody therapeutic libraries via encapsulation or recombinant protein expression strategies. While its underlying mechanism remains to be elucidated, our discovery may provide a promising new avenue to circumvent challenges associated with the effective delivery of therapeutics to the ocular surface.

**Author Contributions:** Conceptualization, W.W., G.W.L., and J.A.M.; Methodology, W.W. and J.A.M.; Formal Analysis, W.W. and C.L.; Investigation, J.A.M.; Resources, G.W.L. and J.A.M.; Data Curation, W.W., C.L., M.P., and J.A.M.; Writing – Original Draft Preparation, W.W.; Writing – Review & Editing, C.L.; Supervision, J.A.M.; Project Administration, G.W.L. and J.A.M.; Funding Acquisition, G.W.L. and J.A.M.

**Funding:** This research was funded by the USAMRMC/TATRC 11262019, NIH R21EB012281, NIH R01EY026635, NIH R01GM114839 to J.A.M., NIH R01EY024327 and NIH R01EY026171 to G.W.L., P30 EY029220 to the USC Ophthalmology Center Core Grant for Vision Research, P30 CA014089 to the USC Norris Comprehensive Cancer Center, University of Southern California (USC), USC Stevens Technology Advancement Grant (TAG), USC Ming Hsieh Institute, USC L.K Whittier Foundation, and an unrestricted departmental grant from Research to Prevent Blindness (RPB). The authors would like to acknowledge the support of the USC School of Pharmacy Translational Research Core for this study. The authors thank Dr. Kraig Scot Bower, M.D. for donating some of the contact lenses used in this study.

**Conflicts of Interest:** J.A.M., G.W.L., and W.W, are inventors on patents describing the delivery of protein therapeutics to the ocular surface by contact lenses related to this work. G.W.L. also holds patents or patents pending on the use of lacritin for treating dry eye, and is cofounder of TearSolutions, Inc, that is currently testing the efficacy of a lacritin synthetic peptide for Sjogren's syndrome dry eye in a phase 2 clinical trial. The funders had no role in the design of the study; in the collection, analyses, or interpretation of data; in the writing of the manuscript, or in the decision to publish the results.

## References

1. Duncan, R. The dawning era of polymer therapeutics. *Nat. Rev. Drug Discov.* **2003**, *2*, 347–360. [CrossRef] [PubMed]
2. Kearney, C.J.; Mooney, D.J. Macroscale delivery systems for molecular and cellular payloads. *Nat. Mater.* **2013**, *12*, 1004–1017. [CrossRef] [PubMed]
3. Langer, R.; Tirrell, D.A. Designing materials for biology and medicine. *Nature* **2004**, *428*, 487–492. [CrossRef]
4. Hubbell, J.A.; Chilkoti, A. Nanomaterials for Drug Delivery. *Science* **2012**, *337*, 303–305. [CrossRef]
5. MacEwan, S.R.; Chilkoti, A. Elastin-like polypeptides: Biomedical applications of tunable biopolymers. *Biopolymers* **2010**, *94*, 60–77. [CrossRef]
6. Chilkoti, A.; Christensen, T.; MacKay, J.A. Stimulus responsive elastin biopolymers: Applications in medicine and biotechnology. *Curr. Opin. Chem. Biol.* **2006**, *10*, 652–657. [CrossRef] [PubMed]
7. Janib, S.M.; Moses, A.S.; MacKay, J.A. Imaging and drug delivery using theranostic nanoparticles. *Adv. Drug Deliv. Rev.* **2010**, *62*, 1052–1063. [CrossRef] [PubMed]
8. Shah, M.; Edman, M.C.; Janga, S.R.; Shi, P.; Dhandhukia, J.; Liu, S.Y.; Louiee, S.G.; Rodgers, K.; MacKay, J.A.; Hamm-Alvarez, S.F. A rapamycin-binding protein polymer nanoparticle shows potent therapeutic activity in suppressing autoimmune dacryoadenitis in a mouse model of Sjogren's syndrome. *J. Control. Release* **2013**, *171*, 269–279. [CrossRef]
9. Wang, W.; Sreekumar, P.G.; Valluripalli, V.; Shi, P.; Wang, J.; Lin, Y.A.; Cui, H.; Kannan, R.; Hinton, D.R.; Mackay, J.A. Protein polymer nanoparticles engineered as chaperones protect against apoptosis in human retinal pigment epithelial cells. *J. Control. Release* **2014**, *191*. [CrossRef]
10. Janib, S.M.; Pastuszka, M.; Aluri, S.; Folchman-Wagner, Z.; Hsueh, P.Y.; Shi, P.; Yi, A.; Cui, H.; Mackay, J.A. A quantitative recipe for engineering protein polymer nanoparticles. *Polym. Chem.* **2014**, *5*, 1614–1625. [CrossRef]
11. Cho, Y.; Zhang, Y.; Christensen, T.; Sagle, L.B.; Chilkoti, A.; Cremer, P.S. Effects of Hofmeister anions on the phase transition temperature of elastin-like polypeptides. *J. Phys. Chem. B* **2008**, *112*, 13765–13771. [CrossRef] [PubMed]

12. Wang, W.; Jashnani, A.; Aluri, S.R.; Gustafson, J.A.; Hsueh, P.Y.; Yarber, F.; McKown, R.L.; Laurie, G.W.; Hamm-Alvarez, S.F.; MacKay, J.A. A thermo-responsive protein treatment for dry eyes. *J. Control. Release* **2015**, *199*, 156–167. [CrossRef]
13. Despanie, J.; Dhandhukia, J.P.; Hamm-Alvarez, S.F.; MacKay, J.A. Elastin-like polypeptides: Therapeutic applications for an emerging class of nanomedicines. *J. Control. Release* **2016**, *240*, 93–108. [CrossRef] [PubMed]
14. Sanghi, S.; Kumar, R.; Lumsden, A.; Dickinson, D.; Klepeis, V.; Trinkaus-Randall, V.; Frierson, H.F., Jr.; Laurie, G.W. cDNA and genomic cloning of lacritin, a novel secretion enhancing factor from the human lacrimal gland. *J. Mol. Biol.* **2001**, *310*, 127–139. [CrossRef]
15. McKown, R.L.; Wang, N.N.; Raab, R.W.; Karnati, R.; Zhang, Y.H.; Williams, P.B.; Laurie, G.W. Lacritin and other new proteins of the lacrimal functional unit. *Exp. Eye Res.* **2009**, *88*, 848–858. [CrossRef]
16. Ma, P.; Beck, S.L.; Raab, R.W.; McKown, R.L.; Coffman, G.L.; Utani, A.; Chirico, W.J.; Rapraeger, A.C.; Laurie, G.W. Heparanase deglycanation of syndecan-1 is required for binding of the epithelial-restricted prosecretory mitogen lacritin. *J. Cell Biol.* **2006**, *174*, 1097–1106. [CrossRef]
17. Urtti, A. Challenges and obstacles of ocular pharmacokinetics and drug delivery. *Adv. Drug Deliv. Rev.* **2006**, *58*, 1131–1135. [CrossRef]
18. Lang, J.C. Ocular Drug-Delivery Conventional Ocular Formulations. *Adv. Drug Deliv. Rev.* **1995**, *16*, 39–43. [CrossRef]
19. Gaudana, R.; Ananthula, H.K.; Parenky, A.; Mitra, A.K. Ocular Drug Delivery. *AAPS. J.* **2010**, *12*, 348–360. [CrossRef] [PubMed]
20. Gaudana, R.; Jwala, J.; Boddu, S.H.S.; Mitra, A.K. Recent Perspectives in Ocular Drug Delivery. *Pharm. Res.* **2009**, *26*, 1197–1216. [CrossRef]
21. Novack, G.D. Ophthalmic Drug Delivery: Development and Regulatory Considerations. *Clin. Pharmacol. Ther.* **2009**, *85*, 539–543. [CrossRef] [PubMed]
22. MSintzel, B.; Bernatchez, S.F.; Tabatabay, C.; Gurny, R. Biomaterials in ophthalmic drug delivery. *Eur. J. Pharm. Biopharm.* **1996**, *42*, 358–374.
23. Guzman-Aranguez, A.; Colligris, B.; Pintor, J. Contact Lenses: Promising Devices for Ocular Drug Delivery. *J. Ocul. Pharmacol. Ther.* **2013**, *29*, 189–199. [CrossRef] [PubMed]
24. Ali, Y.; Lehmussaari, K. Industrial perspective in ocular drug delivery. *Adv. Drug Deliv. Rev.* **2006**, *58*, 1258–1268. [CrossRef]
25. Li, C.C.; Chauhan, A. Ocular transport model for ophthalmic delivery of timolol through p-HEMA contact lenses. *J. Drug Deliv. Sci. Technol.* **2007**, *17*, 69–79. [CrossRef]
26. Kim, J.; Chauhan, A. Dexamethasone transport and ocular delivery from poly(hydroxyethyl methacrylate) gels. *Int. J. Pharm.* **2008**, *353*, 205–222. [CrossRef]
27. Gulsen, D.; Chauhan, A. Dispersion of microemulsion drops in HEMA hydrogel: A potential ophthalmic drug delivery vehicle. *Int. J. Pharm.* **2005**, *292*, 95–117. [CrossRef] [PubMed]
28. Kapoor, Y.; Thomas, J.C.; Tan, G.; John, V.T.; Chauhan, A. Surfactant-laden soft contact lenses for extended delivery of ophthalmic drugs. *Biomaterials* **2009**, *30*, 867–878. [CrossRef]
29. Alvarez-Lorenzo, C.; Yanez, F.; Barreiro-Iglesias, R.; Concheiro, A. Imprinted soft contact lenses as norfloxacin delivery systems. *J. Control. Release* **2006**, *113*, 236–244. [CrossRef] [PubMed]
30. Peng, C.C.; Chauhan, A. Extended cyclosporine delivery by silicone-hydrogel contact lenses. *J. Control. Release* **2011**, *154*, 267–274. [CrossRef]
31. Hiratani, H.; Alvarez-Lorenzo, C. The nature of backbone monomers determines the performance of imprinted soft contact lenses as timolol drug delivery systems. *Biomaterials* **2004**, *25*, 1105–1113. [CrossRef]
32. Jung, H.J.; Abou-Jaoude, M.; Carbia, B.E.; Plummer, C.; Chauhan, A. Glaucoma therapy by extended release of timolol from nanoparticle loaded silicone-hydrogel contact lenses. *J. Control. Release* **2013**, *165*, 82–89. [CrossRef]
33. Ciolino, J.B.; Stefanescu, C.F.; Ross, A.E.; Salvador-Culla, B.; Cortez, P.; Ford, E.M.; Wymbs, K.A.; Sprague, S.L.; Mascoop, D.R.; Rudina, S.S.; Trauger, S.A.; Cade, F.; Kohane, D.S. In vivo performance of a drug-eluting contact lens to treat glaucoma for a month. *Biomaterials* **2014**, *35*, 432–439. [CrossRef]
34. Espandar, L.; Caldwell, D.; Watson, R.; Blanco-Mezquita, T.; Zhang, S.; Bunnell, B. Application of adipose-derived stem cells on scleral contact lens carrier in an animal model of severe acute alkaline burn. *Eye Contact Lens* **2014**, *40*, 243–247. [CrossRef]

35. Leader, B.; Baca, Q.J.; Golan, D.E. Protein therapeutics: A summary and pharmacological classification. *Nat. Rev. Drug Discov.* **2008**, *7*, 21–39. [CrossRef]
36. Frokjaer, S.; Otzen, D.E. Protein drug stability: A formulation challenge. *Nat. Rev. Drug Discov.* **2005**, *4*, 298–306. [CrossRef]
37. Laukkanen, A.; Valtola, L.; Winnik, F.M.; Tenhu, H. Formation of colloidally stable phase separated poly(N-vinylcaprolactam) in water: A study by dynamic light scattering, microcalorimetry, and pressure perturbation calorimetry. *Macromolecules* **2004**, *37*, 2268–2274. [CrossRef]
38. Maeda, Y.; Higuchi, T.; Ikeda, I. Change in hydration state during the coil-globule transition of aqueous solutions of poly(N-isopropylacrylamide) as evidenced by FTIR spectroscopy. *Langmuir* **2000**, *16*, 7503–7509. [CrossRef]
39. Debelle, L.; Alix, A.J.; Jacob, M.P.; Huvenne, J.P.; Berjot, M.; Sombret, B.; Legrand, P. Bovine elastin and kappa-elastin secondary structure determination by optical spectroscopies. *J. Biolog. Chem.* **1995**, *270*, 26099–26103. [CrossRef]
40. Baker, E.N.; Hubbard, R.E. Hydrogen bonding in globular proteins. *Prog. Biophys. Mol. Biol.* **1984**, *44*, 97–179. [CrossRef]
41. Hansen, A.S.; Du, L.; Kjaergaard, H.G. Positively Charged Phosphorus as a Hydrogen Bond Acceptor. *J. Phys. Chem. Lett.* **2014**, *5*, 4225–4231. [CrossRef] [PubMed]
42. Court, J.L.; Redman, R.P.; Wang, J.H.; Leppard, S.W.; Obyrne, V.J.; Small, S.A.; Lewis, A.L.; Jones, S.A.; Stratford, P.W. A novel phosphorylcholine-coated contact lens for extended wear use. *Biomaterials* **2001**, *22*, 3261–3272. [PubMed]
43. Sreekumar, P.G.; Li, Z.; Wang, W.; Spee, C.; Hinton, D.R.; Kannan, R.; MacKay, J.A. Intra-vitreal alphaB crystallin fused to elastin-like polypeptide provides neuroprotection in a mouse model of age-related macular degeneration. *J. Control. Release* **2018**, *283*, 94–104. [CrossRef]
44. Wang, W.; Despanie, J.; Shi, P.; Edman-Woolcott, M.C.; Lin, Y.A.; Cui, H.; Heur, J.M.; Fini, M.E.; Hamm-Alvarez, S.F.; MacKay, J.A. Lacritin-mediated regeneration of the corneal epithelia by protein polymer nanoparticles. *J. Mater. Chem. B* **2014**, *2*, 8131–8141. [CrossRef]
45. Rincon, A.C.; Molina-Martinez, I.T.; de las Heras, B.; Alonso, M.; Bailez, C.; Rodriguez-Cabello, J.C.; Herrero-Vanrell, R. Biocompatibility of elastin-like polymer poly(VPAVG) microparticles: In vitro and in vivo studies. *J. Biomed. Mater. Res. A* **2006**, *78*, 343–351. [CrossRef] [PubMed]
46. George, E.M.; Mahdi, F.; Logue, O.C.; Robinson, G.G.; Bidwell, G.L., 3rd. Corneal Penetrating Elastin-Like Polypeptide Carriers. *J. Ocul. Pharmacol. Ther.* **2016**, *32*, 163–171. [CrossRef]
47. Jung, H.J.; Chauhan, A. Temperature sensitive contact lenses for triggered ophthalmic drug delivery. *Biomaterials* **2012**, *33*, 2289–2300. [CrossRef]
48. Stuart, M.A.C.; Huck, W.T.S.; Genzer, J.; Muller, M.; Ober, C.; Stamm, M.; Sukhorukov, G.B.; Szleifer, I.; Tsukruk, V.V.; Urban, M.; et al. Emerging applications of stimuli-responsive polymer materials. *Nat. Mater.* **2010**, *9*, 101–113. [CrossRef]

© 2019 by the authors. Licensee MDPI, Basel, Switzerland. This article is an open access article distributed under the terms and conditions of the Creative Commons Attribution (CC BY) license (http://creativecommons.org/licenses/by/4.0/).

*Review*

# Ionic Polymethacrylate Based Delivery Systems: Effect of Carrier Topology and Drug Loading

Dorota Neugebauer *, Anna Mielańczyk, Rafał Bielas, Justyna Odrobińska, Maria Kupczak and Katarzyna Niesyto

Faculty of Chemistry, Department of Physical Chemistry and Technology of Polymers, Silesian University of Technology, 44-100 Gliwice, Poland
* Correspondence: dorota.neugebauer@polsl.pl; Tel.: +48-322-371-973

Received: 14 June 2019; Accepted: 12 July 2019; Published: 15 July 2019

**Abstract:** The presented drug delivery polymeric systems (DDS), i.e., conjugates and self-assemblies, based on grafted and star-shaped polymethacrylates have been studied for the last few years in our group. This minireview is focused on the relationship of polymer structure to drug conjugation/entrapment efficiency and release capability. Both graft and linear polymers containing trimethylammonium groups showed the ability to release the pharmaceutical anions by ionic exchange, but in aqueous solution they were also self-assembled into nanoparticles with encapsulated nonionic drugs. Star-shaped polymers functionalized with ionizable amine/carboxylic groups were investigated for drug conjugation via ketimine/amide linkers. However, only the conjugates of polybases were water-soluble, giving opportunity for release studies, whereas the self-assembling polyacidic stars were encapsulated with the model drugs. Depending on the type of drug loading in the polymer matrix, their release rates were ordered as follows: Physical ≥ ionic > covalent. The studies indicated that the well-defined ionic polymethacrylates, including poly(ionic liquid)s, are advantageous for designing macromolecular carriers due to the variety of structural parameters, which are efficient for tuning of drug loading and release behavior in respect to the specific drug interactions.

**Keywords:** polymer carriers; drug delivery; conjugates; self-assemblies; star polymers; graft polymers; poly(ionic liquid)s

---

## 1. Introduction

Conventional drug delivery formulations have significantly contributed to the effectiveness of disease treatment. Nevertheless, there is still a strong need to create modern carriers, including polymers, which are supposed to improve control of pharmacokinetics and biodistribution of "classic" and new drugs, their selective accumulation with reduced side effects, and enhanced effectiveness of therapeutic treatment. The progress in drug delivery has been advanced by the use of polymeric carriers for noninvasive and spatiotemporal release of different therapeutics [1,2]. The functional materials [3] with great biocompatibility, and optional biodegradability [4], are based on the "tailor-made" polymers with the well-defined hydrophilic/hydrophobic balance, particle size, or electric charge distribution. In contrast to linear polymers, the branched topology offers a broader spectrum of structural parameters, such as length and number of grafts/arms, and higher content of reactive groups, which can be used to adjust physicochemical properties responsible for efficiency of drug introduction and delivery. The polymers with sophisticated architectures, like star-shaped [5] and graft copolymers [6,7], are provided, by strategy of macromolecular engineering [8], to design carriers for drug delivery systems (DDS) with programmed activities, including controlled drug release profile, specific targeting to diseased tissues, and prolonged release time.

The nanosized DDS are classified into drug conjugates [9,10] and self-assembling systems [11]. The conjugate formation requires suitable functionalities in the polymer to attach bioactive compounds by covalent bonding, whereas proper hydrophilic–hydrophobic balance in the amphiphilic polymer affords the self-assembling behavior in aqueous solution resulting in micellar and aggregate superstructures with capability of drug entrapment via physical interactions [12]. The amphiphilic polymer conjugates can also be designed to provide dual DDS [13,14], containing two drugs loaded with various strength into the polymer matrix (conjugation vs. encapsulation) in additional respect to both drug and polymer nature. The interesting alternative is drug attachment via ionic bonding, which is weaker than the covalent bond due to electrostatic interactions between ions with opposite charges, but it seems to be more stable than the physical interactions. This variant requires the use of ionic polymers, including poly(ionic liquid)s (PIL) [15], where the counterions can be biologically active.

Both star-shaped [5,16] and graft copolymers [6,17] are convenient for introduction of multiple terminal active/functional groups, which can be used as conjugation sites. Generally, the nonlinear polymers form more stable micelles, which are characterized by longer time release of the drug than that of linear block copolymers [18]. Due to this, the latter have been stabilized by cross-linking in the core and/or shell (e.g., the doughnut shape micelles) [19]. The stability of micelles based on graft or star copolymers has been developed by hydrophobic–hydrophilic block structures of side chains/arms, which yielded amphiphilic core–shell cylindrical brushes [20], miktobrushes [21], or scorpion-like polymers [22]. These macromolecules, with the well-organized hydrophobic inner surrounded by hydrophilic outer layer, exhibited low critical micelle concentrations (CMC) and higher drug loading capacity compared with micelles of linear block copolymers [23]. It has also been reported that the star-shaped copolymers, in comparison to their linear analogs with similar molecular weight and composition, exhibit lower solution viscosity and smaller hydrodynamic radius, which is beneficial in excretion of system after drug release [24]. Additionally, in the case of the presence of polyester segments, which were shorter due to branching into the arms, the lower crystallinity improved control of degradation in correlation with the enhanced drug release [25]. Another advantage of polymers containing acidic units [26] has been indicated by pH activated drug release, i.e., significantly faster release at pH below 7.4 than at neutral pH (37 °C), which was observed for micellar systems of block copolymers grafted with 2-alkanone chains via acid-sensitive linker providing pH-dependent degradation [27]. The pH-dependent systems were also investigated for graft copolymers containing acidic units in the backbone or in the side chains [20], as well as for star copolymers with polyacidic segments in the arms [28]. Moreover, disability of self-assembling for some amphiphilic linear copolymers has been efficiently solved by their grafting onto polymer backbone [29].

In recent years our work was focused on the non-linear polymethacrylates designed for the nanosized DDS, including conjugates and self-assemblies (Figure 1). These studies provided better understanding of the influence of the hydrophobic/hydrophilic content on physicochemical and delivery properties of polymer carriers, which were varied by topology (graft vs. linear and stars vs. miktostars) and architecture (grafting degree, length of backbone and side chains or number and length of arms, core type) to regulate physical entrapment or chemical attachment of a drug. Polymethacrylates with trimethylammonium groups carrying salicylate anions (Sal$^-$), which can be classified as the grafted and linear poly(ionic liquid)s (GPIL1 and LPIL1, respectively), demonstrated the release of pharmaceutical anions by ionic exchange with phosphate ones in buffer solution. Additionally, these polymers, as well as the analogical ones containing Cl$^-$ (GPIL2, LPIL2), were self-assembled into the micellar carriers of non-ionic drugs, such as indomethacin (IMC) or erythromycin (ERY). Star-shaped polymethacrylates with D-glucopyranoside core were functionalized with carboxylic/amine groups (DGL1 and DGL2, respectively) to conjugate doxorubicin (DOX), which might be released by decomposition of hydrolysable covalent bonds. The polyacidic stars with pentaerythritol (PTL1-3) or D-glucopyranoside core (DGL3), including miktoarmed copolymers with extra poly($\varepsilon$-caprolactone) (PCL) arms (DGL4), were also studied for encapsulation of the model drugs (DOX, IMC), which were delivered by polymeric micelles via diffusion process. The naming protocol of the discussed systems, for example GPIL1.1, consists of the symbol corresponding to the polymer group with the first number describing the series of polymers, and the second number identifying the sample in the series. In all these systems, the polymer composition and architecture can be used to adjust drug content and release properties, which were investigated to verify chemical potential of the prepared polymers as the drug carriers.

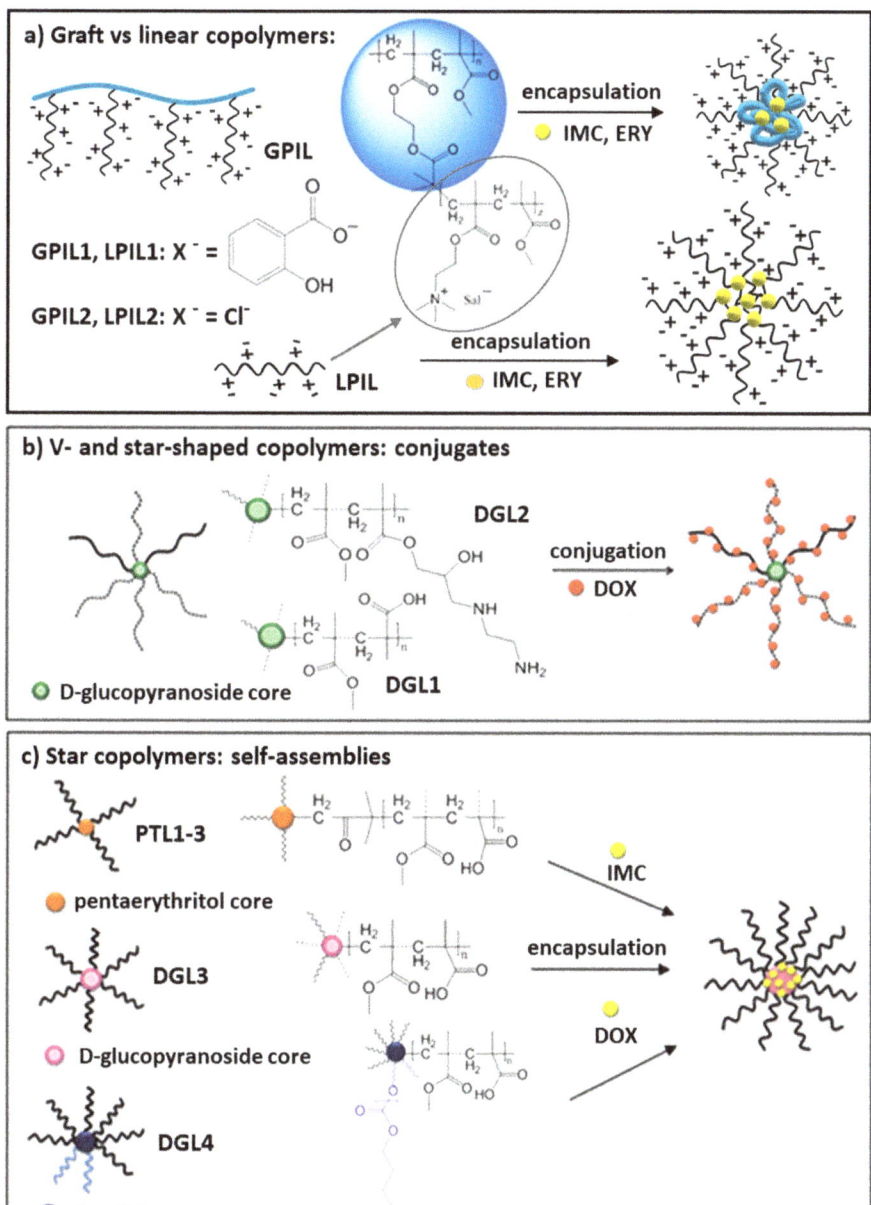

**Figure 1.** Drugs carried by grafted poly(ionic liquid)s (GPIL) (a), and star shaped polymers (DGL, PTL) (b,c), in the form of conjugates and self-assemblies.

## 2. Non-Linear Polymers Containing Ionic Groups in Drug Delivery

In our studies, the atom transfer radical polymerization (ATRP) has been used to obtain the well-defined polymers with various topologies, i.e., grafted copolymers, as well as star copolymers, including V-shaped and miktoarmed structures (Figure 1). Pre-polymerization replacement of chloride anion in monomeric ionic liquid by pharmaceutical one, i.e., salicylate anion (Sal$^-$), led to the design of the polymerized ionic drug carriers (DDS type I with ionically bonded drug). Post-polymerization modifications allowed for the introduction of specific functional groups (carboxyl or amine), which provided the possibility of chemical conjugation of the chosen drug (DDS type II with covalently bonded drug), whereas the induced amphiphilicity supported drug encapsulation (DDS type III with drug entrapped via physical interactions). The amphiphilic nature of macromolecules with diverse topology, including the grafted poly(ionic liquid)s with chloride anions, was beneficial in forming the self-assembling carriers in aqueous solution.

*2.1. Poly(Ionic Liquid) Graft Copolymers (DDS Type I)*

Previously designed by our group, amphiphilic ionic graft copolymers for DDS were based on anionic polyelectrolyte (polyacid) segments grafted from polymethacrylate backbone [30,31], or used as an extension of polyether (polyethylene or polypropylene glycol) side chains [32,33]. In aqueous solutions, depending on nature of side chains, they were self-assembled into different core–shell superstructures with ability for successful encapsulation of IMC [34–36].

The ionic properties are also represented by unique poly(ionic liquid)s, which are made of ionic monomers containing organic cation and organic or inorganic anion [37]. They gained great interest in material science because of macromolecular architectures, which can be tailored by combining both properties of cations and anions [38]. In our recent studies on the amphiphilic graft copolymers, the monomeric ionic liquids, i.e., commercially available (2-(methacryloyloxy)ethyl)-trimethylammonium chloride (ChMACl known as choline methacrylate) and containing pharmaceutical salicylate anion (ChMASal, [39,40]), were grafted from the standard multifunctional ATRP macroinitiators, i.e., poly(methyl methacrylate-*co*-(2-(2-bromoisobutyryloxy)ethyl methacrylate))) (poly(MMA-*co*-BIEM: MI) [41,42] with various contents of bromoester initiating groups, 25–75% (GPIL1 series and GPIL2, Figure 1c, Table 1). The side chains resulted in the use of methyl methacrylate (MMA) as the comonomer. It has been proved that polymers obtained from similar monomers, like phosphorylcholine methacrylate, provided low cytotoxicity [43]. The studies on polymers of ChMA indicated that depending on the grafting density, lengths of backbone, and side chains, the anionic drug content in the resulted cylindrical brushes can be tuned up to 40% weight of the polymer. In aqueous solution, the water-soluble macromolecules formed small superstructures with hydrodynamic diameters ranging from 20 to 60 nm. Low aggregation effect was probably caused by repulsive interactions between ionic moieties in the side chains, which yielded bigger particles with the increase in content of ionic units (and thus salicylate) at the same grafting degree (GPIL1.1 vs. GPIL1.2, GPIL1.3). However, the increase in grafting degree in polymers with the same content of ionic units resulted in particle size reduction (GPIL1.3 vs. GPIL1.4), although this difference was smaller when the ionic drug content was significantly higher due to larger grafting density (GPIL1.3 vs. GPIL1.5).

In the case of amphiphilic graft polymer with chloride anions (GPIL2.1), the therapeutic activity was introduced by self-assembling with encapsulation of nonionic drug, i.e., ERY or IMC. In a similar way, ERY was encapsulated by the salicylate-containing graft polymer (GPIL1.1) to form a dual drug system, but the content of encapsulated ERY was smaller than that for GPIL2.1 (6% vs. 20%). For comparison, the analogous linear poly(ionic liquid)s (LPIL1–2) were synthesized using standard ATRP initiator. i.e., ethyl α-bromoisobutyrate (EBiB) [39,40]. The LPIL systems, both chloride (LPIL2.1) and salicylate (LPIL1.1–1.2) ones in relation to analogical GPIL, exhibited higher drug loading content (DLC$_{LPIL}$ > DLC$_{GPIL}$), but this difference was significantly higher for salicylate systems, whereas the opposite DLC dependency was indicated for IMC encapsulation [44].

Table 1. Characterization of poly(ionic liquid)s GPIL vs. LPIL (DDS type I).

| No. | $n_{sc}$ [a] | $F_{ChMAX}$ [b] (mol %) | $DP_{sc}$ [c] | DG [d] (%) | $M_{n,NMR}$ (g/mol) | $M_{n,SEC}$ [e] (g/mol) | Đ [e] | $D_h$ [f] (nm) | DC (%) | Ref. |
|---|---|---|---|---|---|---|---|---|---|---|
| GPIL1.1 | 75 | 25 | 24 | 22 | 201,500 | 17,680 | 1.04 | 28 | 19 / 6 [g] | |
| GPIL1.2 | 75 | 67 | 12 | 22 | 127,000 | 19,500 | 1.04 | 56 | 32 | [44] |
| GPIL1.3 | 75 | 71 | 28 | 22 | 314,500 | 19,000 | 1.10 | 51 | 36 | |
| GPIL1.4 | 105 | 74 | 39 | 53 | 724,100 | 30,400 | 1.06 | 22 | 38 | |
| GPIL1.5 | 185 | 74 | 43 | 74 | 1,498,500 | 828,700 | 1.28 | 40 | 39 | |
| LPIL1.1 | - | 78 | 187 * | - | 49,200 | 6200 | 1.25 | 293 | 41 / 49 [g] | [39,40] [42] |
| LPIL1.2 | - | 45 | 119 * | - | 23,100 | 8300 | 1.42 | 232 | 32 / 51 [g] | [39,40] [42] |
| GPIL2.1 | 165 | 19 | 26 | 53 | 579,000 | nd | nd | 24 | 20 [g] / 32 [h] | [44] |
| LPIL2.1 | - | 26 | 233 * | - | 28,000 | 10,400 | 1.36 | 149 | 45 [g] / 11 [h] | [42] [42] |

GPIL: grafted poly(ionic liquid)s, where GPIL1: poly(MMA-*co*-(BIEM-*graft*-P(MMA-*co*-ChMASal)), GPIL2: poly(MMA-*co*-(BIEM-*graft*-P(MMA-*co*-ChMACl)); LPIL: linear poly(ionic liquid), where LPIL1: P(MMA-*co*-ChMASal), LPIL2: P(MMA-*co*-ChMACl); [a] number of side chains, [b] content of ionic units in polymer, [c] degree of polymerization of side chains, [d] degree of grafting related to $n_{sc}$ per total DP of backbone, [e] determined in DMF, [f] determined in deionized water, [g] DLC of ERY for the weight ratio of polymer to encapsulated drug P:D = 1:1; [h] DLC of IMC for the weight ratio of P:D = 1:1; * DP of LPIL; nd – not determined.

The ionic exchange was also postulated as the most probable mechanism of drug release. The dialysis experiments in PBS solution (pH = 7.4) indicated facile displacement of salicylate anions by phosphate ones, which represent better capability for coordination of the cations in polymer matrix. The burst release of the ionic drug attached to grafts can be explained by the dense packing character of grafting polymer topology, which intensified the repulsive interactions between negative charges on the aromatic rings of Sal$^-$. Another advantageous ability of these ionic systems was good solubility of the polymer matrix in PBS environment after drug release. In the release studies there was no influence of the content of ionic units in side chains, when the copolymers with the same grafting degree were compared (Figure 2). However, at high grafting density corresponding to 3 grafts per 4 units in the backbone (GPIL1.5), the salicylate release was slightly accelerated.

Comparing the drug release profiles for the grafted poly(ionic liquid)s of ChMA (GPIL) and linear copolymer analogs (LPIL), there was no significant difference. Although, the content of salicylate anions ionically bonded to the polymer matrix was larger in the linear copolymers (Sal$^-$ content 3.4–4.8 mg in 10 mg of linear polymer vs. 1.9–4.0 mg in 10 mg of graft polymer). According to the release studies approximately half of these drug amounts were exchanged and removed from the systems, that is 50% in the linear and 50–60% in the grafted carriers. The advantages for the latter ones were the size of particles, which were formed in aqueous solution. The largest particles of grafted polymers were 5–6 times smaller than the aggregates of linear analogs (56 nm vs. ~250 nm). These results suggested that the nonionic backbone was entangled into globular form with stretched stiff ionic side chains in the shell. Surprisingly, the release of a nonionic drug appeared to be troublesome, especially for the systems based on grafted copolymers (GPIL1.1 and GPIL2.1), where the release was not detected. The LPIL systems provided the miscellaneous release properties because LPIL1.1–1.2 containing Sal anions were able to release ERY, but this effect was not reached for chloride contained LPIL2.1, which supported IMC release (Figure 2). The lack of correlation can be explained by the system complexity related to carrier topology, possible repulsion effect of ionic groups, drug-polymer interactions, and anion bulkiness as responsible factors, which can cooperate providing extraordinary drug release behavior.

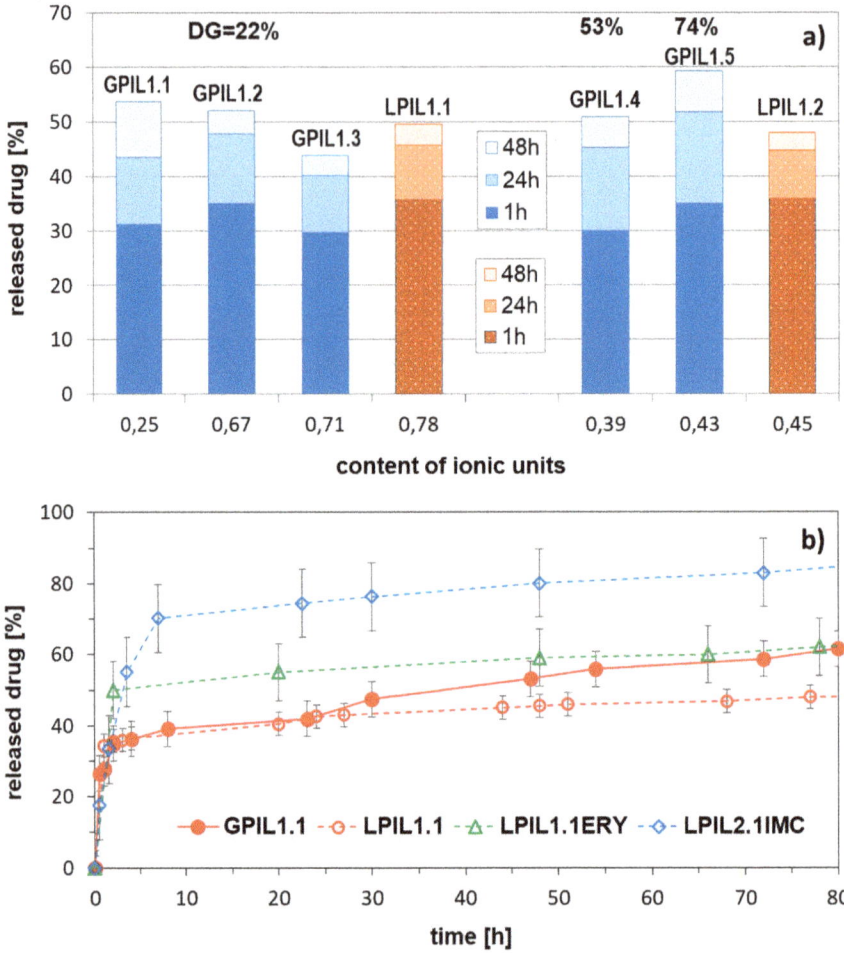

**Figure 2.** Amount of drug released after 48h in correlation with polymer topology (GPIL vs. LPIL), grafting degree (DG = 22–73%), and content of hydrophilic fraction (**a**), and release profiles of salicylate (GPIL1.1, LPIL1.1) and nonionic drugs (ERY, IMC) (**b**) in PBS at pH = 7.4 and 38 °C.

### 2.2. Conjugates of D-Glucopyranoside Based Star Copolymers (DDS Type II)

Star-shaped polymers with cleavable sugar core were obtained using di-, tri-, tetra-, hexakis(2-bromoisobutyrate) of mono-, and diacetal derivatives of D-glucopyranosides (DGL, $f$ = 3–6) [45] as the multifunctional initiators in the controlled ATRP of methacrylates by core first technique. The resulting 2-, 3-, 4-, and 6-armed star copolymers containing protecting *tert*-butyl or reactive glycidyl groups, were modified by acidolysis (into methacrylic acid (MAA)) [46–48] or aminolysis (into 2-hydroxy-3-[(2-aminoethyl)amine]propyl methacrylate units (AmPMA)) [49] to introduce hydrophilic moieties, which in the next step were used as the sites for conjugation. Their contents were controlled by the length and number of arms, efficiency of the modification reaction, and by amount of the functionalized units having pendant carboxylic groups (polyacids, DGL1 series) [50] or amine groups (polybases, DGL2 series) [49,51] in the polyelectrolyte stars (Table 2, Figure 1a).

Previously, the conjugation for these stars has been performed with proper derivatives of fluorescein dye (amine in reaction with polyacids to form amide spacer or isothiocyanate with polybases through thiocarbamide bonding) [48,49], which was more successful for cationic than anionic polyelectrolytes (68% vs. 5%, respectively). Their cellular uptake studies by confocal laser scanning microscopy indicated that the fluorescent star polymers were found in the entire volume of cytoplasm, but the signal intensity received from polybases was stronger in comparison with polyacids, for which less effective internalization was caused probably by electrostatic repulsion with negatively charged cell membrane [49,52].

Table 2. Characterization of star-shaped polyacids DGL1 and polybases DGL2 (DDS type II).

| No. | Polyelectrolytes | | | | | | DOX Conjugates | | | ref. |
|-----|---|---|---|---|---|---|---|---|---|---|
| | $f$ | $F_{\text{h-philic}}$ [a] | $DP_{\text{arm}}$ [b] | $M_{n,\text{NMR}}$ (g/mol) | $M_{n,\text{SEC}}$ [c] (g/mol) | $Đ$ [c] | $D_h$ [d] (nm) | DC (%) | $D_h$ [d] (nm) | |
| DGL1.1 | 4 | 0.56 | 58 | 21,800 | 11,800 | 1.17 | 10 | 5 | insoluble | |
| DGL1.2 | 4 | 0.74 | 68 | 24,200 | 10,900 | 1.28 | 8 | 14 | insoluble | [48,52] |
| DGL1.3 | 6 | 0.51 | 62 | 35,600 | 16,300 | 1.20 | 9 | 6 | insoluble | |
| DGL1.4 | 6 | 0.75 | 54 | 30,100 | insoluble | insoluble | 8 | 19 | insoluble | |
| DGL2.1 | 2 | 0.54 | 51 | 15,800 | nd | nd | 8 | 27 | 8 | |
| DGL2.2 | 3 | 0.49 | 57 | 26,500 | nd | nd | 7 | 28 | 8 | [49,51] |
| DGL2.3 | 4 | 0.53 | 65 | 41,100 | nd | nd | 12 | 17 | 11 | |
| DGL2.4 | 4 | 0.77 | 52 | 38,000 | nd | nd | 12 | 24 | 12 | |

DGL star-shaped polymer with D-glucopyranoside core, where DGL1: s-P(MMA-co-MAA)$_f$, DGL2: s-P(MMA-co-AmPMA)$_f$; [a] content of hydrophilic fraction in the polymer, [b] degree of polymerization of arm, [c] determined in THF, [d] determined in PBS solution 0.4 mg/mL; $f$: number of arms; nd: not determined.

Presented polymer–drug conjugates were prepared from star-shaped polyacids (DGL1.1 – DGL1.4) and polybases (DGL2.1 – DGL2.4) via amide or ketoimine linking DOX, respectively [52,53]. Various polymeric prodrugs with DOX have been investigated by other groups [54–58], to reduce its well-known severe side effects. Our approach was to take advantage of several aspects in the structure of the polymeric carrier, such as sugar-derived biodegradable core, cleavable amphiphilic arms, decreased hydrodynamic volume in solution in comparison with linear analogues, increased effectiveness of the drug protection, and longer time of circulation in the blood stream. Comparing drug conjugation with 4-armed stars based on polybases and polyacids with equimolar content of hydrophilic fraction higher efficiency was observed for formation of ketimine than amide bond (64% at $n_{\text{DOX}}$ = 59 in DGL2.3 vs. 4% at $n_{\text{DOX}}$ = 2 in DGL1.1) similarly to the fluorescein conjugates. Moreover, the conjugation efficiency of polybases decreased with the increase of amine repeating units per arm, whereas in the case of polyacids, the amount of attached drug increased with the number of the arms as well as the content of acidic units.

The release studies were performed only for polymeric prodrugs based on polybases, due to the poor solubility of polyacid-DOX conjugates in water. The lowest amount of drug was released by conjugate, based on 4-armed star (DGL2.3), whereas 3-armed system (DGL2.2) was able to supply release of twice larger drug doses (Figure 3). In acidic conditions (pH 5.0), which are more favorable for hydrolysis of ketimine group than the neutral pH (0.01M PBS, pH 7.4), the drug release occurred faster.

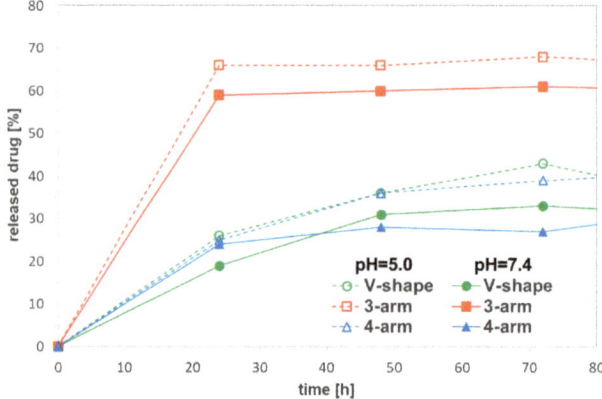

**Figure 3.** DOX release profiles from conjugates of star-shaped copolymers with equimolar compositions and various number of arms, DGL2.1 (V-shaped), DGL2.2 (3-armed), and DGL2.3 (4-armed) in PBS at 37 °C.

### 2.3. Self-Assembling Star-Shaped Copolymers (DDS Type III)

Another group of 4-armed stars was synthesized in a similar way to sugar based stars using a core-first strategy via ATRP initiated by tetrakis(2-bromoisobutyrate) of pentaerythritol (PTL, $f = 4$), and then acidolysis of *tert*-butyl groups in the copolymers to deprotect carboxylic groups (PTL1-3 series, Figure 1b, Table 3) [47]. Combinations of two methacrylates (MAA and MMA, PTL1) or methacrylate and acrylate (MAA and methyl acrylate (MA) as PTL2, MMA and acrylic acid (AA) as PTL3) with various proportions in the arms were investigated to form the self-assembling PTL cored star copolymers as the micellar carriers of IMC.

**Table 3.** Characterization of star-shaped polyacids used for drug encapsulation (DDS type III).

| Polyacids | $f$ | $F_{\text{h-philic}}$ [a] | $DP_{\text{arm}}$ [b] | $M_{n,\text{NMR}}$ (g/mol) | $M_{n,\text{SEC}}$ [c] (g/mol) | $D$ [c] | $D_h$ [d] (nm) | DLE [h] (%) | $D_h$ [d] (nm) | ref. |
|---|---|---|---|---|---|---|---|---|---|---|
| PTL1.1 | | 0.48 | 34 | 15,400 | 11,600 | 1.17 | 147 [e] | 74 | 571 | |
| PTL2.1 | | 0.36 | 63 | 27,800 | 18,200 | 1.32 | 198 [f] | 48 | 628 | |
| PTL2.2 | 4 | 0.70 | 67 | 31,800 | 16,800 | 1.31 | 161 [f] | 6 | 463 | [47] |
| PTL3.1 | | 0.55 | 56 | 27,000 | 17,700 | 1.30 | 198 [f] | 86 | 874 | |
| PTL3.2 | | 0.76 | 39 | 20,200 | 13,600 | 1.28 | 162 [f] | 9 | 579 | |
| PTL3.3 | | 0.98 | 31 | 17,400 | 11,500 | 1.24 | 114 [g] | 7 | 731 | |
| DGL3.1 | | 0.50 | 55 | 32,100 | insoluble | insoluble | 202 | 66 | 531 | |
| DGL3.2 | 6 | 0.75 | 50 | 28,700 | 6000 | 1.33 | 165 | 47 | 321 | [46, 48] |
| DGL3.3 | | 0.97 | 47 | 25,700 | insoluble | insoluble | 180 | 7 | 1165 | |
| DGL4.1 | | 0.48 | 54/19 * | 28,700 | 8400 | 1.57 | 517 | 48 | 705 | |
| DGL4.2 | 8 (6 + 2) * | 0.69 | 44/10 * | 28,900 | 4000 | 1.67 | 252 | 42 | 1169 | [46] |
| DGL4.3 | | 0.92 | 65/10 * | 41,900 | 6600 | 1.44 | 384 | 60 | > 10,000 | |

PTL: Star-shaped polymer with pentaerythritol core, where PTL1: s-P(MMA-co-MAA), PTL2: s-P(MMA-co-AA), PTL3: s-P(MA-co-MAA); DGL: Star-shaped polymer with D-glucopyranoside core, where DGL3: s-P(MMA-co-MAA)$_6$, DGL4: s-P(MMA-co-MAA)$_6$PCL$_2$; [a] content of hydrophilic fraction in the polymer, [b] degree polymerization of arm, [c] determined in THF, [d] determined in PBS solutions 0.4 mg/mL, [e] 0.5 mg/mL, [f] 0.2 mg/mL, [g] 1 mg/mL; [h] for the weight ratio of polymer to encapsulated drug P:D = 1:1 (PTL) and 2:1 (DGL); * P(MMA-co-MAA) and PCL arms, respectively.

The adjustable distribution of acidic units was convenient for controlling the contents of hydrophilic fraction, which affected the efficiency of drug encapsulation and release. The highest drug loading content was obtained for copolymers with equimolar compositions (DLC = 50–90% (50/50)). The copolymers of MMA/MAA and MMA/AA, with comparable amounts of hydrophilic fractions

(PTL1.1 and PTL2.1), exhibited formation of aggregates at the same concentrations, whereas CMC for MA/MAA system (PTL3.1) was twice as high in comparison with MMA copolymers (0.030 vs. 0.017 mg/mL). The rate of drug release for these systems can be summarized by the following order: MMA/MAA < MMA/AA << MA/MAA, which shows strong influence of arm composition. During release studies we have noticed that within 1 h the drug was released faster in neutral conditions than in an acidic environment, whereas after longer time this tendency was reversed (Figure 4). Additionally, it was detected that the reduced drug release can also be forced by increased encapsulation ratio of drug to polymer as it is presented for PTL2.2 (MMA/AA system).

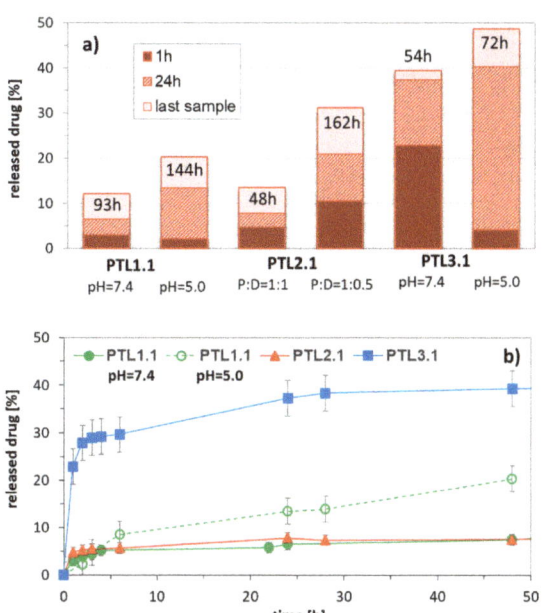

**Figure 4.** Influence of pH and weight ratio of polymer to encapsulated drug on amount of released drug in correlation with composition of 4-armed copolymers of MMA/MAA (PTL1.1), MMA/AA (PTL2.1), and MA/MAA (PTL3.1) with equimolar content of hydrophilic fraction (**a**), and representative release profiles of IMC (**b**) in PBS at 37 °C.

Amphiphilic character of star copolymers bearing hexakis(2-bromoisobutyrate)-dihydroxy-D-(−)-salicin core (DGL3 series, Figure 1b, Table 3) [46,48] gave the opportunity to form more or less globular self-assembling systems in aqueous solution depending on ionization degree of statistically distributed ionizable hydrophilic units in the arms. However, the hydrophilic–hydrophobic balance can also be shifted by introduction of additional hydrophobic arms. In our studies, the miktostars (DGL4 series, Figure 1b, Table 3) [46] were obtained using two unprotected hydroxyl groups in the core of polymethacrylate-based macroinitiator (DGL3) as the initiating sites in the insertion-coordination ring opening polymerization (ROP) catalyzed by tin(II) bis(2-ethylhexanoate). Combined with amphiphilic polymethacrylate arms containing acidic groups, the formed two poly($\varepsilon$-caprolactone) (PCL) arms provided phase-separation in aqueous solution. The yielded micelles with biodegradable core were detected as bigger particles than their amphiphilic polymethacrylate precursors, playing the role of the bifunctional macroinitiators in ROP, for example DGL3.2 vs. DGL4.2.

The self-assembly studies on polyacidic stars DGL3 with a similar length of polymethacrylic arms ($DP_{arm} \sim 50$) has revealed that CMC drastically decreased with the increase in hydrophilic content in the arms (0.172, 0.024, and 0.006 mg/mL, respectively). Comparing 6-armed polymethacrylate stars and their miktoarm analogues (with extra two hydrophobic PCL arms), the aggregation of the latter ones was not dependent on the hydrophilic–hydrophobic ratio (CMC of DGL4.1 = DGL4.2 = 0.030 mg/mL).

The successful aggregation of sugar-cored polyacidic stars encouraged us to provide systems with encapsulated DOX ($\leq$ 65% at polymer/drug ratio = 1:0.5). The drug loading content was reduced with the increase in arm length and the same with the hydrophilic content in the micellar carriers, but comparing stars with the corresponding miktostars, the latter ones seemed to be more promising carriers, especially those with the equimolar compositions (Figure 5). It is also worth noticing that the 6-armed acidic copolymers with similar arm lengths and equimolar compositions were able to entrap physically larger amounts of DOX than it was chemically conjugated (22% in DGL3.1 vs. 9 wt% in DGL1.3 at $DP_{arm}$ = 60), but in more hydrophilic systems this difference was insignificant (16% in DGL3.2 vs. 19 wt% in DGL1.4 at $DP_{arm}$ = 50) (Figure 6).

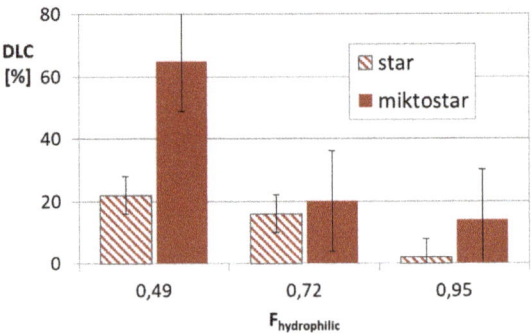

**Figure 5.** Drug content in micellar systems based on amphiphilic star (DGL3 series) and miktostar (DGL4 series) copolymers, where $F_{hydrophilic}$ is average value for comparable pair of star and miktostar.

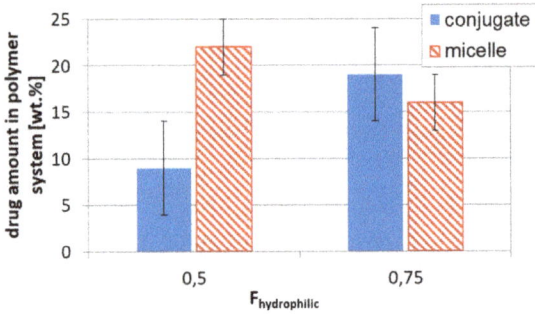

**Figure 6.** Influence of drug loading type, conjugates DGL1 vs. micelles DGL3, on drug content in respect to hydrophilic–hydrophobic balance.

The equimolar system DGL3.1 has also provided the fastest drug release. No significant difference in percentage amount of released drug between star and miktostar analogs was observed at pH 5, but in neutral solution the miktostar with dominating hydrophilic fraction (DGL4.2) was more efficient than the analogous star (DGL3.2), which was in contrast to the more hydrophilic systems DGL4.3 vs. DGL3.3 (Figure 7). Another comparison of drug release from polybase-DOX conjugates (DGL2) and polyacid based micelles loaded with DOX (DGL3, DGL4), indicated in almost all cases that the drug was released faster within first 6 h (for micelles) and 24 h (for polymer-DOX conjugates) at acidic environment than at neutral pH. For example, 60% of DOX was delivered at pH = 5.0 and 44% at pH = 7.4 by DGL4.3, whereas 84% and 56% by DGL2.1, respectively. It is highly probable that the destabilization of self-assembly containing carboxylic groups in the outer layer was activated by the ionized DOX with amine groups, which are protonated in acidic conditions.

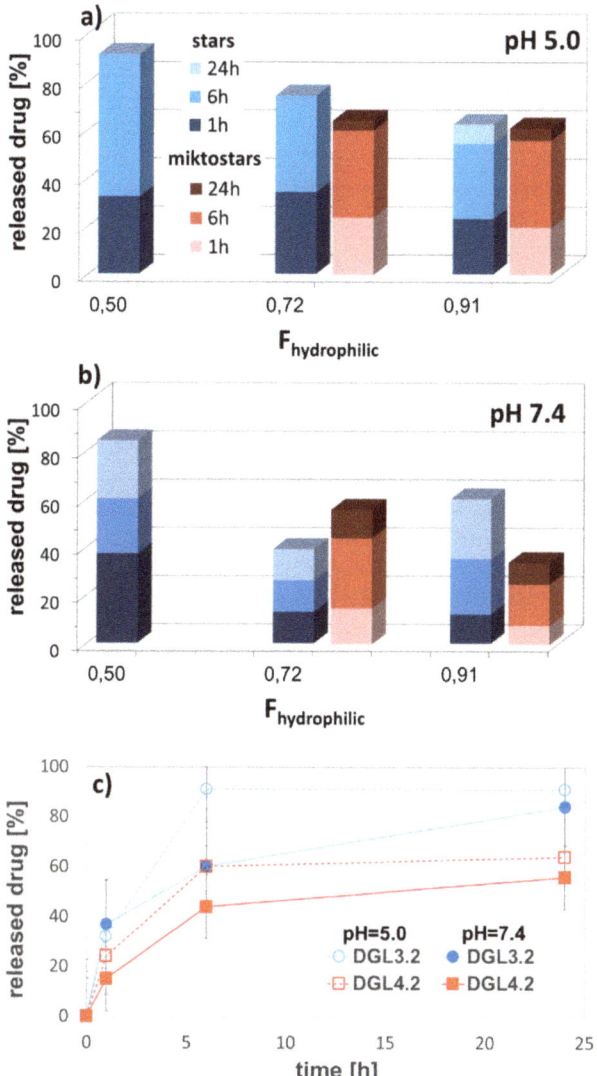

**Figure 7.** Amount of released DOX from self-assembled star DGL3 vs. miktostar DGL4 copolymers in correlation with content of hydrophilic fraction in acidic and neutral conditions (**a,b**), and representative release profiles of encapsulated DOX (**c**) in PBS at 37 °C.

## 3. Drug Distribution and Cytotoxicity

The mathematical models describing the kinetics of drug release suggested the diffusion mechanism. It was confirmed by the good agreement with the Higuchi model represented as the plot of the cumulative amount of released drug against the square root of time (correlation coefficient $R^2$ = 0.90–0.99) for the release of ionic and nonionic drugs (ERY, IMC) from LPIL and GPIL systems [42,44]. Additionally, the kinetics of the nonionic drug release from LPIL1, LPIL2, and GPIL2 systems was also concentration dependent, showing good fit with the first-order kinetic model ($R^2$ = 0.93–0.98) [44], which is expressed by a logarithm of the percentage of drug remaining vs. time. Similarly, the concentration dependent, and diffusion controlled, release of IMC was reported for the self-assembling graft copolymers with PMAA side chains (the first-order kinetic model $R^2$ = 0.9–0.99 and Higuchi model $R^2$ = 0.9–0.97) [34,36], as well as the linear block copolymers PCL-*b*-PMAA, and their three armed stars (the first-order kinetic model $R^2$ = 0.85–0.99 and Higuchi model $R^2$ = 0.91–0.99) [59] or stars with PTL core (the first-order kinetic model $R^2$ = 0.87–0.99 and Higuchi model $R^2$ = 0.9–0.99) [47]. In the case of conjugate systems DGL2-DOX the Korsmeyer–Peppas model, based on the diffusion exponent n, which describes Fickian ($n \leq 0.45$) and non-Fickian ($0.45 < n < 0.89$), the release of drug was applied for verification of release mechanism. According to the "n" values, almost all samples followed the Fickian diffusion ($n < 0.45$, $R^2$ = 0.928–0.999) [51].

In respect to DDS applications, the cytotoxicity of designed copolymers was verified by viability of the selected cell lines. The human bronchial epithelial cells (BEAS-2B) were applied for both linear and graft copolymers with trimethylammonium groups and salicylate counterions, which due to their anti-inflammatory and antibacterial activity potential can be beneficial in the treatment of lung and bronchi diseases [42]. In vitro studies evaluated by MTT assay exhibited very low cytotoxicity towards BEAS-2B as it is shown in Figure 8. Generally, the cells treated with GPIL5 showed slightly lower viability than LPIL1.2. However, at concentration 0.0025 µg/mL both LPIL1.2 and GPIL1.5 stimulated cell growth, which was depicted by cell proliferation at levels of 114% and 110% in comparison to the control, respectively. In the case of polymers functionalized with carboxylic or amine groups, the influence of charged polymer particles on their interactions with cells was evaluated by MTS tests using DOX-resistant breast cancer cells (MCF-7/R), because the star-shaped copolymers with DGL core were prepared for conjugation or encapsulation of anticancer cytostatic DOX. The results for MCF-7/R treated with representative drug-free, 4-armed, star-shaped polyacid DGL1.1 showed significant cytotoxicity in comparison to the polybase DGL2.3 with the same number of arms and hydrophilic content (Figure 8). However, cytotoxicity of polybasic carriers increased with a decrease in the number of arms DGL2.1 > DGL2.2 > DGL2.3 [53], whereas 6-armed polyacids were statistically less cytotoxic than their 4-armed analogs [52]. Moreover, polyacids did not display changes in viability of colon cancer cells (HCT-116) and viability of normal human dermal fibroblasts (NHDF) cells, which stayed at the acceptable level [52]. These results are in contrast to those obtained for polybases, which showed inhibition of HCT-116 cells proliferation and low cytotoxicity toward NHDF cells [49].

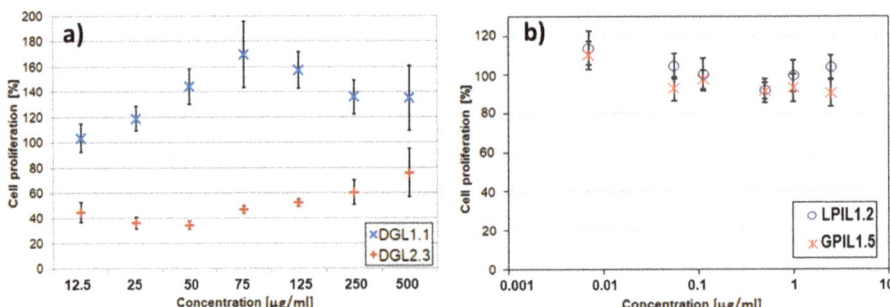

**Figure 8.** In vitro cytotoxicity effect of polymethacrylate carriers containing carboxylic (DGL1.1) or amine groups (DGL2.3) without drug on MCF-7/R cells (**a**), and trimethylammonium groups with salicylate anions (GPIL1.5, LPIL1.2) towards BEAS-2B cells (**b**).

## 4. Experimental

*4.1. Characterization Techniques*

Molecular weights and dispersity indices (Đ) were determined by size exclusion chromatograph (SEC, 1100 Agilent 1260 Infinity) equipped with an isocratic pump, autosampler, degasser, thermostatic box for columns (PLGel 5 mm MIXED-C 300 7.5 mm and pre-column guard 5 mm × 7.5 mm), and differential refractometer MDS RI Detector. Addon Rev. B.01.02 data analysis software (Agilent Technologies) was used for data collecting and processing. The calculation of molecular weight was based on calibration using linear polystyrene standards (580–300,000 g/mol). The measurements were carried out in THF or DMF (HPLC grade) as the solvent at 40 °C with flow rate of 0.8 mL/min.

$^1$H NMR spectra of copolymers in CDCl$_3$, DMSO-d$_6$, or D$_2$O were collected on Varian Inova 600 MHz spectrometer at 25 °C using appropriate internal standard (TMS or TSP).

The hydrodynamic diameters (D$_h$) of particles were measured by dynamic light scattering (DLS) using a Malvern Zetasizer. Samples placed in PMMA cell after appropriate dilution with a solvent (0.2, 0.4, 0.5, or 1 mg/mL) were put in the thermostatted cell compartment of the instrument at 25 °C.

The CMC was measured by fluorescence spectrophotometry (Hitachi F-2500) using pyrene as fluorescence probe. Excitation spectra of pyrene ($\lambda$ = 390 nm) were recorded at constant concentration of pyrene (3.0 × 10$^{-4}$ mol/L) and polymer concentrations in the range of 5 × 10$^{-4}$–1.0 mg/mL. The intensity ratio (I$_{336}$/I$_{332}$) from pyrene excitation spectrum vs. logC (where C is concentration in mg/mL) was plotted, where the cross-over point was estimated as the CMC value.

For the determination of drug content, polymer systems were dissolved in H$_2$O under vigorous vortexing and analyzed using a UV–Vis spectrophotometer (Thermo Scientific Evolution 300) at 480 nm (DOX), 320 nm (IMC). Calibration curves were obtained for drug-H$_2$O solutions with different drugs concentrations. Drug loading content (DLC for micelles), or drug content (DC for conjugates) and drug loading efficiency (DLE) were calculated using the following equations:

$$DLC = \frac{\text{weight of drug loaded into micelle}}{\text{total weight of polymer and loaded drug}} \times 100\% \qquad (1)$$

$$DC = \frac{\text{weight of drug in conjugate}}{\text{weight of drug} - \text{polymer conjugate}} \times 100\% \qquad (2)$$

$$DLE = \frac{\text{weight of drug loaded into micelle}}{\text{weight of drug in feed}} \times 100\%. \qquad (3)$$

In vitro drug release studies were performed in 0.01 M PBS at pH 7.4 or 5.0 (the pH of 0.01 M PBS was subsequently adjusted with 0.1 N HCl to pH 5.0). The lyophilized drug-contained system (2.0 mg) was dissolved in PBS (2.0 mL) and transferred into a dialysis bag (MWCO of 3.5 kDa, Spectrum Laboratories Inc). Then, the dialysis bag was immersed into PBS (20.0 mL) at 37 °C and stirred. At predetermined time intervals, 1 mL of the buffer solution outside the dialysis bag was taken out. UV–vis spectroscopy (Thermo Scientific Evolution 300) was used to determine the amount of released drug by measuring the absorbance maximum (NaSal at 295–298 nm, ERY at 285 nm, IMC at 320 nm, and DOX at 480 nm). Each result is an average of three parallel measurements.

*4.2. Cell Viability Assessment*

In vitro cytotoxicity of the selected copolymers was measured using the MTS or MTT assay (Promega cell proliferation assays) and following cell lines: Bronchial epithelial cells (BEAS-2B from ATCC, Manassas, VA, USA), human colon cancer cells, and human breast cancer cells resistant to DOX (HCT116 and MCF-7/R from ATCC as a kind gift from the Center of Oncology – Maria Sklodowska-Curie Memorial Institute).

Briefly, the selected cells were seeded in a 96-well micro titer plates at a density of 10,000 cells per well and incubated overnight at 37 °C. All cells were grown in DMEMF12 medium (Sigma-Aldrich,

Germany), supplemented with 10% (v/v) inactivated fetal bovine serum (FBS) (EURx, Poland) and 1% antibiotics (10,000 µg/mL of streptomycin and 10,000 units/mL of penicillin) (Sigma-Aldrich, Germany), at 37 °C in humidified atmosphere with 5% $CO_2$. After 24 h of incubation under standard conditions, a series of suspension dilutions were added into wells. The cytotoxicity was evaluated after a predetermined time of incubation (24 or 72 h). The absorbance at 490 nm was measured using a microplate reader (Epoch, Biotek, Winooski, VT, USA). All experiments were performed in quadruplicate, and the relative cell proliferation (%) was expressed as a percentage relative to the untreated control cells (positive control). For more details, see ref. [24,36,43].

## 5. Conclusions

The nonlinear amphiphilic polymers, including star and graft topologies, are attractive nanocarriers for drug delivery because, similarly to linear macromolecules, they are capable of transporting drugs. The bioactive compounds can be effectively introduced into polymer matrix via chemical bonding to form drug-polymer conjugates, including poly(ionic liquid)s with pharmaceutical counterions, or by physical interactions in the self-assemblies. Due to the extra structural parameters related to topology, e.g., number of arms/grafts, the nonlinear polymers offer much broader range of drug loading and release moderation. The reviewed polymeric systems varying with types of drug loading (physical vs. covalent vs. ionic) were represented by nanoparticles with sizes up to 200 nm, with exception of miktostars (up to 500 nm). Although the grafted copolymers were able to carry similar amounts of ionically attached drug as their linear analogs, they are still advantageous systems for delivery due to smaller sizes of nanoparticles. Moreover, the design of DDS combining different strengths of drug entrapment can be strategic for sequential drug release in the combined therapy. Our studies confirmed that knowing the general correlations between structure, and loading/release effect for the obtained polymers, the nanocarrier activity with controllable pharmacokinetic properties can be properly regulated.

**Author Contributions:** Conceptualization, D.N.; methodology, D.N. and A.M.; formal analysis, D.N., A.M., J.O., R.B., K.N.; investigation, J.O., R.B., A.M., K.N.; data curation, J.O., R.B., M.K.; writing—original draft preparation, D.N., A.M., R.B., J.O., M.K.; review & editing, D.N.; visualization, D.N., J.O.; supervision & project administration, D.N.

**Funding:** These studies were financially supported by the National Science Center, grant no. 2017/27/B/ST5/00960.

**Acknowledgments:** The biological experiments were performed in the Biotechnology Center of the Silesian University of Technology in Gliwice using equipment financed by the Silesian Biofarma program supported by POIG.02.03.01-24-099/13 grant. Authors thank Magdalena Skonieczna for help in cytotoxic analyses.

**Conflicts of Interest:** The authors declare no conflicts of interest.

## References

1. Kadajji, V.G.; Betageri, G.V. Water soluble polymers for pharmaceutical applications. *Polymers* **2011**, *3*, 1972–2009. [CrossRef]
2. Felice, B.; Prabhakaran, M.P.; Rodríguez, A.P.; Ramakrishna, S. Drug delivery vehicles on a nano-engineering perspective. *Mater. Sci. Eng. C* **2014**, *41*, 178–195. [CrossRef]
3. Gao, H.; Matyjaszewski, K. Synthesis of functional polymers with controlled architecture by CRP of monomers in the presence of cross-linkers: From stars to gels. *Prog. Polym. Sci.* **2009**, *34*, 317–350. [CrossRef]
4. Xiao, Y.; Yuan, M.; Zhang, J.; Yan, J.; Lang, M. Functional Poly(ε-caprolactone) Based Materials: Preparation, Self-assembly and Application in Drug Delivery. *Curr. Top. Med. Chem.* **2014**, *14*, 781–818. [CrossRef]
5. Ren, J.M.; McKenzie, T.G.; Fu, Q.; Wong, E.H.H.; Xu, J.; An, Z.; Shanmugam, S.; Davis, T.P.; Boyer, C.; Qiao, G.G. Star Polymers. *Chem. Rev.* **2016**, *116*, 6743–6836. [CrossRef] [PubMed]
6. Ito, S.; Goseki, R.; Ishizone, T.; Hirao, A. Synthesis of well-controlled graft polymers by living anionic polymerization towards exact graft polymers. *Polym. Chem.* **2014**, *5*, 5523–5534. [CrossRef]
7. Neugebauer, D. Two decades of molecular brushes by ATRP. *Polymer* **2015**, *72*, 413–421. [CrossRef]

8. Hadjichristidis, N.; Iatrou, H.; Pitsikalis, M.; Mays, J. Macromolecular architectures by living and controlled/living polymerizations. *Prog. Polym. Sci.* **2006**, *31*, 1068–1132. [CrossRef]
9. Chen, W.; Zhang, J.Z.; Hu, J.; Guo, Q.; Yang, D. Preparation of amphiphilic copolymers for covalent loading of paclitaxel for drug delivery system. *J. Polym. Sci. Part A Polym. Chem.* **2014**, *52*, 366–374. [CrossRef]
10. Luo, C.; Sun, J.; Sun, B.; He, Z. Prodrug-based nanoparticulate drug delivery strategies for cancer therapy. *Trends Pharmacol. Sci.* **2014**, *35*, 556–566. [CrossRef]
11. Riess, G. Micellization of block copolymers. *Prog. Polym. Sci.* **2003**, *28*, 1107–1170. [CrossRef]
12. Tang, Z.; He, C.; Tian, H.; Ding, J.; Hsiao, B.S.; Chu, B.; Chen, X. Polymeric nanostructured materials for biomedical applications. *Prog. Polym. Sci.* **2016**, *60*, 86–128. [CrossRef]
13. Zhang, J.; Li, J.; Shi, Z.; Yang, Y.; Xie, X.; Lee, S.M.Y.; Wang, Y.; Leong, K.W.; Chen, M. pH-sensitive polymeric nanoparticles for co-delivery of doxorubicin and curcumin to treat cancer via enhanced pro-apoptotic and anti-angiogenic activities. *Acta Biomater.* **2017**, *58*, 349–364. [CrossRef] [PubMed]
14. Jang, B.; Kwon, H.; Katila, P.; Lee, S.J.; Lee, H. Dual delivery of biological therapeutics for multimodal and synergistic cancer therapies. *Adv. Drug Deliv. Rev.* **2016**, *98*, 113–133. [CrossRef] [PubMed]
15. Egorova, K.S.; Gordeev, E.G.; Ananikov, V.P. Biological Activity of Ionic Liquids and Their Application in Pharmaceutics and Medicine. *Chem. Rev.* **2017**, *117*, 7132–7189. [CrossRef] [PubMed]
16. Deng, Y.; Zhang, S.; Lu, G.; Huang, X. Constructing Well-Defined Star Graft Copolymers. *Polym. Chem.* **2013**, *4*, 1289–1299. [CrossRef]
17. Sheiko, S.S.; Sumerlin, B.S.; Matyjaszewski, K. Cylindrical molecular brushes: Synthesis, characterization, and properties. *Prog. Polym. Sci.* **2008**, *33*, 759–785. [CrossRef]
18. Jie, P.; Venkatraman, S.S.; Min, F.; Freddy, B.Y.C.; Huat, G.L. Micelle-like nanoparticles of star-branched PEO-PLA copolymers as chemotherapeutic carrier. *J. Control. Release* **2005**, *110*, 20–33. [CrossRef]
19. Zhang, Q.; Remsen, E.E.; Wooley, K.L. Shell cross-linked nanoparticles containing hydrolyrically degradable, crystalline core domains. *J. Am. Chem. Soc.* **2000**, *122*, 3642–3651. [CrossRef]
20. Cheng, G.; Böker, A.; Zhang, M.; Krausch, G.; Müller, A.H.E. Amphiphilic cylindrical core-shell brushes via a "grafting from" process using ATRP. *Macromolecules* **2001**, *34*, 6883–6888. [CrossRef]
21. Moughton, A.O.; Sagawa, T.; Gramlich, W.M.; Seo, M.; Lodge, T.P.; Hillmyer, M.A. Synthesis of block polymer miktobrushes. *Polym. Chem.* **2013**, *4*, 166–173. [CrossRef]
22. Djordjevic, J.; Barch, M.; Uhrich, K.E. Polymeric micelles based on amphiphilic scorpion-like macromolecules: Novel carriers for water-insoluble drugs. *Pharm. Res.* **2005**, *22*, 24–32. [CrossRef] [PubMed]
23. Du, J.Z.; Tang, L.Y.; Song, W.J.; Shi, Y.; Wang, J. Evaluation of polymeric micelles from brush polymer with poly(ε-caprolactone)-b-poly(ethylene glycol) side chains as drug carrier. *Biomacromolecules* **2009**, *10*, 2169–2174. [CrossRef] [PubMed]
24. Qiu, L.Y.; Bae, Y.H. Polymer architecture and drug delivery. *Pharm. Res.* **2006**, *23*, 1–30. [CrossRef] [PubMed]
25. Stridsberg, K.M.; Ryner, M.; Albertsson, A.C. Controlled Ring-Opening Polymerization: Polymers with designed Macromolecular Architecture. In *Degradable Aliphatic Polyesters. Advances in Polymer Science*; Springer: Berlin/Heidelberg, Germany, 2002; pp. 41–66.
26. Mori, H.; Müller, A.H.E. New polymeric architectures with (meth)acrylic acid segments. *Prog. Polym. Sci.* **2003**, *28*, 1403–1439. [CrossRef]
27. Lee, H.J.; Bae, Y. Brushed block copolymer micelles with pH-sensitive pendant groups for controlled drug delivery. *Pharm. Res.* **2013**, *30*, 2077–2086. [CrossRef]
28. Gou, P.-F.; Zhu, W.-P.; Zhu, N.; Shen, Z.-Q. Synthesis and Characterization of Novel ResorcinareneCentered Amphiphilic Star-Block Copolymers Consisting of Eight ABA Triblock Arms by Combination of ROP, ATRP, and Click Chemistry. *J. Polym. Sci. Part A Polym. Chem.* **2009**, *47*, 2905–2916. [CrossRef]
29. Qiu, L.Y.; Wu, X.L.; Jin, Y. Doxorubicin-loaded polymeric micelles based on amphiphilic polyphosphazenes with poly(N-isopropylacrylamide-co-*N*,*N*-dimethylacrylamide) and ethyl glycinate as side groups: Synthesis, preparation and in vitro evaluation. *Pharm. Res.* **2009**, *26*, 946–957. [CrossRef]
30. Neugebauer, D.; Bury, K.; Biela, T. Novel hydroxyl-functionalized caprolactone poly(meth)acrylates decorated with tert-butyl groups. *Macromolecules* **2012**, *45*, 4989–4996. [CrossRef]
31. Neugebauer, D.; Bury, K.; Paprotna, M.; Biela, T. Amphiphilic copolymers with poly(meth)acrylic acid chains "grafted from" caprolactone 2-(methacryloyloxy)ethyl ester-based backbone. *Polym. Adv. Technol.* **2013**, *24*, 1094–1101. [CrossRef]

32. Maksym-Bębenek, P.; Biela, T.; Neugebauer, D. Synthesis and investigation of monomodal hydroxy-functionalized PEG methacrylate based copolymers with high polymerization degrees. Modification by "grafting from". *React. Funct. Polym.* **2014**, *82*, 33–40. [CrossRef]
33. Maksym-Bębenek, P.; Biela, T.; Neugebauer, D. Water soluble well-defined acidic graft copolymers based on a poly(propylene glycol) macromonomer. *RSC Adv.* **2015**, *5*, 3627–3635. [CrossRef]
34. Bury, K.; Neugebauer, D. Novel self-assembly graft copolymers as carriers for anti-inflammatory drug delivery. *Int. J. Pharm.* **2014**, *460*, 150–157. [CrossRef]
35. Maksym-Bębenek, P.; Neugebauer, D. Study on Self-Assembled Well-Defined PEG Graft Copolymers as Efficient Drug-Loaded Nanoparticles for Anti-Inflammatory Therapy. *Macromol. Biosci.* **2015**, *15*, 1616–1624. [CrossRef] [PubMed]
36. Maksym, P.; Neugebauer, D. Self-assembling polyether-b-polymethacrylate graft copolymers loaded with indomethacin. *Int. J. Polym. Mater. Polym. Biomater.* **2017**, *66*, 317–325. [CrossRef]
37. Yuan, J.; Antonietti, M. Poly(ionic liquid)s: Polymers expanding classical property profiles. *Polymer* **2011**, *52*, 1469–1482. [CrossRef]
38. Yuan, J.; Mecerreyes, D.; Antonietti, M. Poly(ionic liquid)s: An update. *Prog. Polym. Sci.* **2013**, *38*, 1009–1036. [CrossRef]
39. Bielas, R.; Łukowiec, D.; Neugebauer, D. Drug delivery via anion exchange of salicylate decorating poly(meth)acrylates based on a pharmaceutical ionic liquid. *New J. Chem.* **2017**, *41*, 12801–12807. [CrossRef]
40. Bielas, R.; Mielańczyk, A.; Siewniak, A.; Neugebauer, D. Trimethylammonium-Based Polymethacrylate Ionic Liquids with Tunable Hydrophilicity and Charge Distribution as Carriers of Salicylate Anions. *ACS Sustain. Chem. Eng.* **2016**, *4*, 4181–4191. [CrossRef]
41. Börner, H.G.; Duran, D.; Matyjaszewski, K.; Da Silva, M.; Sheiko, S.S. Synthesis of molecular brushes with gradient in grafting density by atom transfer polymerization. *Macromolecules* **2002**, *35*, 3387–3394. [CrossRef]
42. Bielas, R.; Mielańczyk, A.; Skonieczna, M.; Mielańczyk, Ł.; Neugebauer, D. Choline supported poly(ionic liquid) graft copolymers as novel delivery systems of anionic pharmaceuticals for anti-inflammatory and anti-coagulant therapy. Unpublished.
43. Kania, G.; Kwolek, U.; Nakai, K.; Yusa, S.I.; Bednar, J.; Wójcik, T.; Chłopicki, S.; Skórka, T.; Szuwarzyński, M.; Szczubiałka, K.; et al. Stable polymersomes based on ionic-zwitterionic block copolymers modified with superparamagnetic iron oxide nanoparticles for biomedical applications. *J. Mater. Chem. B* **2015**, *3*, 5523–5531. [CrossRef]
44. Bielas, R.; Siewniak, A.; Skonieczna, M.; Adamiec, M.; Mielańczyk, Ł.; Neugebauer, D. Choline based polymethacrylate matrix with pharmaceutical cations as co-delivery system for antibacterial and anti-inflammatory combined therapy. *J. Mol. Liq.* **2019**, *285*, 114–122. [CrossRef]
45. Neugebauer, D.; Mielańczyk, A.; Waśkiewicz, S.; Biela, T. Epoxy functionalized polymethacrylates based on various multifunctional D-glucopyranoside acetals. *J. Polym. Sci. Part A Polym. Chem.* **2013**, *51*, 2483–2494. [CrossRef]
46. Mielańczyk, A.; Odrobińska, J.; Grządka, S.; Mielańczyk, Ł.; Neugebauer, D. Miktoarm star copolymers from D-(-)-salicin core aggregated into dandelion-like structures as anticancer drug delivery systems: Synthesis, self-assembly and drug release. *Int. J. Pharm.* **2016**, *515*, 515–526. [CrossRef] [PubMed]
47. Neugebauer, D.; Odrobińska, J.; Bielas, R.; Mielańczyk, A. Design of systems based on 4-armed star-shaped polyacids for indomethacin delivery. *New J. Chem.* **2016**, *40*, 10002–10011. [CrossRef]
48. Mielańczyk, A.; Neugebauer, D. Synthesis and characterization of D-(-)-Salicine-based star copolymers containing pendant carboxyl groups with fluorophore dyes. *J. Polym. Sci. Part A Polym. Chem.* **2014**, *52*, 2399–2411. [CrossRef]
49. Mielańczyk, A.; Skonieczna, M.; Bernaczek, K.; Neugebauer, D. Fluorescein nanocarriers based on cationic star copolymers with acetal linked sugar cores. Synthesis and biochemical characterization. *RSC Adv.* **2014**, *4*, 31904–31913. [CrossRef]
50. Mielańczyk, A.; Biela, T.; Neugebauer, D. Synthesis and self-assembly behavior of amphiphilic methyl α-D-glucopyranoside-centered copolymers. *J. Polym. Res.* **2014**, *21*. [CrossRef]
51. Mielańczyk, A.; Neugebauer, D. Designing Drug Conjugates Based on Sugar Decorated V-Shape and Star Polymethacrylates: Influence of Composition and Architecture of Polymeric Carrier. *Bioconjug. Chem.* **2015**, *26*, 2303–2310. [CrossRef] [PubMed]

52. Mielańczyk, A.; Skonieczna, M.; Neugebauer, D. Cellular response to star-shaped polyacids. Solution behavior and conjugation advantages. *Toxicol. Lett.* **2017**, *274*, 42–50. [CrossRef] [PubMed]
53. Mielańczyk, A.; Skonieczna, M.; Mielańczyk, Ł.; Neugebauer, D. In Vitro Evaluation of Doxorubicin Conjugates Based on Sugar Core Nonlinear Polymethacrylates toward Anticancer Drug Delivery. *Bioconjug. Chem.* **2016**, *27*, 893–904. [CrossRef] [PubMed]
54. Muggia, F.M. Doxorubicin-polymer conjugates: Further demonstration of the concept of enhanced permeability and retention. *Clin. Cancer Res.* **1999**, *5*, 7–8. [PubMed]
55. Veronese, F.M.; Schiavon, O.; Pasut, G.; Mendichi, R.; Andersson, L.; Tsirk, A.; Ford, J.; Wu, G.; Kneller, S.; Davies, J.; et al. PEG-doxorubicin conjugates: Influence of polymer structure on drug release, in vitro cytotoxicity, biodistribution, and antitumor activity. *Bioconjug. Chem.* **2005**, *16*, 775–784. [CrossRef] [PubMed]
56. Etrych, T.; Strohalm, J.; Chytil, P.; Černoch, P.; Starovoytova, L.; Pechar, M.; Ulbrich, K. Biodegradable star HPMA polymer conjugates of doxorubicin for passive tumor targeting. *Eur. J. Pharm. Sci.* **2011**, *42*, 527–539. [CrossRef] [PubMed]
57. Gao, A.X.; Liao, L.; Johnson, J.A. Synthesis of acid-labile PEG and PEG-doxorubicin-conjugate nanoparticles via Brush-First ROMP. *ACS Macro Lett.* **2014**, *3*, 854–857. [CrossRef]
58. Sun, Y.; Zhang, J.; Han, J.; Tian, B.; Shi, Y.; Ding, Y.; Wang, L.; Han, J. Galactose-Containing Polymer-DOX Conjugates for Targeting Drug Delivery. *AAPS PharmSciTech* **2017**, *18*, 749–758. [CrossRef] [PubMed]
59. Bury, K.; Du Prez, F.; Neugebauer, D. Self-assembling linear and star shaped Poly(Ĭμ-caprolactone)/poly[(meth)acrylic acid] block copolymers as carriers of indomethacin and quercetin. *Macromol. Biosci.* **2013**, *13*, 1520–1530. [CrossRef]

© 2019 by the authors. Licensee MDPI, Basel, Switzerland. This article is an open access article distributed under the terms and conditions of the Creative Commons Attribution (CC BY) license (http://creativecommons.org/licenses/by/4.0/).

*Review*

# Silk Fibroin as a Functional Biomaterial for Drug and Gene Delivery

**Mhd Anas Tomeh [1], Roja Hadianamrei [1] and Xiubo Zhao [1,2,*]**

[1] Department of Chemical and Biological Engineering, University of Sheffield, Sheffield S1 3JD, UK; matomeh1@sheffield.ac.uk (M.A.T.); rhadianamrei1@sheffield.ac.uk (R.H.)
[2] School of Pharmaceutical Engineering and Life Science, Changzhou University, Changzhou 213164, China
* Correspondence: xiubo.zhao@sheffield.ac.uk; Tel.: +44-0114-222-8256

Received: 25 August 2019; Accepted: 24 September 2019; Published: 26 September 2019

**Abstract:** Silk is a natural polymer with unique physicochemical and mechanical properties which makes it a desirable biomaterial for biomedical and pharmaceutical applications. Silk fibroin (SF) has been widely used for preparation of drug delivery systems due to its biocompatibility, controllable degradability and tunable drug release properties. SF-based drug delivery systems can encapsulate and stabilize various small molecule drugs as well as large biological drugs such as proteins and DNA to enhance their shelf lives and control the release to enhance their circulation time in the blood and thus the duration of action. Understanding the properties of SF and the potential ways of manipulating its structure to modify its physicochemical and mechanical properties allows for preparation of modulated drug delivery systems with desirable efficacies. This review will discuss the properties of SF material and summarize the recent advances of SF-based drug and gene delivery systems. Furthermore, conjugation of the SF to other biomolecules or polymers for tissue-specific drug delivery will also be discussed.

**Keywords:** silk fibroin; drug delivery; gene delivery; controlled release; bioconjugation

## 1. Introduction

Polymeric drug delivery systems have emerged as a new efficient alternative to the conventional formulations to provide a reservoir to the active pharmaceutical ingredients (APIs), improve their physicochemical properties, and overcome some of the major challenges in drug delivery including specific targeting, intracellular transport, and biocompatibility in order to improve the treatment efficiency and life quality of patients [1–4]. An ideal drug delivery system should stabilize the loaded API, allow for modulating its release kinetics and minimize its adverse effects by tissue-specific targeting, especially in the case of highly toxic drugs such as anticancer agents. Silk has been known as a valuable natural material for the fabric industry for centuries, but in the past decades it has attracted immense attention as a promising biopolymer for biomedical and pharmaceutical applications [5–7]. Silk protein possesses a unique combination of properties which is rare among natural polymers. It also enjoys desirable characteristics such as mild aqueous possessing conditions, high biocompatibility and biodegradability, and the ability to enhance the stability of the loaded APIs (e.g., proteins, pDNA, and small molecule drugs) [4,5,8,9]. Moreover, silk fibroin (SF) solution can be processed by various methods to produce different types of delivery systems including hydrogels, films, scaffolds, microspheres, and nanoparticles [10]. SF exists in three different structural forms: Silk I, Silk II and Silk III. Silk I exists in water-soluble form and consists of a high percentage of α-helix domains in addition to random coils [11]. Contrarily, Silk II has mainly β-sheet structure and is more stable and water-insoluble, while Silk III prevails at the water/air interface [12]. The transformation from Silk I to Silk II can be tuned by different methods including organic solvent treatment, physical shear, electromagnetic fields, or chemical processing [13,14]. These properties can be utilized in the pharmaceutical industry for producing micro- and nano particles and nano-fibrils or for coating other pharmaceutical preparations such as liposomes [15,16]. Moreover, the availability of carboxyl and amino groups in the SF allows for bio-functionalization with various biomolecules or ligands which could be used for targeted drug delivery [17]. The two main strategies for functionalizing silk protein are chemical conjugation and genetic modification of silk by chainging the amino acid composition or adding a fragment to obtain a specific function [18]. A large proportion of drug formulations including the vast majority of anticancer drug formulations are prepared for parenteral administration, resulting in direct contact with the blood components. Thus, the drug carriers used in such formulations should not induce any haematological toxicity or immune responses [3], which necessitates the use of biocompatible polymers in the formulation. Furthermore, designing delivery systems for biological drugs such as vaccines and antibodies requires maintaining their physical stability as well as their biological activity, which is more crucial for the controlled release systems [19]. This is mainly due to the higher sensitivity of the biological compounds, especially the protein-based therapeutics, to many of the processing conditions throughout the delivery system preparation compared to small molecule drugs, which limits the processing strategies [20,21]. Hence, loading the biological therapeutics into a compatible polymer can increase their stability and consequently their half-life [22]. Moreover, incorporating the APIs into a natural biocompatible protein such as SF has multiple advantages e.g., preserving the API [23], improving the mechanical properties of the formulation [6], modifying drug release kinetics [24–27], enhancing cell adhesion [16], and compatibility with blood components [25]. The versatility of SF protein processing and formulating methods allows the preparation of a wide range of drug carriers with different sizes and morphologies using unmodified or engineered SF (Figure 1). Unmodified SF carriers have been used to deliver various anticancer drugs such as doxorubicin [28], paclitaxel [29], curcumin [9,30], and cisplatin [31]. In this review, the main strategies for obtaining different SF-based drug delivery systems and the recent methods for generating functionalized SF for controlled or targeted drug and gene delivery will be discussed.

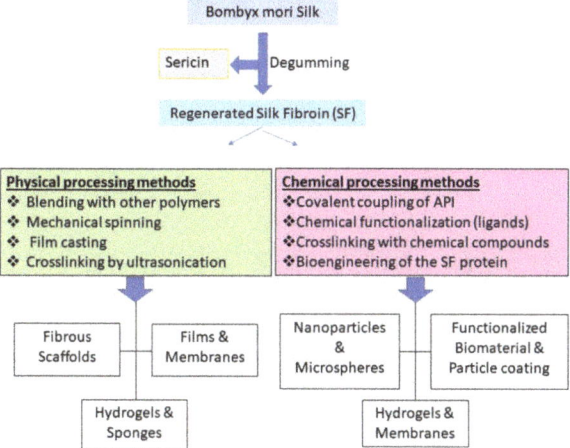

**Figure 1.** A diverse set of physical assemblies and chemical methods for preparing a variety of silk fibroin (SF) formats for pharmaceutical and biomedical applications.

## 2. Physiochemical Properties of Silk Fibroin

SF possesses a unique combination of mechanical and biological properties and exhibits special features of both synthetic and natural polymers [32,33]. Typically, silk represents softness in the clothing industry, but it is considered one of the most robust natural biomaterials due to its tensile strength and modulus [34]. This feature is important for the polymers involved in bone tissue regeneration as mechanical performance of the polymer is of utmost importance for such applications [35]. SF demonstrates excellent stability under high thermal stress (higher than 250 °C) [36].

## 2.1. Biocompatibility

SF is a biocompatible material that has been officially recognized by the Food and Drug Administration (FDA) for the development of a plethora of nanotechnological tools [37]. The biocompatibility of silk has been studied extensively over the past two decades. The majority of the studies have reported excellent biocompatibility and relatively lower immunogenic response in comparison to other common degradable biological polymers in the pharmaceutical industry such as polylactide (PLA), poly(lactic-co-glycolic acid) (PLGA) and collagen [22,38,39]. Cytocompatibility studies on SF formats revealed high compatibility with different cell lines including hepatocytes, osteoblasts, fibroblasts, endothelial cells and mesenchymal stem cells (MSCs) [40,41]. During SF processing, organic solvents such as methanol and hexafluoroisopropanol (HFIP) are used to crosslink SF via inducing structural transformation ($\alpha$ helix to $\beta$ sheet), which has been found responsible for the inflammatory potential of SF formulations [42]. However, mild processing conditions that avoid the use of organic solvents have been used to avoid these inflammatory responses [43]. An evaluation of the levels of lymphocyte activating factor IL-1$\beta$ and inflammatory cyclooxygenase-2 (COX-2) gene expression in relation to SF stimulation did not show any significant differences from those of collagen or PLA, indicating very low immunogenicity [44]. Another study has evaluated the biocompatibility of scaffolds consisting of a combination of calcium polyphosphate (CPP) and SF used for the reconstruction of cartilage and bone defects [45]. The results showed a tangible increase in tissue biocompatibility and osteogenicity of SF-CPP scaffolds in comparison to CPP scaffolds [45].

## 2.2. Mechanical Properties

Mechanical stiffness is a key property of SF-based formulations for pharmaceutical and biomedical applications. The SF material used in tissue engineering, for example, must match the stiffness of the targeted tissue. The stiffness can also affect the stability and degradability of the SF polymer [46]. Many polymers that have been used in drug delivery devices, such as PLGA and collagen, lack sufficient mechanical strength. A common strategy to enhance the mechanical strength of the biopolymers such as collagen is crosslinking. However, the crosslinking reaction could result in undesirable consequences such as cellular toxicity and immunogenicity [47]. SF possesses robust $\beta$-sheet structure which provides excellent mechanical properties without the need for any harsh crosslinking procedures. Based on the $\beta$-sheet content, SF can transform into different formats including liquid, hydrogels, or scaffolds [5]. Measurements of mechanical strength are usually obtained from Young's modulus using nanoindentation techniques [48]. SF exhibits high tensile strength and resistance to compressive force making it a very suitable material for drug delivery and tissue engineering [49]. Moreover, the removal of sericin during the degumming process results in a 50% increase in tensile strength [50] which makes SF more stable during physical pharmaceutical processing.

## 2.3. Stability

The stability of the polymeric materials is one of the most important factors in the production of pharmaceutical formulations. Although biopolymers are preferred to their synthetic counterparts for clinical applications due to their biocompatibility and biodegradability, they must also meet certain stability standards to be considered for utilisation in the pharmaceutical industry. One of the common stability problems of pure SF solution is aggregation or gelation during long-term storage. SF is available in soluble form (with high content of $\alpha$- helix and random coil) and insoluble form (with high content of $\beta$-sheet). Depending on the pharmaceutical preparation, either form should be used and maintained. Storing the soluble SF in highly humid conditions results in a transformation from $\alpha$-helix and random coil to $\beta$-sheet which could lead to gelation and a decrease in the stability of the SF solution [51,52]. SF shows excellent stability under thermal stress compared to other proteins. The best indicator for protein thermal stability is the glass transition temperature ($Tg$) which in the case of SF is affected by its β-sheet content. The $Tg$ of SF films is approximately 175 °C and the protein remains stable up to 250 °C which is desirable for formulation processing. On the other hand, $Tg$ of the frozen SF solution can go to −34 °C [53] which is also advantageous in low temperature pharmaceutical processing. Furthermore, the degree of crystallinity and porosity of SF films are also affected by $Tg$ [54] The increase in β-sheet content of SF causes a transformation from Silk I to Silk II, refelected by a significant change in $Tg$ which changes the degree of crystallinity. Stability of the SF in physiological fluids is another important issue for its biomedical applications. SF could be protected from enzymatic degradation within the body by coating with polymers such as polyethylene glycol (PEG) in order to improve the delivery of the associated drugs to the site of action.

## 2.4. Degradability

Degradability is an important property of biological materials. Although biodegradability is a main advantage of SF in clinical applications, this property makes pure SF particles liable to proteolytic enzymes. The degradation rate of SF can be regulated by modifying the molecular weight, the degree of crystallinity, morphological features, or crosslinking [56]. However, the degree of crystallinity and crosslinking are not the only approaches to stabilize SF against degradation. For example, an in vitro enzymatic degradation experiment revealed that SF sheets slightly transformed from Silk II to Silk I crystalline structure when exposed to collagenase IA. However, when protease XIV was used, the majority of the SF sheets transformed to Silk I leading to a higher degree of crystallinity. Although the degradation time was 15 days in both cases, the degradation rate was significantly lower for protease XIV compared to collagenase IA [57]. Another study reported a predictable loss of mechanical integrity due to SF degradation [58]. Incubation with protease led to an exponential decrease in the SF filament diameter to 66% of the initial diameter after 10 weeks. Gel electrophoresis indicated a decreasing amount of the silk 25 kDa light chain and a shift in the molecular weight of the heavy chain with increasing incubation time with protease XIV [58]. The enzymatic degradation behavior of SF was studied by Wongpinyochit et al. [55] using three different proteases (papain, chymotrypsin, and protease XIV) over a period of 20 days. As shown in Figure 2, the cleavage sites vary from one protease to another, leading to variations in degradation rate. The degradation rate was higher in the presence of chymotrypsin compared to papain and protease XIV but the latter two did not have any significant difference. The degradation rate does not only rely on cleavage sites but also on the enzyme accessibility, SF format, and the secondary structure of the SF [55]. A previous in vivo degradation study in rats found that SF degradation is also related to the phagocytic activity of the cells (fibroblasts) and the presence of sericin with SF [59]. The raw silk (SF and Sericin) has a higher degradation rate than pure SF because sericin promotes degredation by the cells [59].

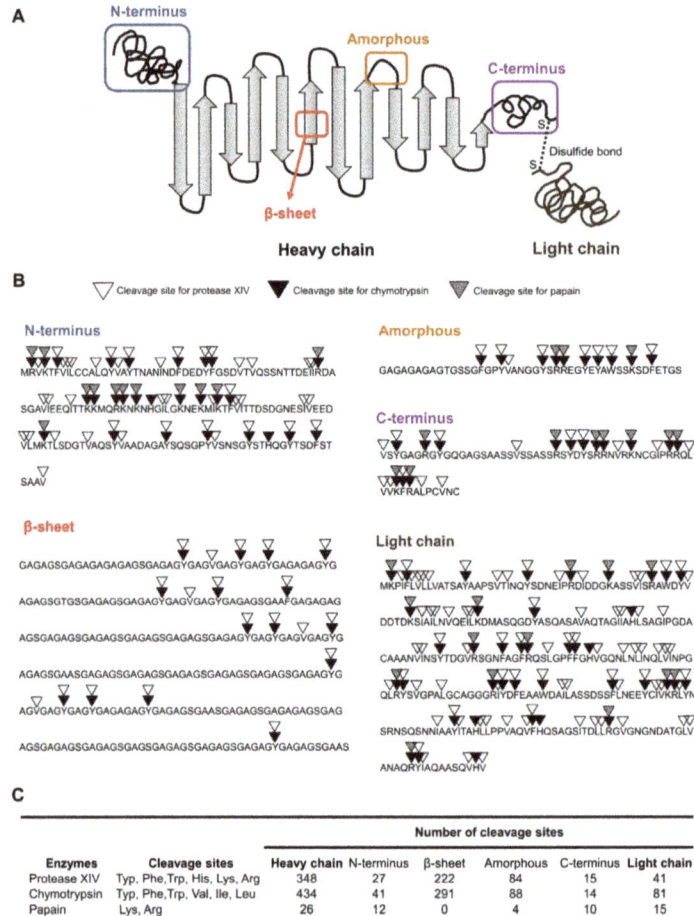

**Figure 2.** Two-dimensional schematic illustration of silk structure including the heavy chain (i.e., N-terminus, crystalline β-sheets, amorphous, and C-terminus) and light chain (**A**). Enzymatic specificities of proteolytic enzymes for the silk sequences (**B**). Number of cleavage sites of proteolytic enzymes on different silk domains (**C**). Reprinted from [55] with permission from American Chemical Society [2018].

## 3. SF-Based Drug Delivery Systems

Delivery of APIs in sustained and controlled release forms is important for many clinical applications. Selection of the particle size, composition, and other features depends on the type of the delivery system and the route of administration. Moreover, using biocompatible and mechanically durable polymers with mild fabrication and processing conditions in such delivery systems is advantageous for preserving the bioactivity of the loaded APIs. As discussed earlier, SF meets all these requirements which makes it a promising candidate for drug delivery [5,60]. SF-based drug delivery systems can be fabricated by different methods, each resulting in a delivery system with unique properties such as modified release kinetics, stability, and other features which could be of benefit in various applications. Various types of SF-based drug delivery systems have been designed including hydrogels, films, micro- and nanoparticles, nanofibers, lyophilized sponges as well as

SF-coated polymeric particles. In the following section, some of the most widely studied SF-based drug delivery systems are reviewed.

*3.1. Hydrogels*

SF aqueous solution was used to generate hydrogels by different methods. The transition from solution to gel can be triggered by physiochemical or chemical processes using natural polymers or synthetic reagents [61]. The physicochemical processes include shearing (spinning), water exclusion via evaporation or osmotic stress, electric field, and heating [62]. The gel form is stabilized because of thermodynamically stable β-sheets which result in a stable gel form in physiological conditions unless extensively degraded by enzymes or oxidative reactions [62]. One recent study used curcumin-loaded gel scaffolds prepared by electrogelation for wound healing [63]. The prepared gel formulation not only improved protein adsorption and sustained the release of curcumin, but also enhanced bacterial growth inhibition by 6-fold against *S. aureus* [63]. Since protein adsorption on substrates is a key factor for cell growth and proliferation, SF gel scaffolds can serve in wound healing by promoting cell proliferation. Sundarakrishnan et al. [64], adapted a chemical approach using horseradish peroxidase (HRP) and hydrogen peroxide to prepare SF hydrogels that were subsequently crosslinked with di-tyrosinase, and loaded with phenol red in order to develop a self-reporting pH system for in vitro environment [64]. Addition of phenol red during di-tyrosine crosslinking resulted in stable entrapment of phenol red within SF hydrogel network due to covalent interactions between phenol red and tyrosine and also prevented leaking [64].

*3.2. Silk Films*

Film preparation from SF has attracted more attention recently due to its huge potential as a biomaterial in pharmaceutical formulations and tissue engineering [65]. SF films can be simply prepared by casting an aqueous SF solution [66]. However, there are other reported SF film preparation techniques such as vertical deposition [67], spin coating [68], centrifugal casting [65], and spin assisted layer-by-layer assembly [69]. Terada et al. [68] investigated the behavior of spin-coated SF films treated with different ethanol concentrations. Alcohol concentrations of 80% or less resulted in a jelly-like hydrogel layer while treatment with more than 90% alcohol provided a rigid film surface. This change in morphology affected the attachment of the fibroblast cells to the SF films. Fibroblasts aggregated on the rigid surface rather than attaching individually to the hydrogel surface [68]. Another study found that blending SF with other polymers such as sodium alginate (SA) before casting the film results in a miscible and transparent film and also induces a structural change in SF [70]. Manipulating the SF/SA blending ratio shifted the SF conformation to the higher β-sheet content. Moreover, mixing SA with SF enhanced water permeability, swelling capacity and tensile strength of SF films [70]. Hence, SF/SA blend can provide unique tunable chataristics that can be benificial in pharmaceutical applications.

*3.3. Silk Particles*

As discussed earlier, the are an increasing number of SF-based systems that have been used for encapsulating APIs and achieving modulated drug delivery among which nanoparticle delivery systems have been studied the most, especially for anticancer drugs. One example of such systems is the lysosomotropic SF nanoparticles designed by Seib et al. [28] for pH-dependent release of the anticancer drug doxorubicin in order to overcome drug resistance. SF nanoparticles are largely employed for controlled release of the loaded drug at the site of action. SF nanoparticles can be fabricated by various methods, for example polyvinyl alcohol (PVA) blends, which are used for fabricating SF spheres with controllable sizes and shapes [71] (Table 1). The determinant factors for drug distribution and encapsulation efficiency in such systems are their charge and lipophilicity. Modifying these factors results in different drug release profiles [72]. Furthermore, addition of PVA results in a tangible improvement in the morphology of the SF-particles [73]. One of the popular methods for fabrication SF particles is the salting-out method. For example, Lammel et al. [74] produced SF particles with controllable sizes ranging from 500 nm to 2 µm using potassium phosphate as the salting out agent. The β-sheet structure and zeta potential of the SF particles were affected by the pH of the potassium phosphate solution [66,74]. In another study conducted by Tian et al. [75] SF nanoparticles were prepared using the salting out method, loaded with a combination of doxorubicin and $Fe_3O_4$ magnetic nanoparticles and driven to the target tissue using an external magnetic field to achieve tissue-specific targeted delivery [75]. It was also found that the entrapment efficiency of doxorubicin can be tuned by changing the concentration of $Fe_3O_4$ in the formulation [75]. However, the size of SF particles produced by potassium phosphate was over 500 nm, which is not ideal for drug delivery. Recently, Song et al. [43] produced magnetic SF nanoparticles (MSNPs) with size range of 90–350 nm by using sodium phosphate. The size and morphology of the MSNPs were governed by the SF concentration, the ionic strength and pH of the salting-out agent (Figure 3) [43]. Compared to potassium phosphate, sodium phosphate produced smaller particles and the size did not increase significantly with increasing SF concentration and ionic strength, providing a promising method to produce smaller particles with high concentration for drug delivery. The size of the particles can be further reduced by increasing the pH of the salting-out agent (Figure 3). Although salting-out and PVA methods are preferred over other methods due to their simplicity and low toxicity, purifying the SF nanoparticles from excess polymers or salting-out agent is required. Therefore, Mitropoulos et al. [76] managed to prepare SF particles with spherical shape using a co-flow capillary device with PVA as the continuous phase and silk solution as the discrete phase. This device allows for generation of SF spheres (2 µm in size) without the requirement for any further purification steps. Moreover, the diameter of the spheres can be simply adjusted by changing the concentration of the polymers, the flow rate, and the molecular weight of the selected polymer [76]. However, the size of the particles produced is not within the desired size range for drug delivery. A more recent study on the microfluidics has used a microfluidic set up (nano-assembler) to produce smaller sizes of SF particles (150–300 nm) by a desolvation method [77] (Table 1). It was found that the characteristics of the SF nanoparticles are controlled by two main factors: Flow rate and flow rate ratio [77]. The use of microfluidic instrument enabled rapid, reproducible and controlled production of SF nanoparticles with desirable sizes for drug delivery. However, solvent residues within the particles and the cost of the equipment should also be taken into account. The properties of the SF particles can also be manipulated by blending with other polymers. For example, Song et al. [9] have recently produced SF nanoparticles blended with different amounts of polyethyleneimine (PEI). The size of the SF nanoparticles was found to increase with increasing SF percentage (Figure 3), while the zeta potential of the particles decreased with increasing SF amount. This allows to fine tune the drug delivery through controlling the size and zeta potential of the particles.

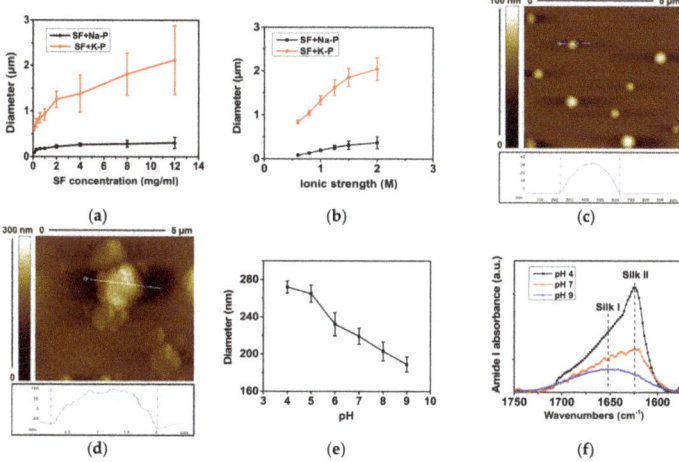

**Figure 3.** Effects of SF concentration, salt, ionic strength and solution pH to the particle size and protein secondary structure. (**a**) Particle diameter as a function of SF concentration when adding SF solutions (concentration from 0.1–12 mg/mL) to sodium phosphate (Na-P) and potassium phosphate (K-P) solutions (both at ionic strength 1.25 M, pH 8) at the volume ratio of 1:5. (**b**) Diameter of SF particles fabricated with K-P or Na-P as a function of their ionic strength. The SF concentration was fixed at 5 mg/mL. (**c**-**d**) AFM images of particles fabricated by adding SF solution (12 mg/mL) in to sodium phosphate (**c**) and potassium phosphate (**d**). Both solutions are at the ionic strength of 1.25 M and pH 8. (**e**) Diameter of SF particles fabricated with Na-P as a function of the Na-P solution pH. (**f**) FTIR spectra of particles produced by 1.25 M sodium phosphate at different pH values. It was found that the use of sodium phosphate, lower ionic strength and higher pH of solution produces smaller SF particles. Reprinted from Reference [43]; with permission from American Chemical Society [2017].

**Table 1.** Preparation teqniques of SF micro- and nanoparticles.

| Preparation Technique | Advantages | Disadvantages | Particle Size |
| --- | --- | --- | --- |
| Self-assembly | Simple and safe procedure Does not require toxic reagents | Sensitive to temperature and vigorous mixing | 100–200 nm [11] |
| Salting out | Low cost method The active ingredient can be loaded during the particle formation | Salting out agent residue Relatively high particle size polydispersity | 100–350 nm [43] 500 nm–2 μm [74] |
| Emulsification | Controllable particle size Low cost method | Organic solvent or surfactant residues | 170 nm [78] |
| Desolvation | Simple and quick method Small particle size Reproduceable technique | Particle aggregation Organic solvent residue | 35–170 nm [79] |
| Electrospraying | High purity particles Very good monodispersity | Requires additional step to insolubilize SF | 59–80 nm [80] 600–1800 nm [72] |
| Microfluidic methods | Rapid procedure Mild operation conditions Controllable particle yield and particle size | Relatively expensive Residual salting agent or organic solvents | 150–300 nm [77] |
| Capillary microdot | Simple procedure | Organic solvents residue | 25–140 nm [81] |
| Freeze drying | Porous particles | Large particle size | 490–940 μm [17] |
| Supercritical fluids | High drug loading | Expensive technique Not easy to operate Requires additional step to insolubilize SF | 50–100 nm [82] |
| PVA Blending method | Time and energy efficient No use of organic solvent | PVA residue | 5–10 μm [71] 300–400 nm [71] |
| Nano-imprinting and inject printing | Tuneable dimensions of different nanostructures | Complicated method Not easy to scale up Not easy to prepare particles | 180 nm–50 μm [83] |

## 4. Applications of Silk Fibroin for Drug and Gene Delivery

Silk has been used as a carrier for delivery of a wide range of therapeutic agents including small molecule drugs [3], biological drugs [84], and genes [85]. For each class of therapeutic agents, different formulations have been designed using various silk processing technologies [86]. One of the main criteria of the SF-based delivery systems is to stabilize the loaded API and manipulate its circulation time to achieve the required therapeutic effect. In addition, the designed formulations are usually optimized to obtain a particular application in drug delivery including stabilising the loaded drug, controlling drug release, and improving cell adhesion [16]. In the following section, an insight into SF applications in drug and gene delivery will be provided a summary of which is presented in Tables 2 and 3.

*4.1. Drug and Gene Stabilization by SF*

One of the main goals of incorporating active ingredients such as small molecules or peptides into SF-based carriers is to stabilise them by different mechanisms including adsorption, covalent interaction, and/or entrapment [87]. Without a stable interaction between the drug and the SF-based carrier to maintain the drug activity, sustained drug release cannot be achieved. Aside from a few exceptions such as growth factors, the majority of the stabilisation approaches rely on entrapping the drug within the SF-matrix or SF-particles in an equally distributed manner [22]. SF-based biomaterials are generally stable to changes in temperature [23], humidity [88], and pH [89]. Therefore, they have been widely studied for enhancing the stability of other materials, for example, encapsulation of antibiotics such as erythromycin, which has very low stability in water. However, porous SF sponges managed to sustain its release and maintain its antimicrobial activity against *Staphilococcus Aureus* for up to 31 days at 37 °C [3]. SF films have also been used for stabilisation of biological compounds. For example, enhanced stability of horseradish peroxidase (HRP) when loaded on SF films or mixed with SF solutions has been reported. The enzymatic activity of SF-loaded HRP was increased by 30–40%, while its half-life showed a tremendous increase from 2 h to 25 days at ambient conditions in comparison to free HRP [90]. A greater improvement in enzymatic activity (80%) was observed in glucose oxidase (GOx) when loaded on SF films [91]. Moreover, SF-loaded GOx demonstrated enhanced thermal and pH stability [92]. Topical application of SF lyogels (gel system in which the pores are filled with both organic and non-organic solvents) containing hydrocortisone in a mouse model of atopic eczema resulted in decreased expression of IgE and enhanced the efficacy of hydrocortisone compared to the commercially available hydrocortisone cream [93]. Moreover, SF lyogels have also been used for stabilising monoclonal antibodies. The lyogels achieved sustained release of IgG1 over 160 days and the release rate was found to be inversely proportional to the SF concentration [19]. In addition to drug stabilisation, SF has also been investigated for DNA preservation in order to protect the DNA from the potential destabilising conditions such as temperature and UV radiation. In a recent study porous cellulose paper was coated with SF and used to preserve the DNA extracted from human dermal fibroblast cells [23]. The results showed that the DNA integrity was maintained for 40 days following 10 h of UV radiation at relatively high temperature (37–40 °C) [23].

*4.2. Controlled Drug Release*

Controlled release drug delivery systems are aimed at releasing the encapsulated API in specified amounts over a specified period of time. One application of such systems is sustained drug release to maintain the therapeutic concentrations of the drug in the blood or site of action for a longer duration which is of great importance for the treatment of chronic diseases. Moreover, sustained drug release reduces the administration frequency and the adverse drug reactions which results in increased patient compliance [22]. Most of the currently available controlled release formulations in the market are composed of synthetic polymers such as PEG and PLGA because they provide desirable pharmacokinetic and pharmacodynamic properties [94]. Although PLGA is approved by FDA as a safe ingredient in pharmaceutical products, the processing requirements might restrict its utilisation in certain controlled release formulations. Therefore, more recently, natural polymers such as SF which offer tunable sustained release kinetics and stabilization of the loaded APIs have gained more attention for use in controlled drug release systems.

One of the unique properties of SF is its ability to undergo diverse structural transformations at the molecular level. The most investigated structural transformation in SF is the change in the ratio of α-helix to β-sheet content. For example, the permeability and release kinetics of the SF films are affected by the percentage of β-sheet structure [7]. The mechanism of controlled release from SF films was studied previously by Hines and Kaplan using different models [95]. The release kinetics of FITC-dextran from methanol-treated and untreated SF films was evaluated as a function of molecular weight of FITC-dextran. The methanol-treated films maintained higher percentage of the loaded FITC-dextran compared to the untreated films which was directly proportional to the molecular weight of FITC-Dextran [95]. In a more recent study, the release profile of the anticancer drug epirubicin from five Heparin-SF films (HEP-SF) treated with methanol (MeOH) or glycerol was investigated and it was found that using different ratios of glycerol in the HEP-SF nanofilm formulation affects the β-sheet content of the nanofilm leading to a modification in the release profile of epirubicin from the nanofilm (Figure 4). This mechanism-causal relationship between SF conformation and release profile also influenced the degree of degradation [7].

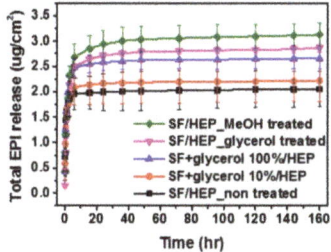

**Figure 4.** Total epirubicin (EPI) release profile from SF nanofilm depending on the ratio of the added glycerol and the solvent treatment. Reprinted from Reference [7] with permission from the American Chemical Society.

In a novel study conducted by Yavuz et al. [27] SF was formulated into insertable discs that can encapsulate either IgG antibody or human immunodeficiency virus (HIV) inhibitor 5P12-RANTES. Three different formulations were prepared by SF layering, water vapor annealing, and methanol treatment. These formulations managed to stabilize the protein cargo and to modify its release profile. High concentrations of IgG were released in a relatively short time from the formulation treated with methanol due to the highly porous structure in comparison to the other two formulations that demonstrated a slower and more controlled release. In the case of 5P12-RANTES, the water vapor annealing showed a sustained release for 31 days and this released protein could inhibit HIV infection in both blood and human colorectal tissue [27].

Controlled release from SF nanoparticles and microspheres has been studied extensively in the past decade. In an attempt to control SF particle features, a recent study conducted by Song et al. [43] demonstrated pH-controlled release of curcumin from SF nanoparticles for up to 20 days with lower pH promoting the release. Moreover, the SF nanoparticles had higher cellular uptake and induced significantly higher growth inhibitory effect in MDA-MB-231 cells compared to curcumin solution (Figure 5).

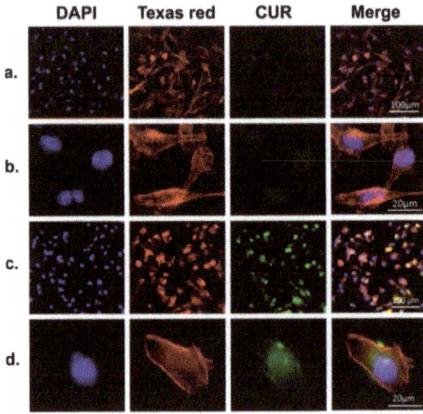

**Figure 5.** Representative microscopic images of MDA-MB-231 cells incubated with free curcumin ((**a**) and (**b**), curcumin amount (10 μg/mL), equivalent to the curcumin amount in CMSPs (curcumin loaded magnetic SF core–shell nanoparticles) and CMSPs ((**c**) and (**d**), 30 μg/mL) for 4 h. The cell nucleus and cytoskeleton were stained with DAPI (blue) and Texas red (red); all images were taken with an AF6000 microscope (Leica). Comparing the images in (**a**) and (**b**) to (**c**) and (**d**), it can be seen that CMSPs significantly improve the cellular uptake of the curcumin. Reprinted from Reference [43] with permission from the American Chemical Society [2017].

Another study developed SF microspheres (2 μm) using DOPC (1,2-dioleoyl-*sn*-glycero-3-phosphocholine) lipid vesicles as a templates [96]. The physically cross-linked β-sheet structure of SF and the residual DOPC in the microspheres played key roles in controlling the release of loaded enzyme (HRP) [96].

In addition to the controlled release systems composed of SF as their main component, SF has also been used as coating to modify the release kinetics of drug delivery systems made of other polymers.

SF, as a biopolymer, can be processed in aqueous conditions and crosslinked by different methods. Therefore, SF solution has been utilised for single or multilayer coating of different pharmaceutical preparations. In a study conducted by Pritchard et al. [97], the adenosine release from SF encapsulated powder reservoirs was evaluated as a function of reservoir coating thickness. The coating thickness was varied by changing the concentration of the silk coating solution and the number of coating layers applied. Increasing the coating thickness or the crystallinity of the SF delayed adenosine burst, decreased average release rate, and increased the duration of release [97]. Eliminating infections by releasing antibiotics such as vancomycin from biodegradable microspheres is a very effective strategy. However, maintaining the antibiotic concentration within the therapeutic window over the required treatment time remains a challenge [98]. A recent study has addressed this challenge by coating vancomycin-loaded poly(ε-caprolactone) (PCL) microspheres with SF to reduce the burst release of vancomycin [98]. The PCL microspheres were prepared by double emulsion ($W_1$/O/$W_2$) solvent evaporation/extraction process and the coating was performed by suspending the microspheres in the SF coating solutions (0.1%, 0.5% or 1%). Methanol was used to induce SF transformation from α-helix to β-sheet [98]. The microspheres coated with 0.1% SF showed smooth surface and presented a better release profile. By increasing the SF concentration, more cracks and defects were detected (Figure 6) [98].

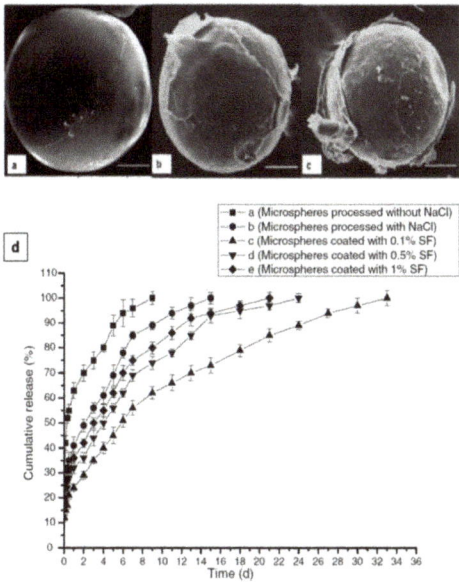

**Figure 6.** Scanning electron microscope (SEM)images of microspheres coated with 0.1% SF (**A**), 0.5% SF (**B**) and 1% SF (**C**). Microsphere coated with 0.1% SF showed a smooth surface without defects or cracks. Microspheres coated with 0.5% and 1% SF showed poor coating comprised of numerous surface defects and cracks in the film structure (bar size 50 mm). In vitro release of vancomycin from microspheres coated with different concentration of SF (**D**). Reprinted from [98] with permission from Taylor & Francis Online [2011].

Many studies have focused on improving the efficiency of the current nanoformulations and increasing the bioavailability of the loaded drugs by using adhesive excipients. For example, ocular drug delivery is a very challenging task and requires highly optimized formulations due to the unique environment in the eye. Because of the low ocular bioavailability, the frequency of applying the eye drops is usually high which can cause cellular damage at the ocular surface [99]. Increasing the residence time of drug on the eye surface will not only improve the efficiency of the therapy but also reduce the frequency of administration. A study presented by Dong et al. [16] prepared SF-coated liposomes loaded with ibuprofen for ocular delivery. The SF-coated liposomes exhibited better cell adhesion in human corneal epithelial cells (HCEC) compared to the conventional liposomes. Moreover, the drug release and permeation rates could be tuned by adjusting the concentration of the SF [16]. Another study also used SF coating on emodin-loaded liposomes (SF-ELP) to enhance keloids cell adhesion [100]. This study showed a selective targeting of keloids cells in comparison to normal cells which was achieved through interaction between SF and the cell in contrast to ligand targeting which is achieved by binding to specific receptors on the cell surface. SF-coating also limited Brownian motion and increased the probability of the nanoparticles attaching to the cell surface [100].

Table 2. SF-based drug delivery systems.

| Type of Drug Delivery System | Associated API | Results | References |
|---|---|---|---|
| SF sponges | Erythromycin | Sustained drug release and prolonged antimicrobial activity against Staphilococcus Aureus | [3] |
| SF films | Horseradish peroxidase (HRP) | Enhanced stability | [36] |
| | Glucose oxidase (GOx) | Increased enzymatic activity | [91] |
| | FITC-dextran | Controlled drug release | [95] |
| | Epirubicin | Controlled drug release | [7] |
| SF lyogels | Hydrocortisone IgG | Enhanced efficacy Enhanced stability and sustained release | [84] |
| Insertable SF discs | IgG and HIV inhibitor 5P12-RANTES | Enhanced stability and modified release profile | [27] |
| SF nanoparticles | Curcumin | Modified release profile and enhanced cellular uptake | [43] |
| SF microspheres | Horseradish peroxidase (HRP) | Modified the release profile | [96] |
| SF-coated PCL microspheres | Vancomycin | Modified the release profile | [98] |
| SF-coated liposomes | Ibuprofen | Enhanced adhesion to human corneal epithelial cells, tunable drug release | [16] |
| | Emodin | Selective targeting of keloid cells | [100] |

*4.3. Gene Delivery*

Gene delivery is defined as the introduction of genetic material including DNA and RNA into the targeted cells to regulate the expression of particular genes or direct the synthesis of specific proteins in order to treat disorders caused by dysregulation or malfunction of those proteins [101]. Using plasmid DNA (pDNA) is more common than RNA in gene delivery, especially to cancer cells [38,101]. Despite that more than 2600 clinical trials on gene therapy have been carried out in 38 countries since the early 1990s, none of these therapies have been granted FDA approval [102]. One of the main reasons why the gene-based formulations have failed to find their way to the market is that the majority of these formulations use viruses as gene vectors. Viral vectors usually give rise to concerns regarding systemic toxicity and immune response [38]. Therefore, the recent studies have shifted their focus toward non-viral vectors using biocompatible polymeric formulations. Non-viral gene delivery systems are typically designed using cationic lipids such as DOTAP (1,2-bis(oleoyloxy)-3-(trimethylammonio)propane) or positively charged synthetic polymers such as polyethyleneimine to provide a safer alternative to viral vectors [103]. However, transfection efficiency, target specificity and cytotoxicity of the common non-viral vectors remain obstacles to overcome. In addition to biocompatibility, SF has demonstrated DNase resistance and high transfection efficiency. These properties make SF a preferable polymeric vector for gene delivery [104]. Using genetic engineering, SF can be modified to gain more functions that suit the desired application. For example, modifying silk by adding PLL (poly(L-lysine)) sequences resulted in higher transfection efficiency of pDNA in human embryonic kidney (HEK) cells [105]. A later study has used SF bioengineered with PLL domains to interact with both pDNA and tumor-homing peptide (THP) for targeted pDNA delivery [85]. Even though silk based gene delivery vectors have shown high transfection efficiency and an acceptable degree of specificity [85], higher specificity is required for more effective targeted cancer therapy. An improved silk-based gene delivery system has been developed by adding F3 and Lyp1 peptides to recombinant silk proteins with a relatively high content of THP (25 mol%, 3.4 kDa/13.6 kDa) [103]. F3 peptide is capable of specifically binding to MDA-MB-435 cancer cells while Lyp1 binds specifically to tumor lymphatics and can also induce cell death in MDA-MB-435 cells [106–109]. The designed system achieved specific delivery of pDNA to the tumorigenic cells [103].

Another approach to capture DNA is using cationic polymers such as polyethyleneimine (PEI) within the silk formulation [9]. Luo et al. [4] developed cationic SF scaffolds by coating PEI on SF scaffold for delivery of vascular endothelial growth factor 165/angiopoietin-1 coexpression plasmid DNA (pDNA & VEGF165–Ang-1). PEI coating converted the surface charge from negative to positive through amidation reaction between the spermine and the carboxyl groups in the side chains of SF. The positive surface charge allowed the SF scaffold to form complexes with pDNA which demonstrated higher transfection efficiency and lower cytotoxicity than PEI/DNA complexes [4]. SF coated PEI/DNA complexes were also used to transfect HEK 293 and human colorectal carcinoma HCT 116 cells, and the system exhibited higher selectivity for HCT 116 compared to HEK 293 [110].

Recently, Song et al. [9] investigated the delivery of oligodeoxynucleotides (ODNs) to MDA-MB-231 breast cancer cells. The addition of SF to the nanoparticle formulation not only increased the cellular uptake of ODN (70%) but also significantly reduced its cytotoxicity (Table 3) [9]. Cell-penetrating peptides (CCPs) have also been used in SF-based gene delivery systems due to their ability to penetrate or destabilize cellular membranes. When designing a non-viral carrier for gene delivery, CCPs are considered among the preferred components to facilitate clathrin-dependent endocytosis of the particles [111,112]. Functionalization of silk protein with ppTG1 which is a CPP increased its transfection efficiency for pDNA in HEK 293 cells compared to unmodified silk (Table 3) [113].

**Table 3.** SF-based formulations for gene delivery.

| Formulation | Gene | Cell line | Reference |
|---|---|---|---|
| Recombinant silk–elastin-like polymer hydrogels (SELPs) | Adenovirus Ad[1]–CMV[2]–LacZ[3] | Head and neck cancer in mice | [114] |
| | pDNA[4] (pRL[5]-CMV-luc[6]) | NA | [115] |
| | Ad–Luc–HSVtk[7] | Head and neck cancer in mice | [116] |
| 3D porous scaffold | Adenovirus Ad-BMP7[8] | Human BMSCs | [117] |
| Bioengineered silk films | pDNA (GFP[9]) | Human HEK cells | [105] |
| Spermine modified SF | pDNA and VEGF165–Ang-1[10] | In vivo-rat | [4] |
| SF-Coated PEI/DNA Complexes | pDNA (GFP) | HEK 293 and HCT 116 cells | [110] |
| SF layer-by-layer assembled microcapsules | pDNA-Cy5[11] | NIH/3T3 fibroblasts | [118] |
| Bioengineered silk–polylysine–ppTG1 nanoparticles | pDNA | Human HEK and MDA-MB-435 cells | [113] |
| Magnetic-SF/polyethyleneimine core-shell nanoparticles | c-Myc[12] antisense ODNs[13] | MDA-MB-231 cells | [9] |

[1] Adenovirus; [2] cytomegalovirus promoter gene; [3] beta galactosidase reporter gene; [4] plasmid DNA; [5] renilla luciferase; [6] luciferase reporter gene; [7] herpes simplex virus thymidine kinase gene; [8] bone morphogenic protein; [9] green fluorescent protein; [10] vascular endothelial growth factor and angiopoietin-1; [11] fluorescent probe; [12] MYC Proto-Oncogene; [13] oligodeoxynucleotides.

## 5. Modification of SF for Enhanced Delivery

### 5.1. SF Bioconjugates

There are many protein-based drugs that have shown a very short half-life in the body. In order to enhance their in vivo stability, an approach has been designed to utilize SF by forming bioconjugates.

A covalent bond between the protein or enzyme and SF can be formed by the cross-linking reagents [5]. SF consists of 18 different amino acids among which 10% are polar amino acids such as serine and lysine with hydroxyl and amino groups in their side chains. These functional groups in SF can be covalently conjugated to polar groups in other proteins such as insulin using bifunctional reagent glutaraldehyde [119]. SF-insulin (SF-Ins) bioconjugate not only demonstrated higher in vitro stability than bovine serum albumin-insulin (BSA-Ins) conjugate, but also prolonged the pharmacological activity 3.5 times in comparison to native insulin [119]. Covalent conjugation of growth factor BMP-2 to SF using carbodiimide chemistry preserved BMP-2 activity and also reduced its degradation rate due to reduction in its unfolding rate as well as protecting it from proteases [120]. Immobilization of enzymes on silk particles has also been studied recently to enhance the catalytic efficiency of enzymes by improving enzymatic stability. SF has several active amino groups that have the potential for covalent binding to several enzymes to immobilize them (Figure 7) such as catalase immobilization on SF particles via tyrosinase crosslinking [121]. SF films have been also used to immobilize antibodies such as mouse IgG simply through the conformational transition to fabricate biocompatible biosensors [122]. The immobilization was achieved by slowly drying concentrated SF solution to reach the semisolid state and then blending it with antibody solution before complete drying. It was found that more antibody was immobilized on the surface of SF film by controlling the conformational changes during the drying process in comparison to covalent methods [122]. These results indicate that SF can be functionalized with antibodies with or without crosslinking agents, offering a wide range of biomedical applications.

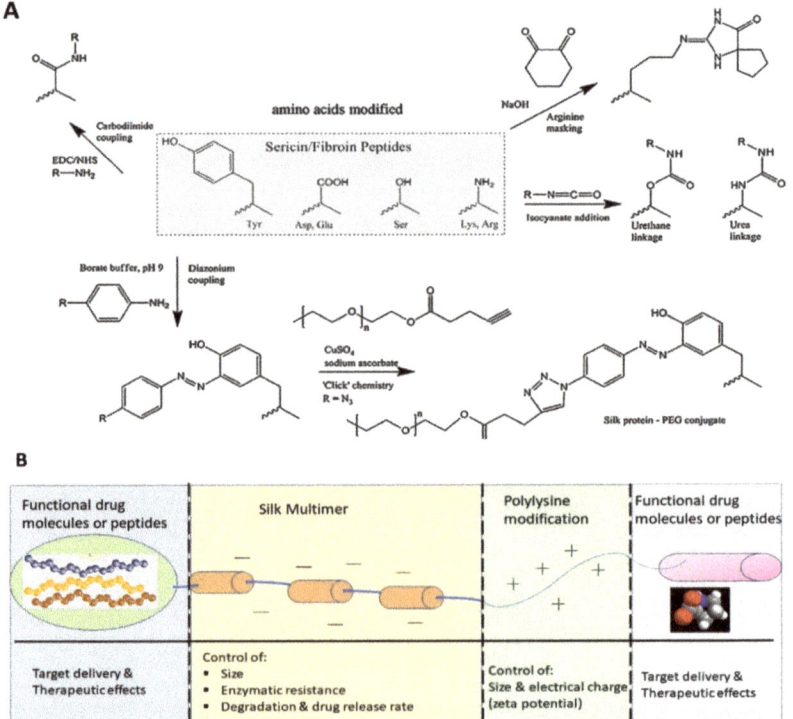

**Figure 7.** (**A**) Possible routes toward chemical modification of amino acids of silk proteins. Reprinted from [50] with permission from Elsevier. [2014].(**B**) The new properties obtained by SF when functionalised or modified in different positions.

*5.2. Functionalization of SF with Ligands*

One of the fundamental advantages of nanoparticles is the greater surface area to the volume ratio compared to larger particles. This property is essential for encapsulating the APIs such as anticancer agents and delivering them to the site of action. Moreover, the loaded APIs must be delivered at a proper concentration to cause the required effect on the target cells and minimise the damage to other cells [123]. However, recent studies have found that engineered particles with the optimum size and shape are limited to less than 1% tumor tissue accumulation [124]. Therefore, decorating polymeric nanocarriers with a targeting molecule has emerged as an effective approach to increase the specificity of the nanoparticles for the targeted cell lines [125]. As mentioned earlier SF has several active amino groups (Figure 7) which can be used for binding to other macromolecules [126]. For example, Arg-Gly-Asp (RGD) sequence that acts as a ligand for cell surface integrin receptors can be linked to SF particles to enhance their attachment to the certain cancer cells that overexpress integrins [127]. In a similar fashion, due to the overexpression of folate receptors (FR) in a wide range of tumor cells, modifying the surface of the silk nanoparticles with folate could be used as a tumor-targeting strategy [128]. Folate-conjugated SF particles (SF-FA) were used to enhance targeted delivery of doxorubicin (DOX) to human breast adenocarcinoma cell line (MDA-MB-231) [127]. Folate decoration on silk particle not only increased the retention of the nanoparticles at the tumor site but also promoted cellular uptake of the particles [129]. DOX incorporated in SF-FA nanoparticles demonstrated 3-folds higher cytotoxic activity in comparison to free DOX *in vitro*. Moreover, conjugation of folate to SF nanoparticles changed their cellular uptake mechanism from passive diffusion (free DOX) to endocytosis [126]. Another example of specific targeting using functionalized SF is modification of SF with human epidermal growth factor receptor 2 (Her2) which is overexpressed in 30% of breast carcinomas for targeted drug delivery to breast cancer cells [130]. An alternative functionalization approach involves using tumor-specific ligands such as nucleic acid sequences like CpG-siRNA [131].

## 6. Conclusions

Silk is a versatile biomaterial with great potential for drug and gene delivery applications. SF has been used as a naturally derived biopolymer for development of various types of drug delivery systems including hydrogels, SF films, microparticles and nanoparticles using a variety of fabrication methods. Each of these SF-based systems have shown promising features for different biomedical applications. SF micro- and nanoparticles have been used for delivery of different types of drugs such as curcumin, doxorubicin and ibuprofen as well as pDNA to various types of cells in a time-specific or site-specific manner. SF films have been used for controlled release of drugs such as dextran, epirubicin, and biological agents such as IgG and HIV inhibitor 5P12-RANTES. In addition, they have been used to stabilise biological agents such as horse radish peroxidase (HRP), glucose oxidase, vaccines and monoclonal antibodies in order to enhance their shelf life. Moreover, conjugation of SF to biomolecules such as insulin and BMP-2 has been employed as a strategy to sustain their release and prolong their biological activity. Functionalisation of SF with biological recognition elements such as RGD sequence, folate and Her2 has been used for tissue-specific drug delivery. In addition to the SF-based drug delivery systems, SF has also been used for coating the surface of polymeric microparticles and liposomes in order to modify their release kinetics or enhance their cell adhesion. However, despite all the advances made in the fabrication of SF-based constructs for biomedical and pharmaceutical applications, there is still lack of sufficient studies on the applications of bioengineered or structurally modified silk for tissue-specific targeted drug and gene delivery. Also, modifying the physichochemical and mechanical properties of the SF through miximng it with other natural or synthetic polymers in order to develop tailored SF-based biomaterials is another area to be further explored.

**Funding:** This research was funded by EPSRC, grant numbers EP/N007174/1 and EP/N023579/1; The Royal Society, grant numbers RG160662; and Jiangsu specially appointed professors program. R.H. was funded by the University of Sheffield studentship.

**Conflicts of Interest:** The authors declare no conflict of interest.

**References**

1. Liechty, W.B.; Kryscio, D.R.; Slaughter, B.V.; Peppas, N.A. Polymers for drug delivery systems. *Annu. Rev. Chem. Biomol. Eng.* **2010**, *1*, 149–173. [CrossRef] [PubMed]
2. Torchilin, V.P. Multifunctional, stimuli-sensitive nanoparticulate systems for drug delivery. *Nat. Rev. Drug Discov.* **2014**, *13*, 813. [CrossRef] [PubMed]
3. Pritchard, E.M.; Valentin, T.; Panilaitis, B.; Omenetto, F.; Kaplan, D.L. Antibiotic-Releasing Silk Biomaterials for Infection Prevention and Treatment. *Adv. Funct. Mater.* **2013**, *23*, 854–861. [CrossRef] [PubMed]
4. Luo, Z.; Li, J.; Qu, J.; Sheng, W.; Yang, J.; Li, M. Cationized Bombyx mori silk fibroin as a delivery carrier of the VEGF165–Ang-1 coexpression plasmid for dermal tissue regeneration. *J. Mater. Chem. B* **2019**, *7*, 80–94. [CrossRef]
5. Wenk, E.; Merkle, H.P.; Meinel, L. Silk fibroin as a vehicle for drug delivery applications. *J. Control. Release* **2011**, *150*, 128–141. [CrossRef] [PubMed]
6. Kim, S.Y.; Naskar, D.; Kundu, S.C.; Bishop, D.P.; Doble, P.A.; Boddy, A.V.; Chan, H.-K.; Wall, I.B.; Chrzanowski, W. Formulation of Biologically-Inspired Silk-Based Drug Carriers for Pulmonary Delivery Targeted for Lung Cancer. *Sci. Rep.* **2015**, *5*, 11878. [CrossRef] [PubMed]
7. Choi, M.; Choi, D.; Hong, J. Multilayered Controlled Drug Release Silk Fibroin Nanofilm by Manipulating Secondary Structure. *Biomacromolecules* **2018**, *19*, 3096–3103. [CrossRef] [PubMed]
8. Li, A.B.; Kluge, J.A.; Guziewicz, N.A.; Omenetto, F.G.; Kaplan, D.L. Silk-based stabilization of biomacromolecules. *J. Control. Release* **2015**, *219*, 416–430. [CrossRef] [PubMed]
9. Song, W.; Gregory, D.A.; Al-janabi, H.; Muthana, M.; Cai, Z.; Zhao, X. Magnetic-silk/polyethyleneimine core-shell nanoparticles for targeted gene delivery into human breast cancer cells. *Int. J. Pharm.* **2019**, *555*, 322–336. [CrossRef]
10. Seib, F.P. Reverse-engineered silk hydrogels for cell and drug delivery. *Ther. Deliv.* **2018**, *9*, 469–487. [CrossRef]
11. Jin, H.-J.; Kaplan, D.L. Mechanism of silk processing in insects and spiders. *Nature* **2003**, *424*, 1057–1061. [CrossRef] [PubMed]
12. Crivelli, B.; Perteghella, S.; Bari, E.; Sorrenti, M.; Tripodo, G.; Chlapanidas, T.; Torre, M.L. Silk nanoparticles: From inert supports to bioactive natural carriers for drug delivery. *Soft Matter.* **2018**, *14*, 546–557. [CrossRef] [PubMed]
13. Rnjak-Kovacina, J.; Wray, L.S.; Burke, K.A.; Torregrosa, T.; Golinski, J.M.; Huang, W.; Kaplan, D.L. Lyophilized Silk Sponges: A Versatile Biomaterial Platform for Soft Tissue Engineering. *ACS Biomater. Sci. Eng.* **2015**, *1*, 260–270. [CrossRef] [PubMed]
14. Leisk, G.G.; Lo, T.J.; Yucel, T.; Lu, Q.; Kaplan, D.L. Electrogelation for protein adhesives. *Adv. Mater.* **2010**, *22*, 711–715. [CrossRef] [PubMed]
15. Lu, Q.; Zhu, H.; Zhang, C.; Zhang, F.; Zhang, B.; Kaplan, D.L. Silk self-assembly mechanisms and control from thermodynamics to kinetics. *Biomacromolecules* **2012**, *13*, 826–832. [CrossRef] [PubMed]
16. Dong, Y.; Dong, P.; Huang, D.; Mei, L.; Xia, Y.; Wang, Z.; Pan, X.; Li, G.; Wu, C. Fabrication and characterization of silk fibroin-coated liposomes for ocular drug delivery. *Eur. J. Pharm. Biopharm.* **2015**, *91*, 82–90. [CrossRef] [PubMed]
17. Vepari, C.; Kaplan, D.L. Silk as a Biomaterial. *Prog. Polym. Sci.* **2007**, *32*, 991–1007. [CrossRef]
18. Deptuch, T.; Dams-Kozlowska, H. Silk Materials Functionalized via Genetic Engineering for Biomedical Applications. *Materials* **2017**, *10*, 1417. [CrossRef]
19. Guziewicz, N.A.; Massetti, A.J.; Perez-Ramirez, B.J.; Kaplan, D.L. Mechanisms of monoclonal antibody stabilization and release from silk biomaterials. *Biomaterials* **2013**, *34*, 7766–7775. [CrossRef]
20. Manning, M.C.; Chou, D.K.; Murphy, B.M.; Payne, R.W.; Katayama, D.S. Stability of protein pharmaceuticals: An update. *Pharm. Res.* **2010**, *27*, 544–575. [CrossRef]
21. Hawe, A.; Wiggenhorn, M.; van de Weert, M.; Garbe, J.H.O.; Mahler, H.-C.; Jiskoot, W. Forced Degradation of Therapeutic Proteins. *J. Pharm. Sci.* **2012**, *101*, 895–913. [CrossRef] [PubMed]
22. Yucel, T.; Lovett, M.L.; Kaplan, D.L. Silk-based biomaterials for sustained drug delivery. *J. Control. Release* **2014**, *190*, 381–397. [CrossRef]

23. Liu, Y.; Zheng, Z.; Gong, H.; Liu, M.; Guo, S.; Li, G.; Wang, X.; Kaplan, D.L. DNA preservation in silk. *Biomater. Sci.* **2017**, *5*, 1279–1292. [CrossRef] [PubMed]
24. Subia, B.; Kundu, S.C. Drug loading and release on tumor cells using silk fibroin-albumin nanoparticles as carriers. *Nanotechnology* **2013**, *24*, 035103. [CrossRef] [PubMed]
25. Mehta, A.S.; Singh, B.K.; Singh, N.; Archana, D.; Snigdha, K.; Harniman, R.; Rahatekar, S.S.; Tewari, R.P.; Dutta, P.K. Chitosan silk-based three-dimensional scaffolds containing gentamicin-encapsulated calcium alginate beads for drug administration and blood compatibility. *J. Biomater. Appl.* **2015**, *29*, 1314–1325. [CrossRef] [PubMed]
26. Elia, R.; Newhide, D.R.; Pedevillano, P.D.; Reiss, G.R.; Firpo, M.A.; Hsu, E.W.; Kaplan, D.L.; Prestwich, G.D.; Peattie, R.A. Silk-hyaluronan-based composite hydrogels: A novel, securable vehicle for drug delivery. *J. Biomater. Appl.* **2013**, *27*, 749–762. [CrossRef]
27. Yavuz, B.; Morgan, J.L.; Herrera, C.; Harrington, K.; Perez-Ramirez, B.; LiWang, P.J.; Kaplan, D.L. Sustained release silk fibroin discs: Antibody and protein delivery for HIV prevention. *J. Control. Release* **2019**, *301*, 1–12. [CrossRef]
28. Seib, F.P.; Jones, G.T.; Rnjak-Kovacina, J.; Lin, Y.; Kaplan, D.L. pH-Dependent Anticancer Drug Release from Silk Nanoparticles. *Adv. Healthc. Mater.* **2013**, *2*, 1606–1611. [CrossRef] [PubMed]
29. Chen, M.; Shao, Z.; Chen, X. Paclitaxel-loaded silk fibroin nanospheres. *J. Biomed. Mater. Res. Part A* **2012**, *100A*, 203–210. [CrossRef] [PubMed]
30. Tomeh, M.A.; Hadianamrei, R.; Zhao, X. A review of curcumin and its derivatives as anticancer agents. *Int. J. Mol. Sci.* **2019**, *20*, 1033. [CrossRef]
31. Qu, J.; Liu, Y.; Yu, Y.; Li, J.; Luo, J.; Li, M. Silk fibroin nanoparticles prepared by electrospray as controlled release carriers of cisplatin. *Mater. Sci. Eng. C* **2014**, *44*, 166–174. [CrossRef] [PubMed]
32. Yao, D.; Dong, S.; Lu, Q.; Hu, X.; Kaplan, D.L.; Zhang, B.; Zhu, H. Salt-leached silk scaffolds with tunable mechanical properties. *Biomacromolecules* **2012**, *13*, 3723–3729. [CrossRef] [PubMed]
33. Theodora, C.; Sara, P.; Silvio, F.; Alessandra, B.; Giuseppe, T.; Barbara, V.; Barbara, C.; Sabrina, R.; Silvia, D.; Stefania, P. Platelet lysate and adipose mesenchymal stromal cells on silk fibroin nonwoven mats for wound healing. *J. Appl. Polym. Sci.* **2016**, *133*. [CrossRef]
34. Kundu, B.; Rajkhowa, R.; Kundu, S.C.; Wang, X. Silk fibroin biomaterials for tissue regenerations. *Adv. Drug Deliv. Rev.* **2013**, *65*, 457–470. [CrossRef] [PubMed]
35. Luo, K.; Yang, Y.; Shao, Z. Physically Crosslinked Biocompatible Silk-Fibroin-Based Hydrogels with High Mechanical Performance. *Adv. Funct. Mater.* **2016**, *26*, 872–880. [CrossRef]
36. Lu, Q.; Hu, X.; Wang, X.; Kluge, J.A.; Lu, S.; Cebe, P.; Kaplan, D.L. Water-insoluble silk films with silk I structure. *Acta. Biomater.* **2010**, *6*, 1380–1387. [CrossRef] [PubMed]
37. Melke, J.; Midha, S.; Ghosh, S.; Ito, K.; Hofmann, S. Silk fibroin as biomaterial for bone tissue engineering. *Acta. Biomater.* **2016**, *31*, 1–16. [CrossRef] [PubMed]
38. Numata, K.; Hamasaki, J.; Subramanian, B.; Kaplan, D.L. Gene delivery mediated by recombinant silk proteins containing cationic and cell binding motifs. *J. Control. Release* **2010**, *146*, 136–143. [CrossRef] [PubMed]
39. Meinel, L.; Karageorgiou, V.; Hofmann, S.; Fajardo, R.; Snyder, B.; Li, C.; Zichner, L.; Langer, R.; Vunjak-Novakovic, G.; Kaplan, D.L. Engineering bone-like tissue in vitro using human bone marrow stem cells and silk scaffolds. *J. Biomed. Mater. Res. A* **2004**, *71*, 25–34. [CrossRef]
40. Jin, H.J.; Chen, J.; Karageorgiou, V.; Altman, G.H.; Kaplan, D.L. Human bone marrow stromal cell responses on electrospun silk fibroin mats. *Biomaterials* **2004**, *25*, 1039–1047. [CrossRef]
41. Meinel, L.; Hofmann, S.; Betz, O.; Fajardo, R.; Merkle, H.P.; Langer, R.; Evans, C.H.; Vunjak-Novakovic, G.; Kaplan, D.L. Osteogenesis by human mesenchymal stem cells cultured on silk biomaterials: Comparison of adenovirus mediated gene transfer and protein delivery of BMP-2. *Biomaterials* **2006**, *27*, 4993–5002. [CrossRef] [PubMed]
42. Meinel, L.; Hofmann, S.; Karageorgiou, V.; Kirker-Head, C.; McCool, J.; Gronowicz, G.; Zichner, L.; Langer, R.; Vunjak-Novakovic, G.; Kaplan, D.L. The inflammatory responses to silk films in vitro and in vivo. *Biomaterials* **2005**, *26*, 147–155. [CrossRef] [PubMed]
43. Song, W.; Muthana, M.; Mukherjee, J.; Falconer, R.J.; Biggs, C.A.; Zhao, X. Magnetic-Silk Core–Shell Nanoparticles as Potential Carriers for Targeted Delivery of Curcumin into Human Breast Cancer Cells. *ACS Biomater. Sci. Eng.* **2017**, *3*, 1027–1038. [CrossRef]

44. Panilaitis, B.; Altman, G.H.; Chen, J.; Jin, H.J.; Karageorgiou, V.; Kaplan, D.L. Macrophage responses to silk. *Biomaterials* **2003**, *24*, 3079–3085. [CrossRef]
45. Xie, H.; Wang, J.; He, Y.; Gu, Z.; Xu, J.; Li, L.; Ye, Q. Biocompatibility and safety evaluation of a silk fibroin-doped calcium polyphosphate scaffold copolymer in vitro and in vivo. *RSC Adv.* **2017**, *7*, 46036–46044. [CrossRef]
46. Zhao, S.; Chen, Y.; Partlow, B.P.; Golding, A.S.; Tseng, P.; Coburn, J.; Applegate, M.B.; Moreau, J.E.; Omenetto, F.G.; Kaplan, D.L. Bio-functionalized silk hydrogel microfluidic systems. *Biomaterials* **2016**, *93*, 60–70. [CrossRef] [PubMed]
47. Bhrany, A.D.; Lien, C.J.; Beckstead, B.L.; Futran, N.D.; Muni, N.H.; Giachelli, C.M.; Ratner, B.D. Crosslinking of an oesophagus acellular matrix tissue scaffold. *J. Tissue Eng. Regen. Med.* **2008**, *2*, 365–372. [CrossRef] [PubMed]
48. Kurland, N.E.; Drira, Z.; Yadavalli, V.K. Measurement of nanomechanical properties of biomolecules using atomic force microscopy. *Micron* **2012**, *43*, 116–128. [CrossRef] [PubMed]
49. Altman, G.H.; Horan, R.L.; Lu, H.H.; Moreau, J.; Martin, I.; Richmond, J.C.; Kaplan, D.L. Silk matrix for tissue engineered anterior cruciate ligaments. *Biomaterials* **2002**, *23*, 4131–4141. [CrossRef]
50. Kundu, B.; Kurland, N.E.; Bano, S.; Patra, C.; Engel, F.B.; Yadavalli, V.K.; Kundu, S.C. Silk proteins for biomedical applications: Bioengineering perspectives. *Prog. Polym. Sci.* **2014**, *39*, 251–267. [CrossRef]
51. Min, B.M.; Jeong, L.; Lee, K.Y.; Park, W.H. Regenerated silk fibroin nanofibers: Water vapor-induced structural changes and their effects on the behavior of normal human cells. *Macromol. Biosci.* **2006**, *6*, 285–292. [CrossRef] [PubMed]
52. Jin, H.J.; Park, J.; Karageorgiou, V.; Kim, U.J.; Valluzzi, R.; Cebe, P.; Kaplan, D.L. Water-Stable Silk Films with Reduced β-Sheet Content. *Adv. Funct. Mater.* **2005**, *15*, 1241–1247. [CrossRef]
53. Guan, J.; Wang, Y.; Mortimer, B.; Holland, C.; Shao, Z.; Porter, D.; Vollrath, F. Glass transitions in native silk fibres studied by dynamic mechanical thermal analysis. *Soft Matter.* **2016**, *12*, 5926–5936. [CrossRef] [PubMed]
54. Nazarov, R.; Jin, H.-J.; Kaplan, D.L. Porous 3-D Scaffolds from Regenerated Silk Fibroin. *Biomacromolecules* **2004**, *5*, 718–726. [CrossRef] [PubMed]
55. Wongpinyochit, T.; Johnston, B.F.; Seib, F.P. Degradation Behavior of Silk Nanoparticles—Enzyme Responsiveness. *ACS Biomater. Sci. Eng.* **2018**, *4*, 942–951. [CrossRef]
56. You, R.; Zhang, Y.; Liu, Y.; Liu, G.; Li, M. The degradation behavior of silk fibroin derived from different ionic liquid solvents. *Nat. Sci.* **2013**, *5*, 10. [CrossRef]
57. Li, M.; Ogiso, M.; Minoura, N. Enzymatic degradation behavior of porous silk fibroin sheets. *Biomaterials* **2003**, *24*, 357–365. [CrossRef]
58. Horan, R.L.; Antle, K.; Collette, A.L.; Wang, Y.; Huang, J.; Moreau, J.E.; Volloch, V.; Kaplan, D.L.; Altman, G.H. In vitro degradation of silk fibroin. *Biomaterials* **2005**, *26*, 3385–3393. [CrossRef]
59. Liu, B.; Song, Y.-W.; Jin, L.; Wang, Z.-J.; Pu, D.-Y.; Lin, S.-Q.; Zhou, C.; You, H.-J.; Ma, Y.; Li, J.-M.; et al. Silk structure and degradation. *Colloids and Surfaces B. Biointerfaces* **2015**, *131*, 122–128. [CrossRef]
60. Winkler, S.; Wilson, D.; Kaplan, D.L. Controlling beta-sheet assembly in genetically engineered silk by enzymatic phosphorylation/dephosphorylation. *Biochemistry* **2000**, *39*, 12739–12746. [CrossRef]
61. Chirila, T.V.; Suzuki, S.; Papolla, C. A comparative investigation of Bombyx mori silk fibroin hydrogels generated by chemical and enzymatic cross-linking. *Biotechnol. Appl. Biochem.* **2017**, *64*, 771–781. [CrossRef] [PubMed]
62. Matsumoto, A.; Chen, J.; Collette, A.L.; Kim, U.-J.; Altman, G.H.; Cebe, P.; Kaplan, D.L. Mechanisms of Silk Fibroin Sol–Gel Transitions. *J. Phys. Chem. B* **2006**, *110*, 21630–21638. [CrossRef] [PubMed]
63. Karahaliloğlu, Z. Curcumin-loaded silk fibroin e-gel scaffolds for wound healing applications. *Mater. Technol.* **2018**, *33*, 276–287. [CrossRef]
64. Sundarakrishnan, A.; Herrero Acero, E.; Coburn, J.; Chwalek, K.; Partlow, B.; Kaplan, D.L. Phenol red-silk tyrosine cross-linked hydrogels. *Acta. Biomater.* **2016**, *42*, 102–113. [CrossRef] [PubMed]
65. Lee, M.C.; Kim, D.-K.; Lee, O.J.; Kim, J.-H.; Ju, H.W.; Lee, J.M.; Moon, B.M.; Park, H.J.; Kim, D.W.; Kim, S.H.; et al. Fabrication of silk fibroin film using centrifugal casting technique for corneal tissue engineering. *J. Biomed. Mater. Res. Part B* **2016**, *104*, 508–514. [CrossRef]
66. Qi, Y.; Wang, H.; Wei, K.; Yang, Y.; Zheng, R.Y.; Kim, I.S.; Zhang, K.Q. A Review of Structure Construction of Silk Fibroin Biomaterials from Single Structures to Multi-Level Structures. *Int. J. Mol. Sci.* **2017**, *18*, 237. [CrossRef] [PubMed]

67. Sagnella, A.; Pistone, A.; Bonetti, S.; Donnadio, A.; Saracino, E.; Nocchetti, M.; Dionigi, C.; Ruani, G.; Muccini, M.; Posati, T.; et al. Effect of different fabrication methods on the chemo-physical properties of silk fibroin films and on their interaction with neural cells. *RSC Adv.* **2016**, *6*, 9304–9314. [CrossRef]
68. Terada, D.; Yokoyama, Y.; Hattori, S.; Kobayashi, H.; Tamada, Y. The outermost surface properties of silk fibroin films reflect ethanol-treatment conditions used in biomaterial preparation. *Mater. Sci. Eng. C* **2016**, *58*, 119–126. [CrossRef]
69. Jiang, C.; Wang, X.; Gunawidjaja, R.; Lin, Y.H.; Gupta, M.K.; Kaplan, D.L.; Naik, R.R.; Tsukruk, V.V. Mechanical Properties of Robust Ultrathin Silk Fibroin Films. *Adv. Funct. Mater.* **2007**, *17*, 2229–2237. [CrossRef]
70. De Moraes, M.A.; Silva, M.F.; Weska, R.F.; Beppu, M.M. Silk fibroin and sodium alginate blend: Miscibility and physical characteristics. *Mater. Sci. Eng. C* **2014**, *40*, 85–91. [CrossRef]
71. Wang, X.; Yucel, T.; Lu, Q.; Hu, X.; Kaplan, D.L. Silk nanospheres and microspheres from silk/pva blend films for drug delivery. *Biomaterials* **2010**, *31*, 1025–1035. [CrossRef] [PubMed]
72. Cao, Y.; Liu, F.; Chen, Y.; Yu, T.; Lou, D.; Guo, Y.; Li, P.; Wang, Z.; Ran, H. Drug release from core-shell PVA/silk fibroin nanoparticles fabricated by one-step electrospraying. *Sci. Rep.* **2017**, *7*, 11913. [CrossRef] [PubMed]
73. Shi, P.; Goh, J.C.H. Self-assembled silk fibroin particles: Tunable size and appearance. *Powder Technol.* **2012**, *215–216*, 85–90. [CrossRef]
74. Lammel, A.S.; Hu, X.; Park, S.H.; Kaplan, D.L.; Scheibel, T.R. Controlling silk fibroin particle features for drug delivery. *Biomaterials* **2010**, *31*, 4583–4591. [CrossRef] [PubMed]
75. Tian, Y.; Jiang, X.; Chen, X.; Shao, Z.; Yang, W. Doxorubicin-Loaded Magnetic Silk Fibroin Nanoparticles for Targeted Therapy of Multidrug-Resistant Cancer. *Adv. Mater.* **2014**, *26*, 7393–7398. [CrossRef] [PubMed]
76. Mitropoulos, A.N.; Perotto, G.; Kim, S.; Marelli, B.; Kaplan, D.L.; Omenetto, F.G. Synthesis of silk fibroin micro- and submicron spheres using a co-flow capillary device. *Adv. Mater.* **2014**, *26*, 1105–1110. [CrossRef] [PubMed]
77. Wongpinyochit, T.; Totten, J.D.; Johnston, B.F.; Seib, F.P. Microfluidic-assisted silk nanoparticle tuning. *Nanoscale Adv.* **2019**, *1*, 873–883. [CrossRef]
78. Myung, S.J.; Kim, H.-S.; Kim, Y.; Chen, P.; Jin, H.-J. Fluorescent silk fibroin nanoparticles prepared using a reverse microemulsion. *Macromol. Res.* **2008**, *16*, 604–608. [CrossRef]
79. Wongpinyochit, T.; Uhlmann, P.; Urquhart, A.J.; Seib, F.P. PEGylated Silk Nanoparticles for Anticancer Drug Delivery. *Biomacromolecules* **2015**, *16*, 3712–3722. [CrossRef]
80. Gholami, A.; Tavanai, H.; Moradi, A.R. Production of fibroin nanopowder through electrospraying. *J. Nanoparticle Res.* **2011**, *13*, 2089–2098. [CrossRef]
81. Gupta, V.; Aseh, A.; Ríos, C.N.; Aggarwal, B.B.; Mathur, A.B. Fabrication and characterization of silk fibroin-derived curcumin nanoparticles for cancer therapy. *Int. J. Nanomed.* **2009**, *4*, 115–122. [CrossRef]
82. Zhao, Z.; Xie, M.; Li, Y.; Chen, A.; Li, G.; Zhang, J.; Hu, H.; Wang, X.; Li, S. Formation of curcumin nanoparticles via solution-enhanced dispersion by supercritical $CO_2$. *Int. J. Nanomed.* **2015**, *10*, 3171–3181. [CrossRef] [PubMed]
83. Perry, H.; Gopinath, A.; Kaplan, D.L.; Dal Negro, L.; Omenetto, F.G. Nano- and Micropatterning of Optically Transparent, Mechanically Robust, Biocompatible Silk Fibroin Films. *Adv. Mater.* **2008**, *20*, 3070–3072. [CrossRef]
84. Guziewicz, N.; Best, A.; Perez-Ramirez, B.; Kaplan, D.L. Lyophilized silk fibroin hydrogels for the sustained local delivery of therapeutic monoclonal antibodies. *Biomaterials* **2011**, *32*, 2642–2650. [CrossRef] [PubMed]
85. Numata, K.; Reagan, M.R.; Goldstein, R.H.; Rosenblatt, M.; Kaplan, D.L. Spider Silk-Based Gene Carriers for Tumor Cell-Specific Delivery. *Bioconjugate Chem.* **2011**, *22*, 1605–1610. [CrossRef] [PubMed]
86. Meinel, L.; Kaplan, D.L. Silk constructs for delivery of musculoskeletal therapeutics. *Adv. Drug Deliv. Rev.* **2012**, *64*, 1111–1122. [CrossRef] [PubMed]
87. Pritchard, E.M.; Dennis, P.B.; Omenetto, F.; Naik, R.R.; Kaplan, D.L. Review physical and chemical aspects of stabilization of compounds in silk. *Biopolymers* **2012**, *97*, 479–498. [CrossRef] [PubMed]
88. Zhang, X.; Berghe, I.V.; Wyeth, P. Heat and moisture promoted deterioration of raw silk estimated by amino acid analysis. *J. Cult. Herit.* **2011**, *12*, 408–411. [CrossRef]
89. Zong, X.H.; Zhou, P.; Shao, Z.Z.; Chen, S.M.; Chen, X.; Hu, B.W.; Deng, F.; Yao, W.H. Effect of pH and copper(II) on the conformation transitions of silk fibroin based on EPR, NMR, and Raman spectroscopy. *Biochemistry* **2004**, *43*, 11932–11941. [CrossRef]

90. Lu, S.-Z.; Wang, X.-Q.; Uppal, N.; Kaplan, D.L.; Li, M.-Z. Stabilization of horseradish peroxidase in silk materials. *Front. Mater. Sci. China* **2009**, *3*, 367. [CrossRef]
91. Demura, M.; Asakura, T.; Kuroo, T. Immobilization of biocatalysts with bombyx mori silk fibroin by several kinds of physical treatment and its application to glucose sensors. *Biosensors* **1989**, *4*, 361–372. [CrossRef]
92. Kuzuhara, A.; Asakura, T.; Tomoda, R.; Matsunaga, T. Use of silk fibroin for enzyme membrane. *J. Biotechnol.* **1987**, *5*, 199–207. [CrossRef]
93. Ng, S.-F.; Anuwi, N.-A.; Tengku-Ahmad, T.-N. Topical Lyogel Containing Corticosteroid Decreases IgE Expression and Enhances the Therapeutic Efficacy Against Atopic Eczema. *AAPS Pharm. Sci. Tech.* **2015**, *16*, 656–663. [CrossRef] [PubMed]
94. Mao, S.; Guo, C.; Shi, Y.; Li, L.C. Recent advances in polymeric microspheres for parenteral drug delivery – part 1. *Expert Opin. Drug Deliv.* **2012**, *9*, 1161–1176. [CrossRef] [PubMed]
95. Hines, D.J.; Kaplan, D.L. Mechanisms of Controlled Release from Silk Fibroin Films. *Biomacromolecules* **2011**, *12*, 804–812. [CrossRef] [PubMed]
96. Wang, X.; Wenk, E.; Matsumoto, A.; Meinel, L.; Li, C.; Kaplan, D.L. Silk microspheres for encapsulation and controlled release. *J. Control. Release* **2007**, *117*, 360–370. [CrossRef] [PubMed]
97. Pritchard, E.M.; Szybala, C.; Boison, D.; Kaplan, D.L. Silk fibroin encapsulated powder reservoirs for sustained release of adenosine. *J. Control. Release* **2010**, *144*, 159–167. [CrossRef] [PubMed]
98. Zhou, J.; Fang, T.; Wen, J.; Shao, Z.; Dong, J. Silk coating on poly(ε-caprolactone) microspheres for the delayed release of vancomycin. *J. Microencapsul.* **2011**, *28*, 99–107. [CrossRef]
99. Ali, M.; Byrne, M.E. Challenges and solutions in topical ocular drug-delivery systems. *Expert Rev. Clin. Pharmacol.* **2008**, *1*, 145–161. [CrossRef]
100. Gobin, A.S.; Rhea, R.; Newman, R.A.; Mathur, A.B. Silk-fibroin-coated liposomes for long-term and targeted drug delivery. *Int. J. Nanomed.* **2006**, *1*, 81–87. [CrossRef]
101. Lam, J.K.W.; Chow, M.Y.T.; Zhang, Y.; Leung, S.W.S. siRNA Versus miRNA as Therapeutics for Gene Silencing. *Mol. Ther. Nucleic Acids* **2015**, *4*, e252. [CrossRef] [PubMed]
102. Ginn, S.L.; Amaya, A.K.; Alexander, I.E.; Edelstein, M.; Abedi, M.R. Gene therapy clinical trials worldwide to 2017: An update. *J. Gene Med.* **2018**, *20*, e3015. [CrossRef] [PubMed]
103. Numata, K.; Mieszawska-Czajkowska, A.J.; Kvenvold, L.A.; Kaplan, D.L. Silk-Based Nanocomplexes with Tumor-Homing Peptides for Tumor-Specific Gene Delivery. *Macromol. Biosci.* **2012**, *12*, 75–82. [CrossRef]
104. Numata, K.; Kaplan, D.L. Silk-based delivery systems of bioactive molecules. *Adv. Drug Deliv. Rev.* **2010**, *62*, 1497–1508. [CrossRef]
105. Numata, K.; Subramanian, B.; Currie, H.A.; Kaplan, D.L. Bioengineered silk protein-based gene delivery systems. *Biomaterials* **2009**, *30*, 5775–5784. [CrossRef] [PubMed]
106. Christian, S.; Pilch, J.; Akerman, M.E.; Porkka, K.; Laakkonen, P.; Ruoslahti, E. Nucleolin expressed at the cell surface is a marker of endothelial cells in angiogenic blood vessels. *J. Cell Biol.* **2003**, *163*, 871–878. [CrossRef] [PubMed]
107. Porkka, K.; Laakkonen, P.; Hoffman, J.A.; Bernasconi, M.; Ruoslahti, E. A fragment of the HMGN2 protein homes to the nuclei of tumor cells and tumor endothelial cells in vivo. *Proc. Natl. Acad. Sci. USA* **2002**, *99*, 7444–7449. [CrossRef] [PubMed]
108. Laakkonen, P.; Akerman, M.E.; Biliran, H.; Yang, M.; Ferrer, F.; Karpanen, T.; Hoffman, R.M.; Ruoslahti, E. Antitumor activity of a homing peptide that targets tumor lymphatics and tumor cells. *Proc. Natl. Acad. Sci. USA* **2004**, *101*, 9381–9386. [CrossRef]
109. Laakkonen, P.; Porkka, K.; Hoffman, J.A.; Ruoslahti, E. A tumor-homing peptide with a targeting specificity related to lymphatic vessels. *Nat. Med.* **2002**, *8*, 751–755. [CrossRef]
110. Liu, Y.; You, R.; Liu, G.; Li, X.; Sheng, W.; Yang, J.; Li, M. Antheraea pernyi Silk Fibroin-Coated PEI/DNA Complexes for Targeted Gene Delivery in HEK 293 and HCT 116 Cells. *Int. J. Mol. Sci.* **2014**, *15*, 7049. [CrossRef]
111. Richard, J.P.; Melikov, K.; Vives, E.; Ramos, C.; Verbeure, B.; Gait, M.J.; Chernomordik, L.V.; Lebleu, B. Cell-penetrating peptides. A reevaluation of the mechanism of cellular uptake. *J. Biol. Chem.* **2003**, *278*, 585–590. [CrossRef]
112. Holm, T.; Johansson, H.; Lundberg, P.; Pooga, M.; Lindgren, M.; Langel, U. Studying the uptake of cell-penetrating peptides. *Nat. Protoc.* **2006**, *1*, 1001–1005. [CrossRef] [PubMed]

113. Numata, K.; Kaplan, D.L. Silk-based gene carriers with cell membrane destabilizing peptides. *Biomacromolecules* **2010**, *11*, 3189–3195. [CrossRef]
114. Gustafson, J.; Greish, K.; Frandsen, J.; Cappello, J.; Ghandehari, H. Silk-elastinlike recombinant polymers for gene therapy of head and neck cancer: From molecular definition to controlled gene expression. *J. Control. Release* **2009**, *140*, 256–261. [CrossRef] [PubMed]
115. Hwang, D.; Moolchandani, V.; Dandu, R.; Haider, M.; Cappello, J.; Ghandehari, H. Influence of polymer structure and biodegradation on DNA release from silk-elastinlike protein polymer hydrogels. *Int. J. Pharm.* **2009**, *368*, 215–219. [CrossRef] [PubMed]
116. Price, R.; Gustafson, J.; Greish, K.; Cappello, J.; McGill, L.; Ghandehari, H. Comparison of silk-elastinlike protein polymer hydrogel and poloxamer in matrix-mediated gene delivery. *Int. J. Pharm.* **2012**, *427*, 97–104. [CrossRef]
117. Zhang, Y.; Wu, C.; Luo, T.; Li, S.; Cheng, X.; Miron, R.J. Synthesis and inflammatory response of a novel silk fibroin scaffold containing BMP7 adenovirus for bone regeneration. *Bone* **2012**, *51*, 704–713. [CrossRef]
118. Li, L.; Puhl, S.; Meinel, L.; Germershaus, O. Silk fibroin layer-by-layer microcapsules for localized gene delivery. *Biomaterials* **2014**, *35*, 7929–7939. [CrossRef]
119. Zhang, Y.-Q.; Ma, Y.; Xia, Y.-Y.; Shen, W.-D.; Mao, J.-P.; Zha, X.-M.; Shirai, K.; Kiguchi, K. Synthesis of silk fibroin-insulin bioconjugates and their characterization and activities in vivo. *J. Biomed. Mater. Res. Part B* **2006**, *79B*, 275–283. [CrossRef]
120. Karageorgiou, V.; Meinel, L.; Hofmann, S.; Malhotra, A.; Volloch, V.; Kaplan, D. Bone morphogenetic protein-2 decorated silk fibroin films induce osteogenic differentiation of human bone marrow stromal cells. *J. Biomed. Mater. Res. A* **2004**, *71*, 528–537. [CrossRef]
121. Wang, P.; Qi, C.; Yu, Y.; Yuan, J.; Cui, L.; Tang, G.; Wang, Q.; Fan, X. Covalent Immobilization of Catalase onto Regenerated Silk Fibroins via Tyrosinase-Catalyzed Cross-Linking. *Appl. Biochem. Biotechnol.* **2015**, *177*, 472–485. [CrossRef]
122. Lu, Q.; Wang, X.; Zhu, H.; Kaplan, D.L. Surface immobilization of antibody on silk fibroin through conformational transition. *Acta. Biomater.* **2011**, *7*, 2782–2786. [CrossRef]
123. Cao, Y.; Gu, Y.; Ma, H.; Bai, J.; Liu, L.; Zhao, P.; He, H. Self-assembled nanoparticle drug delivery systems from galactosylated polysaccharide–doxorubicin conjugate loaded doxorubicin. *Int. J. Biol. Macromol.* **2010**, *46*, 245–249. [CrossRef]
124. Wilhelm, S.; Tavares, A.J.; Dai, Q.; Ohta, S.; Audet, J.; Dvorak, H.F.; Chan, W.C.W. Analysis of nanoparticle delivery to tumours. *Nat. Rev. Mater.* **2016**, *1*, 16014. [CrossRef]
125. Gobin, A.S.; Butler, C.E.; Mathur, A.B. Repair and regeneration of the abdominal wall musculofascial defect using silk fibroin-chitosan blend. *Tissue Eng.* **2006**, *12*, 3383–3394. [CrossRef]
126. Subia, B.; Talukdar, S.; Kundu, S.C.; Chandra, S. Folate conjugated silk fibroin nanocarriers for targeted drug delivery. *Integr. Biol.* **2013**, *6*, 203–214. [CrossRef]
127. Patra, C.; Talukdar, S.; Novoyatleva, T.; Velagala, S.R.; Muhlfeld, C.; Kundu, B.; Kundu, S.C.; Engel, F.B. Silk protein fibroin from Antheraea mylitta for cardiac tissue engineering. *Biomaterials* **2012**, *33*, 2673–2680. [CrossRef]
128. Torchilin, V.P. Recent advances with liposomes as pharmaceutical carriers. *Nat. Rev. Drug Discov.* **2005**, *4*, 145. [CrossRef]
129. Zhang, Y.; Hong, H.; Myklejord, D.V.; Cai, W. Molecular imaging with SERS-active nanoparticles. *Small* **2011**, *7*, 3261–3269. [CrossRef]
130. Witton, C.J.; Reeves, J.R.; Going, J.J.; Cooke, T.G.; Bartlett, J.M.S. Expression of the HER1-4 family of receptor tyrosine kinases in breast cancer. *J. Pathol.* **2003**, *200*, 290–297. [CrossRef]
131. Kozlowska, A.K.; Florczak, A.; Smialek, M.; Dondajewska, E.; Mackiewicz, A.; Kortylewski, M.; Dams-Kozlowska, H. Functionalized bioengineered spider silk spheres improve nuclease resistance and activity of oligonucleotide therapeutics providing a strategy for cancer treatment. *Acta. Biomater.* **2017**, *59*, 221–233. [CrossRef]

© 2019 by the authors. Licensee MDPI, Basel, Switzerland. This article is an open access article distributed under the terms and conditions of the Creative Commons Attribution (CC BY) license (http://creativecommons.org/licenses/by/4.0/).

*Review*

# Injectable Hydrogels for Cancer Therapy over the Last Decade

Giuseppe Cirillo, Umile Gianfranco Spizzirri *, Manuela Curcio, Fiore Pasquale Nicoletta and Francesca Iemma

Department of Pharmacy, Health and Nutritional Sciences, University of Calabria, 87036 Rende (CS), Italy; giuseppe.cirillo@unical.it (C.C.); manuela.curcio@unical.it (M.C.); fiore.nicoletta@unical.it (F.P.N.); francesca.iemma@unical.it (F.I.)
* Correspondence: g.spizzirri@unical.it; Tel.: +39-0984493298

Received: 30 July 2019; Accepted: 17 September 2019; Published: 19 September 2019

**Abstract:** The interest in injectable hydrogels for cancer treatment has been significantly growing over the last decade, due to the availability of a wide range of starting polymer structures with tailored features and high chemical versatility. Many research groups are working on the development of highly engineered injectable delivery vehicle systems suitable for combined chemo-and radio-therapy, as well as thermal and photo-thermal ablation, with the aim of finding out effective solutions to overcome the current obstacles of conventional therapeutic protocols. Within this work, we have reviewed and discussed the most recent injectable hydrogel systems, focusing on the structure and properties of the starting polymers, which are mainly classified into natural or synthetic sources. Moreover, mapping the research landscape of the fabrication strategies, the main outcome of each system is discussed in light of possible clinical applications.

**Keywords:** injectable hydrogels; drug delivery; anticancer activity; natural polymers; synthetic polymers; stimuli-responsive materials

## 1. Introduction

Injectable hydrogels can be defined as three-dimensional hydrophilic polymeric networks with a very high affinity for body fluids that may be delivered into body through a catheter or by direct injection with a syringe [1]. Injectable hydrogels have been proposed in the biomedical field as a platform for tissue engineering, as well as for the delivery of therapeutics (Figure 1) [2–4].

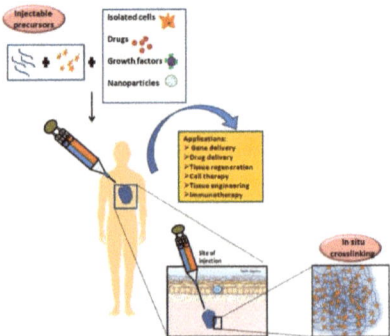

**Figure 1.** Application of injectable hydrogel systems in biomedical field. Reproduced with permission from [3]. Elsevier, [2018].

A gelling mechanism allows injectable hydrogels to be classified into chemically and physically cross-linked hydrogels [5].

Chemical intermolecular cross-linking can be created by the generation of new covalent bonds between polymer chains via photo- or thermo-irradiation [6], or by specific reaction mechanisms involving Schiff's base formation [7], Diels–Alder cycloaddition [8], Michael-type addition [9], and azide–alkyne (CuAAC) click chemistry [10,11]. The encapsulation of suitable therapeutic agents within the gels during hydrogel formation allows the preparation of three-dimensional structures able to act as a platform for controlled drug delivery or tissue engineering [12]. Chemical hydrogels possess higher mechanical strength (due to high stable crosslink points [13]), longer physical stability, and a prolonged degradation period [14]. Nevertheless, in vivo applications appear reduced due to some potential toxic agents, such as cross-linking monomers, photo-initiators, organic solvents, or catalyzers [2]. Non-covalent bonds such as hydrophobic interactions [15], hydrogen bonding [16], ion cross-linking [17], and host-guest interactions [18] can be exploited in the formation of injectable physical hydrogels. Usually, in the synthesis of this kind of structure, the required mild reaction conditions avoid the generation of any toxic by-products. Furthermore, organic solvents, cross-linking catalysts, or photo-initiation processes are not required during the gelation process [2]. On the contrary, physical hydrogels suffer from some drawbacks compared with the chemically cross-linked formulations, particularly related to bond stability and poor mechanical properties [19].

The mechanical properties of injectable hydrogels are a critical parameter for its function and applications, with the nature of gel being evident by a storage modulus G' higher than the corresponding values of the loss modulus G" [20,21]. The resulting mechanical properties of any injectable hydrogels should be adequate to withstand the deformations occurring in the body [22]. The viscosity of the polymer solution is an important parameter that should also be considered in the case of injectable matrices: Precursor aqueous solutions should possess sufficiently low viscosity, or at least adequate shear-thinning properties, to allow for easy injection [23–25]. This requirement makes molecular weight control, polymer architecture, as well as chemical composition, very important parameters to be controlled in the design of an effective hydrogel system, which should also allow a homogeneous drug dispersion before the gelation of the cross-linked structure [26]. The U.S. Food and Drug Administration (FDA) fixed the upper limit for any injectable solutions to 0.05 Pa s [27]. Upon gelation, a rapid increase in this value was observed, followed by a leveling off over time [28]. The mechanical properties of the whole hydrogel are strictly dependent on another important structural parameter, such as the porosity (e.g., the space between cross-links). An increased concentration or cross-linking density would enhance the mechanical strength, thus promoting the integrity duration of the hydrogels [29]. Nevertheless, this would determine the hydrogel's porosity to be significantly reduced, limiting the movement of nutrients and solutions for either the growth of the cell in tissue engineering applications or the modulation of the release profile in drug delivery [30]. Thus, a valuable balance between these parameters should be achieved.

Clinical applications of injectable hydrogels require some fundamental mandatories, such as biodegradability, biocompatibility, stability, non-toxicity, and suitable mechanical and viscoelastic properties. A biocompatible injectable hydrogel should be non-carcinogenic, non-toxic, and should not induce any chronic or adverse physiological response after its degradation. To develop systems with high biocompatibility towards tissues, cells, and body fluids, natural polymers are more suitable than synthetic cross-linked structures due to their subunits, which are more similar to the natural extracellular matrix [31]. Gradual degradation of the hydrogel into biocompatible by-products should also be considered because of their possible accumulation that could generate adverse effects. Usually, carbohydrates, peptides, and nucleic acids naturally degrade in non-toxic by-products [31]. Among the different applications of injectable hydrogels, cancer therapy is one of the most widely explored [32]. The treatment of cancer by systemic chemotherapeutic procedure, indeed, often determines a high level of cytotoxicity [33] and, to overcome this inconvenience, intratumor delivery of therapeutics employing injectable hydrogels can provide a controlled and targeted release within the tumor site [34].

Here, we have reviewed the synthesis and the application of different injectable hydrogels proposed as drug delivery systems for the local delivery of chemotherapeutics. Additionally, stimuli-responsive release of anticancer agents have been treated by the analysis of thermo-, pH-, photo-, or multi-sensitive drug delivery systems, as well as active targeting hydrogels [35]. Based on the main component of the polymer network, herein we have classified the injectable hydrogels reviewed and discussed as synthetic or natural systems. For each class of materials, a summarizing table containing information about composition, carrier and delivery properties, as well as cancer models employed in either in vivo or in vitro experiments has been introduced. Moreover, when available, data about studies in health models have given information about side toxicity and pharmacokinetic profiles. Finally, injectable hydrogels containing nanoparticle systems as functional additive to control the releasing rate have been defined as composite materials, while N/S hybrid hydrogels refer to the simultaneous presence of natural and synthetic polymers within the same polymer structure.

## 2. Synthetic Injectable Hydrogels

### 2.1. Polyphosphazenes

Polyphosphazenes (PPZs) are a class of hybrid organic–inorganic macromolecules consisting in a linear or branched skeletal structure of repeating phosphorus and nitrogen atoms with alternating single and double bonds [36]. Each phosphorus atom is linked to two organic side groups, ranging from alkyl and aryl moieties to amino acids (Figure 2) [37].

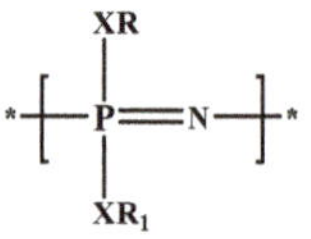

**Figure 2.** Representation of Polyphosphazenes. X = O, NH; R and $R_1$ = Alkyl, Aryl, amino acid.

PPZs are obtained via different synthetic routes, with most of the biologically-relevant materials being prepared by a ring-opening polymerization, followed by macromolecular substitution reactions [38]. Either the modification of organic side groups and their ratios, or the attachment of multiple different side groups to the same backbone, allow the preparation of a wide range of PPZs, with finely tuned physical and mechanical properties [39]. The interest in PPZs as materials for the formulation of injectable hydrogels is related to the ability of their aqueous solutions to undergo reversible sol–gel transitions depending on the temperature. In fact, PPZs are in the sol state at room temperature (or below), but they gelate at body temperature. Such transition is tunable by adjusting the balance of hydrophobic to hydrophilic substituents [40]. Furthermore, a growing number of hydrolytically-sensitive PPZ hydrogels have been designed, with negligible toxicity arising from the degradation of by-products generally consisting of $H_3PO_4$, ammonium, and free organic side groups [41]. On the contrary, the employment of cyclic PPZ architecture should be accurately investigated, because such derivatives are characterized by a relatively long time of degradation which can reduce the biomedical applicability [42]. Although a large number of PPZ polymers have not found commercial success [43], several research groups have developed different types of PPZ injectable hydrogels (Table 1). PPZ-based hydrogels were successfully tested for the delivery of cytotoxic drugs or sRNA to solid tumors, both in vitro and in vivo [40,44–50]. They proved the ability of these systems to extend the release profiles overtime [47] with no-toxicity on healthy mice [46,47] and the possibility to confer targeted behavior [50]. A further upgrade of the use of PPZ was proposed in [51], where the injectable hydrogels consisted of a Camptotechin (CPT) prodrug useful for the treatment of lung and colon cancer cell lines. The insertion of metal ferrite superparamagnetic iron oxide nanoparticles within the hydrogel structure was proved to be a suitable strategy for enabling tumor imaging and magnetic hyperthermia ablation [52,53].

Table 1. Composition and anticancer performance of injectable hydrogels based on polyphosphazenes.

| Ref | Composition | | Carrier Properties | | Delivery Properties | | Cancer Model | | | Health Model | |
|---|---|---|---|---|---|---|---|---|---|---|---|
| | Hydrogel (Gelation Process) | Composite Component | Degradation Time (Days) | Smart Responsivity | Bioactive Agent (DL% w/w) | Release Time (Days) | Type | In Vitro | In Vivo | In Vitro | In Vivo |
| [40] | PPZ (Physical – T) | --- | 30 | --- | ME (0.15) | 35 | Breast | MDA-MB-231 | MDA-MB-231 | --- | --- |
| [44] | PPZ (Physical – T) | --- | --- | --- | DOX (10) | 30 | Stomach | HSC44Luc | HSC44Luc | --- | --- |
| [45] | PPZ (Physical – T) | --- | >50 | --- | DOX (0.3–0.6) | 35 | Stomach | SNU-601 | SNU-601 | --- | --- |
| [46] | PPZ (Physical – T) | --- | --- | --- | DTX (10) | 28# | Stomach | 44As3Luc | 44As3Luc | --- | Mice |
| [47] | PPZ (Physical – T) | --- | 10–20 | --- | DTX (1–3) | 10–20 | Stomach Pancreas Liver | SNU-601 AsPC-1 SNU-398 | SNU-601 --- --- | --- | Mice |
| [48] | PPZ (Physical – T) | --- | --- | --- | PTX (0.6–0.9) | 60 | Colon Stomach | DLD-1 SNU-601 | --- SNU-601 | --- | --- |
| [49] | PPZ (Physical – T) | --- | --- | --- | PTX-DOX (0.6) | 60–100# | Stomach | SNU-601 | SNU-601 | --- | --- |
| [50] | PPZ (Physical – T) | --- | --- | --- | sRNA | 30# | Prostate Lung | PC3 A549 | PC3 --- | --- | --- |
| [51] | PPZ (Physical – T) | --- | 12–25 | --- | CPT* (10) | 60 | Colon | DLD-1 HCT-116 HT-29 | HT-29 | --- | --- |
| [52] | PPZ (Physical – T) | CoFe$_2$O$_4$ | 30 | Magnetic | SN-38 (0.8–0.12) | 60 | Glioblastoma | U-87 | U87 | 3T3 | --- |
| [53] | PPZ (Physical – T) | Zn$_{0.47}$Mn$_{0.53}$Fe$_2$O$_4$ | 25 | Magnetic | --- | --- | Glioblastoma | U-87 | U87 | 3T3 | Mice |

* Conjugated to hydrogel; # from in vivo experiments; DL: Drug loading; T: Temperature; CPT: Camptotechin; DOX: Doxorubicin; DTX: Docetaxel; ME: 2-Methoxyestradiol; PPZ: Poly(organophosphazene); PTX: Paclitaxel; SN-38: 7-ethyl-10-hydroxycamptothecin.

## 2.2. Polaxamers

Poloxamers (also known as Pluronics) are tri-block amphiphilic polymers consisting of poly(ethylene oxide)-poly(propyleneoxide)-poly(ethylene oxide) (PEO-PPO-PEO) repeating units [54]. They are non-ionic surfactants, with physical and chemical properties depending on the molecular weight and hydrophilic (PEO) to hydrophobic (PPO) balance (Figure 3) [55].

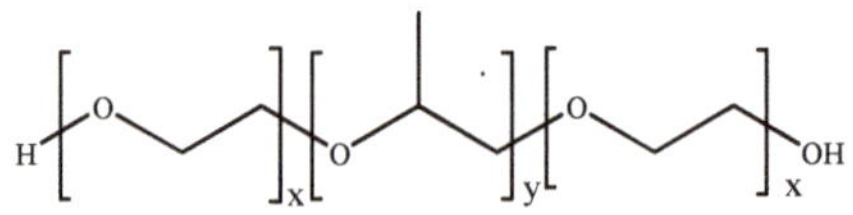

**Figure 3.** Schematic representation of poloxamers. x: 2–130; y: 15–67.

Among others, PF127 (PEO/PPO balance 70/30) is one of the most widely employed poloxamers for biomedical applications due to the ability to form either micellar nanocarriers for lipophilic drugs (due to PPO content) or hydrogel networks upon reverse thermal gelation. PF127 water solutions (>20% by weight) show a low-viscosity state at 4 °C, while semisolid gels are obtained upon heating to room or body temperature, probably due to micellar packing and entanglement [56,57].

To date, PF127 injectable hydrogels (Table 2) have been proposed as delivery vehicles for drug and drug crystals in the treatment of both blood and solid tumors [58,59]. Interestingly, such systems were found to reverse the multi-drug resistance in MCF-7/ADR cells because of the ability to increase the intracellular drug concentration escaping the efflux pumps on the cell membrane [59]. To extend the drug release profiles overtime, nanoparticle carriers (e.g., micelles or polymeric nanoparticles) loaded with the cytotoxic agent were incorporated into the hydrogels [60–62]. This approach allowed a co-delivery of 5-Fluoruracil (5-FU) and Doxorubicin-loaded Poly(D,L-lactide-co-glycolide) nanoparticles (DOX@PLGA) for the in vitro and in vivo treatment of melanoma [61]. When metal nanoparticles (e.g., Cu or Au) were used as loaded nanocarriers, photothermal and hyperthermia effects were achieved (Figure 4) [62,63].

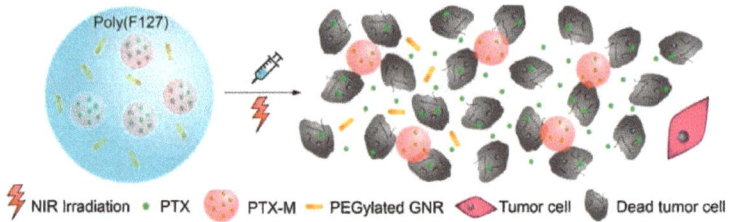

**Figure 4.** Schematic representation of the PTX-NPs/AuNRs/gel-mediated photothermal–chemotherapy. PTX: Paclitaxel; GNR: Gold NanoRods; NIR: Near InfraRed. Adapted with permission from [62]. Elsevier, [2016].

Despite the advantageous features of poloxamers, these polymers suffer from weak mucoadhesivity, poor mechanical properties, and short residence time due to the easily dissolution at the action site [64]. To overcome these drawbacks, PF127 was mixed with different polymers from synthetic (polyacrylic acid (PAA) or α-Tocopheryl Polyethylene glycol 1000 Succinate (TPGS)) [65,66] or natural (Hyaluronic acid (HA)) [67,68] origin to increase the gel strength [65] and enhance the drug efficiency [66]. Finally, it should be cited the incorporation of cyclodextrins (α-CD) into the hydrogel network for the preparation of effective depot system in cervix and breast cancer treatment [69]. A further improvement consisted in the insertion of graphene oxide (GO) or reduced graphene oxide (rGO) materials, with the formation of hybrid hydrogels with more sustained drug delivery behavior [70].

Table 2. Composition and anticancer performance of injectable hydrogels based on poloxamers.

| Ref | Composition | | Carrier Properties | | Delivery Properties | | Cancer Model | | | Health Model | |
|---|---|---|---|---|---|---|---|---|---|---|---|
| | Hydrogel (Gelation Process) | Composite Component | Degradation Time (Days) | Smart Responsivity | Bioactive Agent (DL% w/w) | Release Time (Days) | Type | In Vitro | In Vivo | In Vitro | In Vivo |
| [58] | PF127 (Physical – T) | --- | --- | pH | MLX (7.5) | 1 | Leukemia | K562 HL60 | --- | --- | --- |
| [59] | PF127 (Physical – T) | --- | --- | --- | PTX (2.0) LAP (4.0) | 4 | Breast | MCF7 MCF7-ADR | --- | BT 474 | Mice |
| [60] | PF127 (Physical – T) | MPEG-PCL | --- | --- | Q* (7.0) | 9 | Ovary | SK-OV-3 | SK-OV-3 | --- | --- |
| [61] | PF127 (Physical – T) | PLGA | --- | --- | 5-FU (2.0) DOX* (2.0) | 35 | Melanoma | B16F10 | B16F10 | --- | --- |
| [62] | PF127 (Physical – T) | OCS MPEG-AuNRs | --- | --- | PTX* (34.8) | 18# | Liver | HepG2 | HepG2 | --- | --- |
| [63] | PF127 (Physical – T) | PVP | --- | --- | $Cu_2MnS_2$* | --- | Murine breast | 4T1 | 4T1 | --- | --- |
| [65] | PF127/PAA (Physical – I,T) | --- | --- | --- | OXA (2.3) | 1 | Colon | SW480 | --- | IEC-6 | --- |
| [66] | PF127/TPGS (Physical – T) | --- | --- | Temperature | DTX (5.0) | 3 | Liver | SMMC-7721/RT | SMMC-7721/RT | --- | --- |
| [69] | PF127/β-CD (Physical – T) | --- | --- | Temperature pH | CUR (10) | 2 | Cervix Breast | HeLa MCF-7 | --- | L929 | --- |
| [70] | PF127/α-CD (Physical – T) | GO | 8 | --- | DOX (6.0) CPT (14) | 8 6 | --- | --- | --- | --- | --- |
| | PF127/α-CD (Physical – T) | rGO | | | | | | | | | |

* Loaded into composite component; # from in vivo experiments; DL: Drug loading; I: Ionic; T: Temperature; 5-FU: 5-Fluoruracil; CD: Cyclodextrin; AuNRs: Gold nanorods; CPT: Camptotechin; CUR: Curcumin; DOX: Doxorubicin; DTX: Docetaxel; GO: Graphene oxide; rGO: Reduced graphene oxide; LAP: Lapatinib; MLX: Meloxicam; MPEG: Monomethoxy poly(ethylene glycol); OCS: N-octyl chitosan; OXA: Oxaliplatin; PAA: Poly(acrylic acid); PCL: Poly(ε-caprolactone); PF: Pluronic F; PLGA: Poly(lactide-co-glycolide); PTX: Paclitaxel; PVP: Polyvinylpyrrolidone; Q: Quercetin; TPGS: α-Tocopheryl polyethylene glycol 1000 succinate.

## 2.3. Polyesters

During the last decades, thermosensitive in-situ gels of amphiphilic copolymers based on biodegradable polyesters and polyethylene glycol (PEG) have represented a suitable alternative in the intratumoral delivery of hydrophobic therapeutics [71], allowing to recover high drug concentration at the tumor site while overcoming, at same time, the limitations usually associated with the systemic administration of these drugs [72]. The advantages of this class of polymers arise from the possibility to ensure both a physical targeting to the cancer site and a controlled/sustained delivery of hydrophobic drugs [73], as well as from their high biodegradability which allows the obtainment of stimuli responsive and biocompatible delivery platforms [74]. On the other hand, the main drawback of such materials is that their acidic degradation by-products significantly influence the pH value of the surrounding media, with potential limitations in biomedical applications [75].

Different biodegradable polymers have been proposed for the development of injectable hydrogels, each showing peculiar features and biological performances (Table 3). The structures of the main polyesters employed to this regard are sketched in Figure 5.

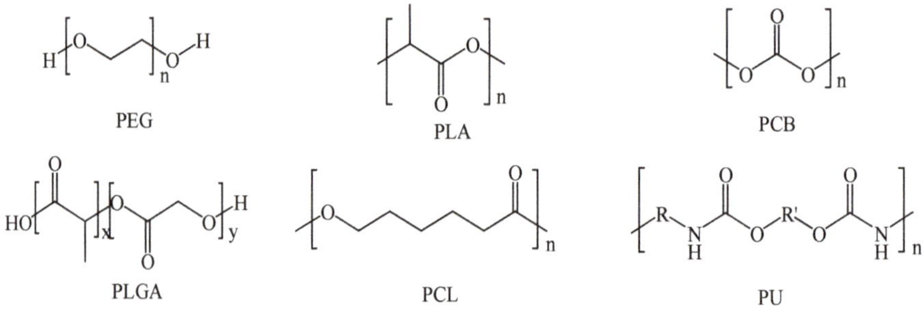

**Figure 5.** Schematic representation of poly(ethylene glycol) (PEG) and main biodegradable polyesters. PLA: Polylactide; PCB: Polycarbonate; PLGA: Poly(lactide-co-glycolide); PCL: Poly(ε-caprolactone); PU: Poly(urethane).

Biodegradable poly(D,L-lactide)-poly(ethylene glycol)-poly(D,L-lactide) (PLA-PEG-PLA) amphiphilic triblock copolymer showed the ability to self-assembly in aqueous medium into core-shell micelles, forming a physical network when exposed to the body temperature [76]. Injectable thermo-sensitive PLA–PEG–PLA for the local delivery of Gemcitabile (GEM) and Cisplatin (CisPt) was employed to promote synergistic combination therapy against pancreatic cancer [77]. Alternatively, poly(D,L-lactide) PLA was combined with pluronic L (PL) moieties in the preparation of three-block hydrogels (PLA–PL–PLA) proposed for intraperitoneal therapy of colon cancer [78,79]. This amphiphilic copolymer displayed thermosensitive behavior freely flowing at lower temperatures but turning into gel at body temperature. D,L-lactic (LA) acid oligomer combined with methoxy poly(ethylene glycol) and poly(octadecanedioic anhydride) was employed in the preparation of thermosensitive amphiphilic triblock copolymer suitable for local cancer chemotherapy. In particular, paclitaxel (PTX) loaded into LA oligomer nanoparticles could be stored as freeze-dried powders, and easily re-dispersed into aqueous medium at ambient temperature, forming a hydrogel in the injection site [80].

Poly(D,L-lactide-co-glycolide) (PLGA) and PEG triblock copolymer (PLGA–PEG–PLGA) hydrogels were synthesized via ring-opening polymerization of D,L-lactide (LA) and glycolide (GA) in the presence of PEG and Tin (II) 2-ethylhexanoate as macroinitiator and catalyst, respectively. Thermo-induced gelation of amphiphilic PLGA–PEG–PLGA can be related to the micellar aggregation as a consequence of the increase in the hydrophobic interactions between the PLGA moieties and the partial dehydration of the PEG chains [81,82]. Literature data indicates that the transition temperatures of PLGA–PEG–PLGA gels were in the range 10–40 °C for a polymer concentration of 15-20% wt [83]. Copolymer concentration influenced sol–gel transition temperature, because the formation of the micellar aggregation network was simplified when the concentration of the polymer increased [84]. PLGA–PEG–PLGA gel was proposed as a carrier of topotecan (TPC), DOX, CisPt, and methotrexate (MTX), and employed for the treatment of osteosarcoma in in vivo experiments (Figure 6) [85,86].

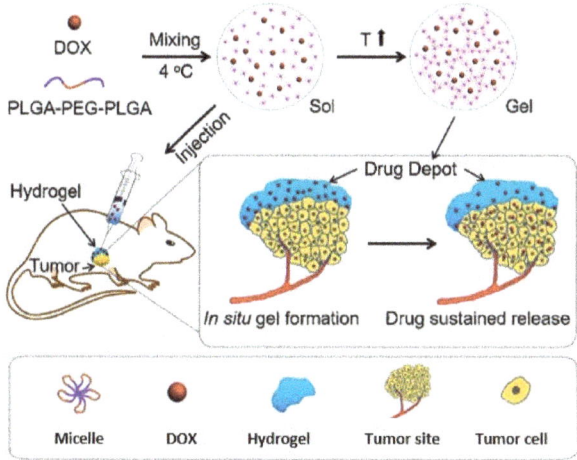

**Figure 6.** Schematic illustration of localized hydrogel formation and drug release. Adapted with permission from [86]; Elsevier, [2018].

Injectable thermosensitive hydrogel can be loaded with either drug or drug-loaded nanoparticles [87]. In particular, the interaction of ionic drugs with specific surfactants has been exploited to achieve sustained release of 2-methoxyestradiol (ME) and Cytarabine (CYT) in the therapy against leukemia and breast cancer, respectively [88,89]. Additionally, drug-loaded particles entrapped in a PLGA–PEG–PLGA hydrogel have been proposed as dual-stimuli responsive drug delivery systems combining the pH-responsivity of the nanoparticles with the temperature response of the PEGylated polyester gels [90,91]. In addition, in a modern scheduled treatment, sustained co-delivery of DOX and sRNA@Poly(ethyleneimine)-Lysine (PEI-Lys) complexes displayed significant synergistic effects in promoting the PLK1 silencing, tumor apoptosis, and cell cycle regulation of osteosarcoma cells [92].

In the pharmaceutical and biomedical fields, the sustained release of both hydrophobic and hydrophilic drugs from a single release device represents a newsworthy challenge, exhibiting different clinical survival advantages compared with the single drug treatment. To this regard, a strategy to realize the synchronous, sustained co-delivery of hydrophilic CisPt and hydrophobic PTX in one injectable device was achieved by synthesis of a Pt(IV) prodrug based on MPEG–PLGA, able to self-assemble in a core-corona micelle showing hydrophobic inner cores where PTX can be incorporated [93].

Finally, a promising strategy involved the use of cytokine-carrying thermosensitive MethoxyPEG (MPEG)–PLGA hydrogels followed by injection of vaccine vectors loading antigens [94]. This device provides a sustained release profile of granulocyte-macrophage colony-stimulating factor, able to facilitate proliferation, recruitment, and maturation of dendritic cells and macrophages at the site of inoculation, providing an efficient tool proposed in the melanoma therapy.

ε-Caprolactone was employed in the synthesis of amphiphilic block copolymers bearing PEG pendants. Different injectable Poly(ε-caprolactone) (PCL)-based nanocomposite hydrogels with multicomponent compatibility were proposed for the sustainable release of therapeutics, such as PTX, Camptotechin (CPT), 5-FU, and DOX. Three-block copolymers (PEG–PCL–PEG) were prepared by ring-opening polymerization in presence of Tin(II) 2-ethylhexanoate as macroinitiator [73,95–97]. Alternatively, PCL–PEG diblock [98] and PCL–PEG–PCL copolymers [99–102] were synthesized in the presence of 1,4,8-trioxa[4.6]spiro-9-undecanone to obtain a modified PCL able to undergo PEGylation reaction.

A MPEG–b–PCL copolymer diblock was proposed in the synthesis of supramolecular hydrogels by combination with α-CD to achieve an injectable delivery system for the release of PTX, DOX, and CisPt in lung and bladder tumors [32,103]. In these systems, α-CD were selectively inserted onto the linear polymer chains, and the resulted supramolecular complex aggregated in packed columns, mainly formed by host–guest interactions or π–π stacking between polymeric chains [104]. These systems have attracted special interest because of their favorable properties, such as thixotropy and reversibility, with their in situ encapsulation characteristics able to prolong the retention time in cancers, reducing side effects [105]. In another system, the coordination between platinum(II) atoms and carboxylic groups of poly-(acrylic acid) (PAA) blocks induced poly(ethylene glycol)–b–poly-(acrylic acid) (PEG–b–PAA) self-assembly into micelles, with the supramolecular hydrogels eventually formed by the addition of α-CD [106]. Different supramolecular hydrogels based on PEG block polymers (e.g., nucleobase (adenine/thymine)-terminated PEG) were tested for the buccal delivery of DOX in in vivo mouse models [107]. Folic acid (FA)-modified cationic and amphiphilic MPEG–PCL–PEI–FA was proposed as supramolecular system able to form polyplexes with anionic plasmid for sustained gene delivery effectively inhibiting in vivo tumor growth [108].

Drug delivery systems based on PEG–PCL–PEG were loaded with 5-FU and PXT and tested in in vivo experiments for the treatment of colon and breast tumors, respectively [73,95]. Another promising injectable hydrogel for in situ gel-forming controlled drug delivery systems is based on PCL–PEG–PCL, due to several benefits, such as prolonged drug release, sol–gel transition around the body temperature, and ease of handling, being in a solid state at room temperature [109]. In situ gelling materials based on PCL–PEG–PCL loaded with PTX and CPT were proposed as drug delivery systems against breast and gastro-intestinal cancers, with excellent results in both in vivo and in vitro experiments [96,97]. However, the preparation of the anticancer-gel formulations require high temperatures or extended times, which are unsuitable for formulations containing unstable drugs [110]. Moreover, strong hydrophobicity and high crystallinity of PCL units confer to PCL–PEG–PCL a slow degradation rate, which is not always desirable.

To address this concern, chemical modification of PCL allowed the synthesis of new polymeric systems with improved properties. In particular, PCL modified with cyclic ether pendant groups, i.e., poly(ε-caprolactone-co-1,4,8-trioxa[4.6]spiro-9-undecanone)-poly(ethyleneglycol)-poly(ε-caprolactone-co-1,4,8-trioxa[4.6]spiro-9-undeca-none), were prepared [111]. The insertion of cyclic ether pendant groups into PCL units was performed by copolymerization of 1,4,8-trioxa[4.6]spiro-9-undecanone with PCL, and the resulting macromer showed modified gelation performances as a consequence of the changing of PCL crystallization properties. By this approach, injectable carriers for DOX and PXT were obtained and proposed for the treatment of breast and liver cancers [99–101,112].

Methoxy poly(ethylene glycol)–b–poly(ε-caprolactone-co-1,4,8-trioxa[4.6]spiro-9-undecanone) (PEG–PCL) diblock copolymer was employed to prepare host–guest inclusion injectable nanocomposite devices based on surface-modified gold nanorods, PTX/PEG–PCL nanoparticles, and α-cyclodextrin [98]. A single local injection of this hydrogel allowed to deliver abundant PTX/PEG–PCL nanoparticles and gold nanorods at the target site, developing remarkable anticancer activity and photothermal effect. Alternatively, the coupling of PTX/PEG–PCL with α-CD allowed the synthesis of supramolecular hydrogels based on the hydrophobic aggregation of pseudorotaxane between cyclodextrins and block copolymers [113].

Co-delivery of anticancer agents and radiosensitizer isotopes was exploited in the design of innovative drug delivery systems able to combine the effects of chemo- and radio-therapy with reduction of the damage to normal tissue and improved therapeutic efficiency [114]. Specifically, PEG–PCL-based hydrogels were employed in the preparation of multifunctional devices for the delivery of DOX and β-emitter species, such as iodine-131 and rhenium-188, for the treatment of the hepatocellular carcinoma [102,115]. Finally, an advanced system involving linear copolymer formed by poly(ε-caprolactone) was proposed for transcatheter arterial chemoembolization, a technique based on the combination of chemotherapeutic efficacy from delivered anticancer drugs and a blockage of tumor feeding vessels with an embolic material [116]. Specifically, sulfamethazine-based anionic pH-sensitive block PCL copolymer was fabricated by free radical polymerization [117]. Aqueous solutions of the synthesized copolymer underwent a sol-to-gel phase transition upon lowering the environmental pH, and created a gel region able to cover the physiological conditions and low pH environments typical of the tumor site.

Polyurethane (PU) derivatives, such as poly(amino ester urethane) (PAEU) block copolymers, were employed as drug delivery systems, thanks to their ability to form electrostatic interactions and hydrogen bonds with bioactive molecules, and to exhibit sol–gel phase transition after injection into the body. PAEU copolymers were proposed for the fabrication of injectable radiopaque embolic materials, based on a mixtures of an aqueous copolymer solution and Lipiodol, a commercial long-lasting X-ray contrast agent [118]. In particular, exploiting the influence of pH and temperature on the self-assembly capacity of this polymeric material, a dual drug delivery system was proposed as a carrier for the regional release of DOX in the liver compartment. Additionally, target-specific release of CisPt was proposed by incorporation of CisPt chondroitin sulfate-based nanogels into pH- and temperature-responsive PEG–PAEU hydrogels [119]. In this case, ionic interactions, under physiological conditions, between the tertiary amine and sulfate groups allowed to form hydrogel networks able to selectively bind a receptor specifically expressed on cancer cells [120].

Linear copolymers obtained by suitable mixing of polyester monomers were used to synthesize injectable hydrogels with tailored properties due to their specific hydrophobic/hydrophilic balance. To this regards, PCLA–PEG–PCLA triblock copolymer was synthesized using a ring-opening copolymerization involving ε-Caprolactone and LA, in the presence of PEG and Tin(II) 2-ethylhexanoate. In particular, amphiphilic copolymer was conjugated with heparin to construct non-anticoagulant heparin prodrugs loaded in thermosensitive hydrogel for anti-metastasis treatment [121] and as a GEM carrier for the treatment of pancreatic cancer [122]. Moreover, PCLA–PEG–PCLA copolymer was modified via polyaddition polymerization with sulfamethazine, acting as anionic pH-sensitive moiety, to synthesize a dual stimuli responsive polymeric system, proposed for the DOX release in liver cancer [123]. Finally, injectable pentablock copolymer hydrogels PEG–PCL–PLA–PCL–PEG, with different ratios of PCL and PLA, were proposed as single-shot sustained release of vaccines. Specifically, vaccine was encapsulated into PLGA nanoparticles and incorporated in the thermoresponsive hydrogels in order to modulate gelation temperature and minimize burst release of antigen and adjuvants in the treatment of melanoma [124]. Nevertheless, the synthetic strategies involving lactide, glycolide, or ε-caprolactone derivatives to generate a temperature-sensitive and biodegradable polymeric backbone suffered from the lack of chemical functionality in the parent aliphatic polyesters that makes it difficult to modify the polymeric chains.

A valuable alternative way exploited the employment of methyltrimethylcarbonate (PCB), cyclic carbonates derived from 2,2-bis(methylol) propionic acid (bis-MPA), as synthon for functional biodegradable monomers [16]. Ring-opening polymerization, followed by N,N′-dicyclohexylcarbodiimide-mediated condensation, was the synthetic strategy proposed to prepare hydrophilic/hydrophobic PEG-functionalized cyclic carbonate based on 2,2-bis(methylol)propionic Acid (bis-MPA) [125]. Micellization provided a physical cross-linked system, displaying a lower critical solution temperature at values near the body temperature that can be suitable for PXT release against hepatic cancer cells. A different protocol involved the formation of a biodegradable polymeric biomaterial consisting of PEG and a polycarbonate of dihydroxyacetone (pDHA), proposed for the prevention of the seroma post-operative complications following ablative breast cancer surgery [126]. Vitamins E and D-functionalized polycarbonates were proposed as a hydrophobic block in the synthesis of three-block copolymers able to form physically cross-linked injectable hydrogels for local and sustained delivery of herceptin in breast cancer treatment [127,128].

Table 3. Composition and anticancer performance of injectable hydrogels based on polyesters.

| Ref | Composition | | Carrier Properties | | Delivery Properties | | Cancer Model | | | Health Model | |
|---|---|---|---|---|---|---|---|---|---|---|---|
| | Hydrogel (Gelation Process) | Composite Component | Degradation Time (Days) | Smart Responsivity | Bioactive Agent (DL% w/w) | Release Time (Days) | Type | In Vitro | In Vitro | In Vitro | In Vitro |
| [77] | PLA-PEG-PLA (Physical – T) | PLA-PEG-PLA | --- | --- | GEM* (10) CisPt* (0.2) | 10 | Pancreas | Bxpc-3 | Bxpc-3 | --- | Mice |
| [78] | PLA-PL64-PLA (Physical – T) | PLA-PL35-PLA | --- | --- | DTX (4.5) LL37* (7.5) | 24 | Colon | HCT 116 | --- | HEK293 | Mice |
| [79] | PLA-PL35-PLA (Physical – T) | PLA-PL35-PLA | --- | --- | OXA DTX* (4.4) | 14 | Murine colon | CT26 | CT26 | 3T3 HEK293 | --- |
| [80] | MPEG-POA-LAO (Physical – T) | MPEG-POA-LAO | --- | --- | PTX* (0.9) | 17# | Breast Cervix | MCF7 HeLa | MCF7 --- | --- | --- |
| [83] | PLGA-PEG-PLGA (Physical – T) | --- | --- | --- | DOX (1.0) CisPt (1.0) MTX (1.0) | 12 | Osteosarcoma | Saos-2 MG-63 | Saos-2 --- | L929 | --- |
| [85] | PLGA-PEG-PLGA (Physical – T) | --- | --- | --- | TPT (1.0) | 5 | Sarcoma | S180 | S180 | --- | --- |
| [86] | PLGA-PEG-PLGA (Physical – T) | --- | 44 | --- | DOX (1.0) | 15 | Osteosarcoma Murine osteosarcoma | Saos-2 K7 | Saos-2 K7 | --- | --- |
| [88] | PLGA-PEG-PLGA (Physical – T) | Vesicles | --- | --- | CYT* (25.4) | 12 | Leukemia | K562 HL-60 | --- | --- | Rabbit |
| [89] | PLGA-PEG-PLGA (Physical – T) | Liposome | --- | --- | ME* (5.6) | 70 | Murine breast | --- | 4T1 mice | --- | --- |
| [90] | PLGA-PEG-PLGA (Physical – T) | Arg Dendrimers / Lys Dendrimers | --- | pH | DOX* (13.6) DOX* (14.3) | 20 | Murine breast | 4T1 | 4T1 mice | 3T3 RAW267 | --- |
| [91] | PLGA-PEG-PLGA (Physical – T) | SLN | --- | pH | ME* (2.05–2.23) | 45 | --- | --- | --- | --- | --- |
| [92] | PLGA-PEG-PLGA (Physical – T) | PEI-Lys | 40 | --- | DOX (1.0) sRNA* | 16# | Osteosarcoma | Saos2 MG-63 | Saos2 --- | --- | --- |
| [93] | MPEG-PLGA (Physical – T) | MPEG-PLGA | --- | --- | PTX (4.0) Pt* (0.8) | 80 | Ovarian | SKOV-3 | SKOV-3 | --- | --- |

Table 3. Cont.

| Ref | Composition | | Carrier Properties | | | Delivery Properties | | Cancer Model | | | Health Model | |
|---|---|---|---|---|---|---|---|---|---|---|---|---|
| | Hydrogel (Gelation Process) | Composite Component | Degradation Time (Days) | Smart Responsivity | | Bioactive Agent (DL% w/w) | Release Time (Days) | Type | In Vitro | In Vitro | In Vitro | In Vitro |
| [94] | MPEG-PLGA (Physical – T) | --- | --- | --- | | GM-CSF (15–25) | 14 | Murine Melanoma | B16 / B16-F10 | B16-F10 | --- | --- |
| [95] | PCL-PEG (Physical – T) | --- | --- | --- | | 5-FU (1.0) | 7 | Colon | CT26 | CT26 | --- | --- |
| [73] | PEG-PCL-PEG (Physical – T) | MPEG-PCL | --- | --- | | PTX* (8.3) | 20 | Breast | 4T1 | 4T1 | --- | --- |
| [96] | PCL-PEG-PCL (Physical – T) | --- | --- | pH | | PTX** (20) | 30 | Liver / Breast | HepG2 / MCF7 | --- | L929 | Mice |
| [97] | PCL-PEG-PCL (Physical – T) | PCL-PEG-PCL | --- | --- | | CPT* (4.1–13.5) | 14 | Colon | CT26 | CT26 | L929 | --- |
| [98] | MPEG-PCL-α-CD (Physical – T) | MPEG-PCL AuNRs | 14 3$ | --- | | PTX* (3.0) | 14 | Murine Breast | 4T1 | 4T1 | --- | --- |
| [100] | PCL-PEG-PCL (Physical – T) | PCL-PEG-PCL | --- | --- | | PTX* (1.25, 2.5) | 45 | Erlich ascites / Ovarian | EAC / --- | EAC / OVCAR-3 | --- | --- |
| [101] | PCL-PEG-PCL (Physical – T) | PCL-PEG-PCL | --- | pH | | DOX* (1.0) | 35 | Liver / Breast | HepG2 / --- | --- / Bcap-37 | --- | --- |
| [102] | PCL-PEG-PCL (Physical – T) | HA | --- | pH | | DOX $^{131}$I* (0.5–2) | 35 | Liver | HepG2 | HepG2 | --- | --- |
| [32] | MPEG-PCL-α-CD (Physical) | MPEG-PCL | --- | --- | | PTX* (18.8) | 3 | Lung | A549 | --- | --- | --- |
| [103] | MPEG-PCL-α-CD (Physical) | MPEG-PCL | --- | pH | | DOX* (15) CisPt* (20) | 8 | Bladder | EJ | --- | HEK293 | --- |
| [106] | MPEG-PAA-α-CD (Physical) | MPEG-PAA | --- | --- | | CisPt* (0.5–1.0) | 4 | Bladder | EJ | --- | HEK293 | --- |
| [107] | APEGA/TPEGT-α-CD (Physical) | --- | --- | --- | | DOX (0.2–0.6) | 11 | Buccal | --- | U14 | L929 | --- |
| [108] | MPEG-PCL-PEI-α-CD (Physical – SE) | --- | --- | --- | | PTX (10) pDNA | 7# | Liver / Lymphoma | HepG2 / Bcl-2 | --- | HEK293 / LO2 | --- |
| [112] | PCL-PEG-PCL/ MPEG-PPFEMA (Physical – T) | --- | --- | --- | | PTX (4.2) DOX (4.2) | 42 | Breast | MCF7 | Bcap-37 | --- | --- |

Table 3. Cont.

| Ref | Composition | | Carrier Properties | | Delivery Properties | | Cancer Model | | | Health Model | |
|---|---|---|---|---|---|---|---|---|---|---|---|
| | Hydrogel (Gelation Process) | Composite Component | Degradation Time (Days) | Smart Responsivity | Bioactive Agent (DL% w/w) | Release Time (Days) | Type | In Vitro | In Vitro | In Vitro | In Vitro |
| [113] | MPEG-PCL-α-CD (Physical) | MPEG-PCL | --- | --- | PTX* (1–3) | 20 | Murine breast | 4T1 | --- | --- | --- |
| [115] | PCL-PEG-PCL (Physical – T) | Liposome | --- | --- | $^{188}$Re DOX* (2.0) | 10 | Murine liver | BNL-Luc | BNL-Luc | --- | --- |
| [117] | PCLMA-PEGMA-SMA (Chemical – RP) | --- | --- | pH | DOX (0.2–0.4) | 27 | Liver | HepG2 | VX2 | L929 | --- |
| [118] | PAEU (Physical – T, pH) | --- | --- | --- | DOX (10) | 14 | Liver | HepG2 | VX2 | L929 | --- |
| [119] | PEG-PAEU (Physical – T) | CHS-Nanogel | 28 | pH | CisPt* (0.2) | 14 | Lung | A549 | --- | 3T3 | Mice |
| [121] | PCLA-PEG-PCLA (Physical – T) | PCLA-PEG-PCLA | 35 | --- | Hep* | 10 | Cervix | HeLa | HeLa | HaCaT | --- |
| [122] | PCLA-PEG-PCLA (Physical – T) | MMT | 56 | --- | GEM* (10) | 7 | Pancreas | --- | Panc-1 | 293T | --- |
| [123] | PCLA-PEG-PUSSM (Physical – T) | --- | 28 | pH | DOX (5.0) | 28 | Liver | HepG2 | VX2 | 293T | --- |
| [124] | PEG-PCL-PLA-PCL-PEG (Physical – T) | PLGA | 30 | --- | OVA (0.8) MPL (0.6) QA (1.1) | 30 | Melanoma | --- | B16 OT-I B16 OT-II | --- | --- |
| [125] | Bis-MPA-PEG (Chemical – ROP) | --- | --- | Temperature | PTX (3.9) | 7 | Liver | HepG2 | --- | --- | --- |
| [126] | PEG-pDHA (Physical) | --- | 1 | --- | --- | --- | Breast | --- | Rat | --- | --- |
| [127] | VitE-PCB-VitE (Physical) | --- | 42 | --- | TZB (4.0) | 16 | | MCF-7 BT474 | --- BT474 | HDF | --- |
| [128] | VitD-PCB-VitD VitE-PCB-VitE (Physical) | --- | --- | --- | TZB (4.0) | 42 | Breast | BT474 | BT474 | --- | Mice |

* Loaded into composite component; # from in vivo experiments; DL: Drug loading; I: Ionic; T: Temperature; RP: Radical polymerization; ROP: Ring-opening polymerization; SE: Solvent evaporation; Arg: Arginine; 5-FU: 5-Fluoruracil; CD: Cyclodextrin; APEGA: Adenine-terminated poly(ethylene glycol); AuNRs: Gold nanorods; CHS: Chondroitin sulfate; CisPt: Cisplatin; CPT: Camptotechin; CYT: Cytarabine; pDHA: Polycarbonate of dihydroxyacetone; pDNA: Plasmid DNA; DOX: Doxorubicin; DTX: Docetaxel; GEM: Gemcitabine; GM-CSF: Granulocyte-macrophage colony-stimulating factor; HA: Hyaluronic acid; Hep: Heparin; LAO: Lactic acid oligomers; Lys: Lysine; ME: 2-Methoxyestradiol; MMT: Montmorillonite; bis-MPA: 2,2-bis(methylol)propionic acid; MPEG: Monomethoxy poly(ethylene glycol); MPL: Monophosphoryl lipid A; MTX: Methotrexate; OVA: Ovalbumin; OXA: Oxaliplatin; PAA: Poly(acrylic acid); PAEU: Poly(β-aminoester urethane); PCB: Polycarbonate; PCL: Poly(ε-caprolactone); PCLA: Poly(ε-caprolactone-co-lactide); PCLMA: Poly(ε-caprolactone) monomethacrylate; PEG: Polyethylene glycol; PEGMA: Methoxypoly(ethylene glycol) monomethacrylate; PEI: Poly(ethylene imine); PL: Pluronic L; PLA: Polylactide; PLGA: Poly(lactide-co-glycolide); POA: Poly(octadecanedioic anhydride); PPFEMA: Poly(2-(perfluorobutyl)ethyl methacrylate); PTX: Paclitaxel; PUSSM: Poly(urethane sulfide-sulfamethazine); QA: Quil A; SLN: Solid lipid nanoparticles; SMA: Sulfamethazine-acrylamide; TPEGT: Thymine-terminated poly(ethylene glycol); TPT: Topotecan; TZB: Trastuzumab; Vit: Vitamin.

## 2.4. Polyacrylates

Photo-induced radical polymerization involving acrylate monomers and/or functionalized macromers represents an alternative to thermal gelation in the preparation of injectable hydrogels able to be self-assemble after injection following a UV-irradiation (Figure 7) [129,130].

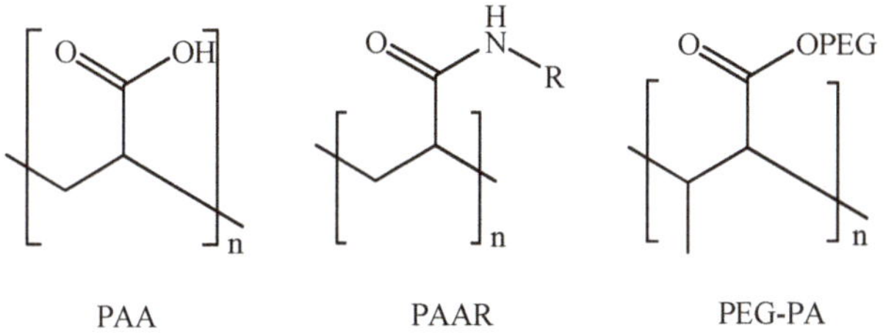

**Figure 7.** Schematic representation of the main acrylate polymers. PAA: Poly(acrylic acid); PAAR: N-alkyl poly(acrylic amide); PEG-PA: PEGylated poly(methacrylic acid).

The main component of this class of materials enclosed PEG acrylate polymers (PEG-PA), which was designed to allow the insertion of PEG properties (e.g., non-cytotoxicity, non-immunogenicity, and ability to reduce opsonization) within a hydrogel network, showing increased drug loading capability and retention time and improved mechanical properties (Table 4) [23,131].

This approach was investigated in the treatment of glioblastoma, employing a system based on polyethylene glycol dimethacrylate (PEGDMA). The photopolymerizable monomer was UV-irradiated in the brain tumor resection bed and employed for the delivery of Temozolomide (TMZ) and Paclitaxel (PTX) [132,133]. This approach could present several advantages, including the killing of the tumor cells that, after the resection of the main primary tumor, could infiltrate the brain tissue and the parenchyma.

Hybrid materials were also prepared by incorporating carbon nanotubes [134] or Zn ferrite nanoparticles [135] for breast cancer treatment by combined DOX/photothermal and thermal ablation therapy, respectively. Injectable hydrogels, proposed for the thermo-responsive delivery of different drug molecules to prostate cancer in vivo, were prepared by radical polymerization of oligo(ethylene glycol) methacrylate (OEGMA) monomers [136]. In another study, PAA was combined with a poly[4-(2,2,6,6-tetramethyl piperidine-N-oxyl)aminomethylstyrene]-b-poly(ethylene glycol)-b-poly[4-(2,2,6,6-tetramethylpiperidine-N-oxyl)aminomethylstyrene] (PMNT–PEG–PMNT) triblock copolymer to obtain a redox-active polyion complex for the local protein therapy of murine colon cancer [137].

A different approach involved the synthesis of specific gold nanorods incorporated into the three-dimensional network achieved by radical polymerization of methacrylated poly-$\beta$-cyclodextrin (MPCD)-based macromer and N-isopropylacrylamide (NIPAAm) as a poly(N-Alkylacrylamide) (PAAR) derivative [138]. The hydrogel, exhibiting near-infrared and pH responsivity, was efficiently loaded by host–guest interactions with adamantane-modified DOX prodrug, and its efficiency was tested in in vitro tests against MCF7 (breast) and HeLa (cervix) cancer cells, and in in vivo experiments carried out in the treatment of murine sarcoma.

Alternatively, thermoresponsive supramolecular poly(N-acryloyl glycinamide-co-acrylamide) (PNAm) hydrogels, bearing polydopamine-coated gold nanoparticles and DOX, were fabricated by radical photopolymerization [139], and proposed as a breast filler. This system, after heating in the sol state, was injected into the cavity of resected breasts, where a rapid gelation occurred during cooling to body temperature.

**Table 4.** Composition and anticancer performance of injectable hydrogels based on polyacrylates.

| Ref | Composition | Carrier Properties | | Delivery Properties | | Cancer Model | | | Health Model | |
|---|---|---|---|---|---|---|---|---|---|---|
| | Hydrogel (Gelation Process) | Composite Component | Degradation Time (Days) | Smart Responsivity | Bioactive Agent (DL% w/w) | Release Time (Days) | Type | In Vitro | In Vivo | In Vitro | In Vivo |
| [132] | PEGDMA (Chemical – RP) | PEG-PCL | --- | --- | TMZ* (0.9–1.3) | 12 | Glioblastoma | --- | U87 | --- | --- |
| [133] | PEGDMA (Chemical – RP) | PLGA | --- | --- | PTX* (4.0) | 7 | Glioblastoma | U87 | U87 | --- | Mice |
| [134] | PEGDMA (Chemical – RP) | $TiO_2$-MWCNT | --- | NIR | DOX* (10) | 3 | Breast | MCF7 | MCF7 | --- | --- |
| [135] | PEGDMA (Physical – T) | $ZnFe_2O_4$ | --- | Magnetic | --- | --- | Breast | --- | 4T1 | --- | --- |
| [136] | p($MEO_2$MA-OEGMA-AA) (Chemical – RP) | --- | --- | --- | BSA (1.6) Epo (1.6) | 3 1.5 | Prostate | --- | PC3 | 3T3 | --- |
| [137] | PMNT–PEG–PMNT/PAA (Physical) | --- | --- | --- | IL-12 (0.07) | 15 | Murine Colon | --- | C26 | --- | --- |
| [138] | p(NIPAAm-MPCD) (Chemical – RP) | AuNRs | --- | pH NIR | DOX (6.6) | 30 | Breast | MCF7 | --- | --- | --- |
| | | | | | | | Cervix | HeLa | --- | | |
| | | | | | | Murine sarcoma | --- | S180 | | |
| [139] | PNAm (Chemical – RP) | PDA-AuRNs | --- | NIR | DOX | 28 | Murine breast | --- | 4T1 | --- | --- |

* Loaded into Composite Component; DL: Drug loading; RP: Radical polymerization; AuNRs: Gold nanorods; BSA: Bovine serum albumin; DOX: Doxorubicin; Epo: Erythropoietin; IL: Interleukin; $MEO_2$MA: Methyl ether methacrylate; MPCD: Methacrylated poly-β-cyclodextrin; MWCNT: Multi-walled carbon nanotubes; NIPAAm: N-isopropyl acrylamide; NIR: Near-infrared; OEGMA: Poly(ethylene glycol) methyl ether methacrylate; PAA: Poly(acrylic acid); PCL: Poly(ε-caprolactone); PDA: Polydopamine; PEG: Polyethylene glycol; PEGDMA: Polyethylen glycole dimethacrylate; PLGA: Poly(lactide-co-glycolide); PMNT: Poly[4-(2,2,6,6-tetramethyl piperidine-N-oxyl)aminomethylstyrene; PNAm: Poly(N-acryloylglycinamide-co-acrylamide); PTX: Paclitaxel; TMZ: Temozolomide.

## 2.5. Synthetic Polypeptide

Polypeptides (Pep) are synthetic protein-mimicking materials particularly attractive for their biocompatibility and biodegradability [140–142]. Another advantage of this class of compounds lies in the great chemical diversity due to the wide number of monomer sources from 21 natural amino acids and their synthetic derivatives (Figure 8).

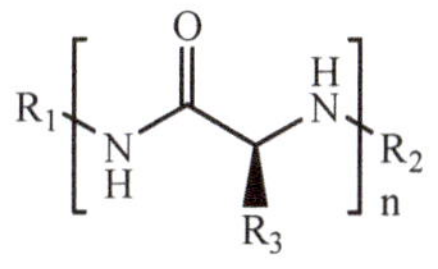

Pep

**Figure 8.** Schematic representation of synthetic polypeptides.

In addition, exploiting intramolecular hydrogen bonds within peptide backbones, polypeptides can adopt ordered secondary structures (i.e., $\alpha$-helix and the $\beta$-sheet) that confer them the self-assembly behavior. Self-assembling polypeptides were employed as starting materials for the preparation of injectable hydrogels (Table 5) [143–145] via gelation processes of their aqueous solutions upon changes in pH, ionic strength, or temperature. The introduction of cytotoxic molecules into the solution led to the encapsulation of bioactive agents for the treatment of different tumors. In detail, ionic gelation was proposed for the preparation of Ce6 carrier system [146] for breast cancer, and stimulation of immune system in health mice [147,148]. Thermo gelation processes were used for the fabrication of injectable hydrogels for TMP-2 [149] and DOX-based therapy [150,151] of breast, cervix, and lung cancers, as well as for DOX or gene (CDN) administration with simultaneous stimulation of immune responses [152,153]. Furthermore, DOX@Liposome formulations were loaded in Pep hydrogels for an Losartan (LST) combination therapy [154]. Another approach for the preparation of starting materials for injectable hydrogels involves the conjugation of peptide moieties to oligoethylene glycol (OEG) [141] or PEG derivatives, with the formation of PEGylated [155–158] or block [15,159,160] copolymers. Such hydrogels were found to be suitable for the preparation of pro-drugs [160–162] and the delivery of different clinically relevant cytotoxic agents, with the possibility to trigger the releasing profile in response to physiological stimuli such as pH [155], temperature [158], and cell redox state [15,160–162], or stimulate the immune system (Figure 9) [157].

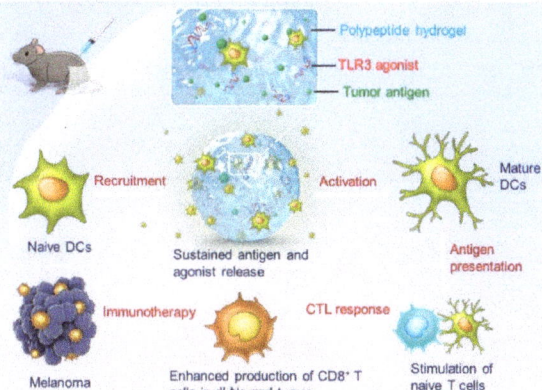

**Figure 9.** In vivo modulation of dendritic cells (DCs) by sustained release of tumor antigens and tumor cell lysates 3 (TLR3) agonist from a polypeptide hydrogel, evoking a strong cytotoxic T-lymphocyte (CTL) response. With permission from [157]; Elsevier, [2018].

Table 5. Composition and anticancer performance of injectable hydrogels based on synthetic polypeptide.

| Ref | Composition | Carrier Properties | | Delivery Properties | | Cancer Model | | | Health Model | |
|---|---|---|---|---|---|---|---|---|---|---|
| | Hydrogel (Gelation Process) | Composite Component | Degradation Time (Days) | Smart Responsivity | Bioactive Agent (DL% w/w) | Release Time (Days) | Type | In Vitro | In Vivo | In Vitro | In Vivo |
| [146] | Fmoc-FF/PLL (Physical – I) | --- | --- | --- | Ce6 (0.4) | 14# | Breast | --- | MCF7 | --- | --- |
| [147] | Fmoc-FF/PLL (Physical – I) | --- | --- | --- | --- | --- | --- | --- | --- | --- | Mice |
| [148] | RGD-PIC (Physical – T) | --- | --- | --- | --- | --- | --- | --- | --- | --- | Mice |
| [149] | AcVES3 (Physical – T) | --- | --- | --- | TIMP-2 (4.0) | 28 | Lung | A549 | --- | --- | --- |
| [150] | FEFFFK (Physical – T) | --- | --- | --- | DOX (0.5) | 20# | Breast | MDA-MB 231 | MDA-MB 231 | --- | --- |
| | | | | | | | Murine breast | 4T1 | 4T1 | | |
| [151] | Nap-GFFYGRGDH$_n$ (n = 0-2) (Physical – T) | --- | --- | pH | DOX** (3.9–12.4) | 7 | Lung | A549 | --- | --- | Mice |
| | | | | | | | Cervix | HeLa | | | |
| | | | | | | | Breast | MCF-7 | | | |
| [152] | K$_2$(SL)$_6$K$_2$ (Physical – T) | --- | --- | --- | CDN (40) | 1 | Murine Oral | MOC2-E6E7 | MOC2-E6E7 | --- | --- |
| [153] | (RADA)$_8$ (Physical – T) | --- | 10 | --- | MEL** DOX (16) | 7 | Murine Melanoma | B16F10 | B16F10 | --- | --- |
| [154] | C$_{16}$-GNNQQNYKD-OH (Physical – T) | Liposome | --- | --- | LST DOX* | 9 | Breast | --- | 4T1 | --- | --- |
| [155] | PEG-PAH (Physical – T) | --- | --- | pH | DOX (1.7) | 2 | Fibrosarcoma | HT1080 | HT1080 | 3T3 | --- |
| [156] | MPEG-(PELG-LG) (Physical – T) | --- | 28 | --- | CisPt (1.0) | 7 | Murine colon | C26 | C26 | --- | --- |
| | | | | | | | Cervix | HeLa | --- | | |
| | | | | | | | Breast | MCF-7 | --- | | |
| [157] | MPEG-PV (Physical – T) | --- | --- | --- | TCL (50) | 21# | Murine Melanoma | --- | B16 | --- | --- |
| [158] | MPEG-PAF (Physical – T) | --- | 28 | Temperature | DOX (6.0) CA4 (6.0) | 28 | Murine cervix | --- | U14 | --- | --- |
| [163] | MPEG-PLD-Arg-α-CD (Physical) | --- | --- | --- | sRNA | 6 | Epithelium | HNE-1 | HNE-1 | 3T3 | --- |
| [159] | PELG-PEG-PELG (Physical – T) | --- | 14 | --- | PTX (6.0) | 21 | Liver | HEPG2 | HEPG2 | --- | Mice |

Table 5. Cont.

| Ref | Composition | Carrier Properties | | Delivery Properties | | Cancer Model | | | Health Model | |
|---|---|---|---|---|---|---|---|---|---|---|
| | Hydrogel (Gelation Process) | Composite Component | Degradation Time (Days) | Smart Responsivity | Bioactive Agent (DL% w/w) | Release Time (Days) | Type | In Vitro | In Vivo | In Vitro | In Vivo |
| [15] | (Me-D-1MT)–PEG–(Me-D-1MT) (Physical – T) | --- | 21 | ROS | PD-1/PD-L1 CTLA-4 | 16# | Murine Melanoma | B16F10 | B16F10 | --- | --- |
| [160] | FFE–EE (Physical – Red) | --- | --- | Redox | SN-38** | 3 | Breast | MD-MBA-231 | --- | --- | --- |
| [161] | KE–EE/AcKE–EE/E–EE/R–EE/S–EE (Physical – Red) | --- | --- | Redox | DXM** TX** | 1 | Liver | HepG2 | --- | --- | --- |

* Loaded to composite component; ** conjugated to hydrogel; # from in vivo experiments; DL: Drug loading; I: Ionic; T: Temperature; Red: Redox; Ac: Acetyl; AcVES3: Ac–VEVSVSVEV<sup>D</sup>PPTEVSVEVEV–NH$_2$; Arg: Arginine; C$_{16}$: Palmitic acid; CA4: Combretastatin CDN: Cyclic dinucleotides; CisPt: Cisplatin; CTLA-4: Cytotoxic T lymphocyte antigen 4; D-1MT: Dextro-1methyl tryptophan; DOX: Doxorubicin; DXM: Dexamethasone; Fmoc-FF: N-fluorenylmethoxycarbonyl diphenylalanine; LG: L-glutamic acid; LST: Losartan; MEL: Melittin; MPEG: Monomethoxy poly(ethylene glycol); Nap: Naphthylacetic acid; PAF: Poly(alanine–phenylalanine); PAH: α,β-polyaspartyl hydrazide; PD-1/PD-L1: Programmed cell death protein 1/programmed cell death-ligand 1; PEG: Polyethylene glycol; PELG: Poly(ethyl-L-glutamate); PIC: Tri-ethylene glycol-substituted polyisocyanopeptide; PLL: Poly-L-lysine; PTX: Paclitaxel; PV: Poly(L-valine); ROS: Reactive oxygen species; SN-38: 7-ethyl-10-hydroxycamptothecin; TIMP-2: Tissue inhibitor of metalloproteinase 2; TLC: Tumor cell lysates; TX: Taxol.

Disulfide bonds were also employed for the preparation of thermo-responsive injectable hydrogels. For example, PEGylated disulfide bond containing poly(L-cysteine) derivative (poly(L-EGx-SS-Cys)) possessed an irreversible thermo-responsive behavior in water, probably ascribed to chemical cross-linking caused by disulfide bond exchange. A thermogel consisting of PEG and poly(L-EG4-SS-Cys) diblock copolymer was used as reduction responsive injectable hydrogel [164]. Physical cross-linking approach was also employed for the preparation of injectable hydrogels with excellent shearing thinning features using $PEG_{44}-NH_2$ as a macroinitiator [165].

## 2.6. Dendrimers and Other Systems

Dendrimers are synthetic branched polymers with a globular structure, nanometric size, and low polydispersity index [166], fabricated via a sequence of reaction steps in which monomer units are added to a Generation 0 core [167]. This class of materials possesses unique features for drug delivery applications, including the high affinity of the inner hydrophobic environment for different drug molecules, the wide number of functional groups suitable for tailored functionalization [167], and the ability to cross the cell membrane via paracellular and endocytosis pathways [168,169]. Different injectable hydrogels based on dendrimers have been proposed in the literature for the treatment of solid cancers (Table 6), mainly consisting in modified PEG [170], poly(amine-ester) [171], and polyamidoamine (PAMAM) (Figure 10) [172,173].

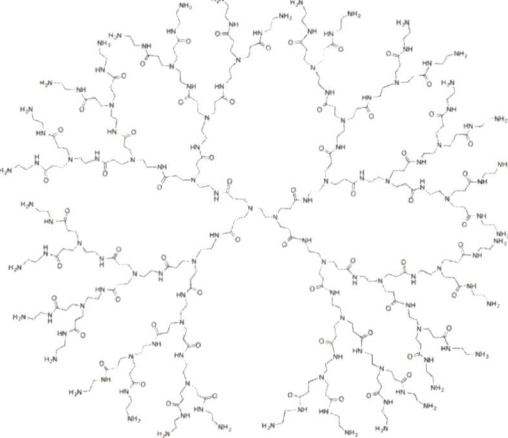

**Figure 10.** Schematic representation of polyamidoamine (PAMAM )dendrimer.

The main component of this class of materials enclosed PEG acrylate polymers (PEG-PA), which was designed to allow the insertion of PEG properties (e.g., non-cytotoxicity). PEG dendrimers were modified by insertion of disulfide bonds [174,175] or boronic acid moieties [176] to confer redox and pH responsivity, respectively. Boronic acid derivatives were also proposed to enhance the pH biodegradability patterns of injectable hydrogels employed for breast cancer treatment in mouse models [177], while the formation of Shiff's base with poly-L-lysine (PL) carried out to an effective MTF and 5-FU delivery system to colon C26 cells [34]. Targeting behavior can be conferred by derivatization with heparin residues [178]. PEGylated PAMAM injectable hydrogels with increased solubility and improved biodistribution characteristics [179] were tested as 5-FU carriers or as pH and redox responsive DOX delivery vehicles for head/neck and cervix cancer treatment, respectively [172,173]. Other examples of injectable hydrogels for cancer therapy consist in lipid nanocapsule-based hydrogels able to cross the blood–brain barrier [180], and in pH responsive PVA/GO hybrids loaded with a CPT-CD complex [181]. The latter systems take the advantages of the peculiar properties of the high biocompatible carbon nanostructures [182–184].

Table 6. Composition and anticancer performance of injectable hydrogels based on dendrimers and other systems.

| Ref | Composition | Carrier Properties | | | Delivery Properties | | | Cancer Model | | Health Model | |
|---|---|---|---|---|---|---|---|---|---|---|---|
| | Hydrogel (Gelation Process) | Composite Component | Degradation Time (Days) | Smart Responsivity | Bioactive Agent (DL% w/w) | Release Time (Days) | Type | In Vitro | In Vivo | In Vitro | In Vivo |
| Dendrimers | | | | | | | | | | | |
| [170] | PGA-MA/PEG-4-SH (Chemical – MR) | --- | --- | --- | TZB (1.0–13) | 42 | Breast | BT474 | BT474 | --- | --- |
| [171] | HPAE (Chemical – RP) | --- | 7 | --- | DOX (0.05) 5-FU (0.5) LC (0.2) | 5 | Breast | MCF7 | --- | L929 | --- |
| [172] | PAMAM(G4)/PEG-PDBCO (Physical) | --- | --- | --- | 5-FU (4) | 0.5 | Head/Neck | HN12 | HN12 | 3T3 | --- |
| [173] | PEGDA–PAMAM (Chemical – MR) | --- | 14-22 1.4-2.0 | pH Redox | DOX (4) | 2 | Cervix | HeLa | --- | --- | --- |
| [174] | PEG-4-SH/PDMA–PEGMA (Physical) | p(AA-co-4-VPBA) | 21 | pH Redox | CA4P (0.05) DOX* (14.3–96.3) | 3 14 | Breast Liver | MCF-7 --- | --- HepG2 | 3T3 | --- |
| [175] | PEG-4-SH/PEG-2-MI (Physical – pH) | --- | 2.7 | Redox | BSA (1.2) | 7 | --- | --- | --- | --- | --- |
| [176] | PEGBA-4/PEGBA-8 (Chemical – RP) | PEGBA-4/PEGBA-8 | --- | pH | PLP* | 20 | Oral | CAL-27 | --- | --- | --- |
| [177] | PBA-PEG (Physical) | --- | --- | pH | DOX (1.2) | 5 | Murine Breast | 4T1 | 4T1 | 3T3 | --- |
| [34] | PEG-4-SH/PLL (Chemical – C) | --- | 14 | pH | MTF (5.0) 5-FU (0.5) | 14 | Colon | C26 | C26 | --- | --- |
| [178] | Hep-PEG-4-SH (Physical) | --- | --- | --- | DOX (0.004–0.08) | 4 | Breast | MCF-7 MDA-MB-231 | MDA-MB-231 | --- | --- |
| Other systems | | | | | | | | | | | |
| [181] | PVA/GO (Physical) | β-CD | --- | pH | CPT (5.0) | 5 | --- | --- | --- | --- | --- |

163

Table 6. Cont.

| Ref | Composition | Carrier Properties | | | Delivery Properties | | Cancer Model | | | Health Model | |
|---|---|---|---|---|---|---|---|---|---|---|---|
| | Hydrogel (Gelation Process) | Composite Component | Degradation Time (Days) | Smart Responsivity | Bioactive Agent (DL% w/w) | Release Time (Days) | Type | In Vitro | In Vivo | In Vitro | In Vivo |
| [180] | GemC$_{12}$LNC (Physical – PI) | --- | --- | --- | --- | --- | Glioblastoma | U251 | --- | --- | --- |
| | | | | | | | | T98-G | --- | | |
| | | | | | | | | 9L-LacZ | --- | | |
| | | | | | | | | U-87 | U87 | | |

\* Loaded into composite component; DL: Drug loading; MR: Michael reaction; RP: Radical polymerization; 4-VPBA: 4-vinylboronic Acid; 5-FU: 5-Fluoruracil; CD: Cyclodextrin; AA: Acrylic acid; BSA: Bovine serum albumin; CA4P: Combrestatin A4 phosphate; DOX: Doxorubicin; GO: Graphene oxide; Hep: Heparin; HPAE: Hyperbranched poly(amine-ester); LC: Leucovorin calcium; LNC: Lipod nanocapsule; MPEG: Monomethoxy poly(ethylene glycol); MTF: Metformin; PAMAM: Polyamidoamine; PBA: Phenylboronic acid; pDMA: Poly(3,4-dihydroxyphenethyl)-methacrylamide; PEG-4-SH: 4-arm PEG; PEGBA-4: 4-arm PEG-boronic acid; PEGBA-8: 8-arm PEG-boronic acid; PEGDA: PEG-based diacrylate; PGA-MA: Maleimide-modified c-polyglutamic acid; PEGMA: Methoxypoly(ethylene glycol) monomethacrylate; PEG-2-MI: Maleimide-functionalized linear PEG; PLD-Arg: Arginine-poly(L-lysine) dendron; PLL: Poly-L-lysine; PLP: Polyphenols mixture; PVA: Polyvinyl alcohol.

## 3. Natural Polymers

### 3.1. Polysaccharides

Polysaccharides are widely employed for the fabrication of injectable hydrogels, owing to their outstanding advantages consisting in water affinity, biocompatibility, biodegradability, non-immunogenicity, and non-fouling features. Furthermore, the presence of multiple chemical functionalities (e.g., acid, amine, hydroxyl, and aldehyde groups) allows easy chemical modifications with the obtainment of a plethora of biomedical devices. They exert biological activities such as cell recruiting, cell adhesion, and modulation of the inflammatory process, and the pharmacokinetic profiles can be tailored by choosing the appropriate molecular weight distribution [185,186].

Polysaccharides are obtained from renewable plant and animal sources, including algae (e.g., dextran, alginate), plants (e.g., cellulose, agarose), microbes (e.g., dextran, gellan gum), and animals (e.g., hyaluronic acid, chitosan). In this review, when polysaccharides are mixed with synthetic polymers to further modify their physical, mechanical, and chemical properties, the resulting systems are referred as N/S hybrids.

Chitosan (CS, Figure 11), the *N*-deacetylated derivative of chitin, is a biomaterial with a wide range of biomedical applications due to its high biocompatibility and biodegradability.

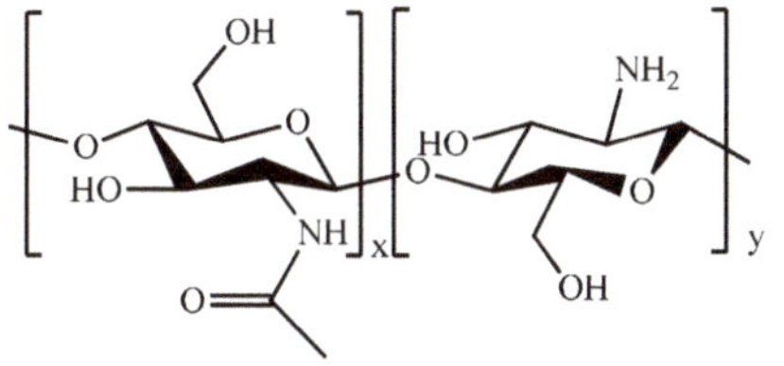

**Figure 11.** Schematic representation of chitosan (CS).

In addition, the wound-healing, anti-tumor, and antimicrobial activities, make CS an ideal starting material for designing pharmaceutical injectable formulations (Table 7) [187–189]. A CS prodrug of a photosensitizing agent was used as base material to obtain an injectable pH-responsive hydrogel to be used in breast cancer and melanoma therapy [190], whereas the chemical cross-linking of CS with β-GP was proposed in several research works as a valuable strategy to obtain thermo-responsive materials for the treatment of a number of cancer diseases. In more detail, CS/β-GP systems were either employed as platforms for the release of antineoplastic drugs [191–194] or loaded with nanoparticles bearing the anticancer agent, in order to obtain a more sustained drug release in the site of interest [195–198]. Other applications involved the possibility to combine chemo- and radio-therapy [195,199], and produce local hyperthermia for different types of cancer [200–202]. Thermal gelation of CS in the presence of G carried out to injectable hydrogels for the treatment of breast cancer [203], while mixed polysaccharide hydrogels, including CS-ALG [204] and CS-HA-NIPAAm [205,206] complexes, were designed to produce targeted delivery of anti-VEGF antibody [204], as well as pH-responsive systems for the DOX [205] and DOX@GO [206] vectorization to colon and breast cancer, respectively. Injectable hydrogels were also prepared using CS hydrophilic derivatives [207]; for example, CS modified with glycol moieties was covalently linked with PEG to obtain hydrogel materials for the release of self-healing [208] and photosensitizing [209] agents. In another approach, DOX@PLGA nanoparticles were inserted into the hydrogel structure, together with magnetic nanoparticles, to raise a more sustained release profile combined with magnetic ablation of breast cancer [210]. Furthermore, supramolecular hydrogels composed of GCS, PF127, and α-CD were proposed as DOX delivery platforms in the treatment of liver carcinomas [211]. Different modifications involved the bonding of hydroxybutyl [212], hydroxypropyl [7], carboxymethyl [213,214], and carboxyethyl [215–217] groups.

Table 7. Composition and anticancer performance of injectable hydrogels based on chitosan.

| Ref | Composition | | Carrier Properties | | | Delivery Properties | | Cancer Model | | | Health Model | |
|---|---|---|---|---|---|---|---|---|---|---|---|---|
| | Hydrogel (Gelation Process) | Composite Component | Degradation Time (Days) | Smart Responsivity | Bioactive Agent (DL% w/w) | Release Time (Days) | Type | In Vitro | In Vivo | In Vitro | In Vivo |
| | | | | | Naturals | | | | | | | |
| [190] | CS (Chemical – C) | TA-ZnPc | 1 | Light pH | TA-ZnPc (6.0) | 8 | Breast / Melanoma | MDA-MB-231 / A435 | --- | --- | --- |
| [191] | CS/β-GP (Physical – T) | --- | --- | --- | CisPt (1.0) | 15 | Colon / Breast | HCT-116 / MCF7 | --- | --- | --- |
| [193] | CS/β-GP/HA (Physical – I) | --- | --- | pH | DOX (0.016–0.033) | 5 | Cervix | HeLa | --- | --- | --- |
| [194] | CS/β-GP/CNT (Physical – T) | --- | 21 | --- | MTX | 7 | Breast | MCF7 | --- | 3T3 | --- |
| [196] | CS/β-GP (Physical – T) | MPEG | --- | --- | MEL** (4.6–10.6) | 4 | --- | --- | --- | --- | --- |
| [198] | CS/β-GP (Physical – T) | Liposome | --- | pH | TPT* (0.97) | 2 | Murine Liver | --- | H22 | --- | --- |
| [197] | CS/β-GP (Physical – T) | Liposome | --- | --- | DOX* (4.5) | 7 | Ovarian | A2780 | --- | --- | --- |
| [195] | CS/β-GP (Physical – T) | Liposome | --- | --- | DOX* (0.2) $^{188}$Re* | 21 | Murine Breast | --- | 4T1 | --- | --- |
| [199] | CS/β-GP (Physical – T) | Sn | --- | --- | DOX (0.025) $^{188}$Re | 2 | Liver | N1-S1 | N1-S1 | --- | --- |
| [200] | CS/β-GP (Physical – T) | Fe$_3$O$_4$ | 48 | Magnetic | BCG | --- | Bladder | --- | Mice | --- | --- |
| [201] | CS/β-GP (Physical – T) ALG (Physical – T) | F$_3$O$_4$ | --- | Magnetic | --- | --- | Breast / Ovarian / Glioblastoma / Colon | --- | SK-BR-3 / SKOV-3 / LN229 / Col12 / T380 | --- | --- |
| [202] | CS/β-GP (Physical – T) | GO/PEI-Fe$_3$O$_4$ | --- | pH Magnetic | DOX* (200) | 0.5 | Breast / Murine Sarcoma | MCF7 / --- | --- / S180 | --- | --- |
| [203] | CS/G (Physical – T) | --- | 21 | --- | MTX (0.0125) | 7.5 | Breast | MCF7 | --- | --- | --- |

Table 7. Cont.

| Ref | Composition | | Carrier Properties | | Delivery Properties | | Cancer Model | | | Health Model | |
|---|---|---|---|---|---|---|---|---|---|---|---|
| | Hydrogel (Gelation Process) | Composite Component | Degradation Time (Days) | Smart Responsivity | Bioactive Agent (DL% w/w) | Release Time (Days) | Type | In Vitro | In Vivo | In Vitro | In Vivo |
| [204] | CS-ALG (Physical – I) | --- | 31 | --- | anti-VEGF (0.018) | --- | --- | --- | --- | HUVECs | --- |
| [212] | HBCS (Physical) | --- | 45 | --- | DOX (2.5–10) | 3 | Murine Breast | 4-T1 | --- | HUVEC | --- |
| [214] | CMCS-oxALG (Chemical – C) | --- | --- | --- | HDBP | --- | Liver / Murine Liver | Bel-7402 / H-22 | --- / H-22 | L02 | --- |
| [216] | CECS-oxALG (Chemical – C) | MGM | 14 | Magnetic | 5-FU* (4.0) | 35 | --- | --- | --- | --- | --- |
| [217] | CECS/HA (Chemical – C) | --- | 10 | pH | DOX (0.3) | 3.5 | Cervix | HeLa | --- | --- | --- |
| [218] | CS-oxDEX (Chemical – C) | PF127 | 10 | pH Redox | 5-FU (2.5) CUR* (7.6) | 10 | Cervix | HeLa | --- | --- | --- |
| [219] | PBCS-oxDEX (Physical – T) | --- | --- | pH Glucose | DOX (1.0) | 0.5 | --- | --- | --- | L929 | --- |
| [220] | CS-DA/oxPLN (Chemical – C) | --- | --- | pH | DOX (0.01–0.32) AMX (0.5) | 2.5 1.5 | Colon | HCT116 | --- | --- | --- |
| [221] | SCS-oxCS (Chemical – C) | --- | 11 | pH | DOX (3.0) FeGl (5.0) | 6 2 | --- | --- | --- | MSC | --- |
| [222] | SCS-oxALG (Physical) | --- | --- | pH | DOX** (7.6) | 2 | Breast | MCF7 | MDA-MB-231 | --- | --- |
| [223] | GTMACS/ePC/LA (Physical) | --- | --- | --- | DTX | --- | --- | --- | --- | --- | mice |
| [224] | CS-CAT (Physical – I) | --- | --- | --- | DOX DTX (2.5) | 18 | Murine Lung / Murine Breast | --- / 4T1 | LLC / 4T1 | C212 | --- |
| [225] | CS-TRIPOD (Chemical – C) | --- | --- | Light pH | TPP** | 12 | Breast / Liver | MCF7 / HepG2 | --- | --- | --- |

Table 7. Cont.

| Ref | Composition | | Carrier Properties | | Delivery Properties | | | Cancer Model | | | Health Model | |
|---|---|---|---|---|---|---|---|---|---|---|---|---|
| | Hydrogel (Gelation Process) | Composite Component | Degradation Time (Days) | Smart Responsivity | Bioactive Agent (DL% w/w) | Release Time (Days) | Type | In Vitro | In Vivo | In Vitro | In Vivo |
| | | | | | N/S Hybrids | | | | | | | |
| [192] | CS/β-GP/NIPAAm-IA (Physical – T) | --- | --- | pH Thermo | DOX (3.0) | 8 | Breast | MCF7 | --- | --- | --- |
| [205] | CS-HA-NIPAAm (Chemical – C) | --- | 40 | pH | DOX (10) | 12 | Murine Colon | CT-26 | CT-26 | --- | --- |
| [206] | CS-HA-NIPAAm (Physical – T) | GO | 60 | pH | DOX* (14.20) | 9 | Breast | MCF7 | MCF7 | --- | --- |
| [207] | GCS/GMA (Chemical – IRD) | --- | --- | --- | DOX (1.0) | 7 | Breast | MCF7 | MCF7 | --- | --- |
| [208] | GCS-PEG (Physical – T) | --- | --- | --- | CRB (2.0) | 0.25 | --- | --- | --- | --- | mice |
| [209] | GCS-PEG (Physical – T) | --- | --- | --- | TMPyP (0.05–0.2) | 7# | Cervix | U14 | U14 | --- | --- |
| [210] | GCS-PEG (Chemical – C) | PLGA-F$_3$O$_4$ | --- | Magnetic | DTX* (9.0) | 30 | Breast | MDA-MB-23MDA-MB-231 | --- | --- | Mice |
| [211] | GCS/PF127/α-CD (Physical) | --- | 11 | pH | DOX (1.0–5.0) | 8 | Liver Murine Liver | HepG2 --- | --- H22 | --- | --- |
| [7] | PPLG-HPCS-PPLL (Physical – I) | oxDEX | 21 | --- | DOX* (22.1) IL-2* (8.3) IFN-γ* (8.7) | 24# | Breast Cervix Lung | MCF7 HeLa A549 | --- | --- | --- |
| [213] | CMCS-NIPAAm (Chemical – RP) | --- | --- | pH Thermo | 5-FU (6.2–8.9) | 2 | Breast Cervix | MCF-7 HeLa | --- | L929 | --- |
| [215] | CECS-PEG (Chemical – C) | --- | 8 | pH | DOX | 7.5 | Liver | HepG2 | --- | L929 | --- |
| [226] | TCS-PEGDMA (Chemical – MR) | STC | --- | Enzyme | CUR (3.8) LSZ* | 7 0.5 | Liver | HepG2 | HepG2 | --- | --- |

* Loaded in composite component; ** conjugated to composite component; # from in vivo experiments; C: Condensation; I: Ionic; IRD: Irradiation; MR: Michael reaction; RP: Radical polymerizarion; T: Temperature; 5-FU: 5-Fluorouracil; CD: Cyclodextrin; β-GP: β-Glycerophosphate; ALG: Alginate; AMX: Amoxicillin; BCG: Bacillus Calmette–Guérin; CAT: Catechol; CECS: Carboxyethyl chitosan; CisPt: Cisplatin; CMCS: Carboxymethyl chitosan; CNT: Carbon nanotubes; CRB: Carbazochrome; CS: Chitosan; CUR: Curcumin; DA: Dihydrocaffeic acid; DOX: Doxorubicin; DTX: Docetaxel; ePC: Egg phosphatidylcholine; FeG1: Non-hormonal contraceptive; G: Graphene; GCS: Glycol chitosan; GMA: Glycidyl methacrylate; GO: Graphene oxide; GTMACS: Glycidyltrimethylammonium chitosan; HA: Hyaluronic acid; HBCS: Hydroxybutyl chitosan; HDBP: Hydrogel degradation by-product; HPCS: Hydroxypropyl chitosan; IA: Itaconic acid; IL: Interleukin; IFN: Interferon; LA: Lauric aldehyde; LSZ: Lysozyme; MEL: Melphalan; MGM: Magnetic gelatin microspheres; MPEG: Monomethoxy poly(ethylene glycol); MTX: Methotrexate; NIPAAm: N-isopropyl acrylamide; oxALG: Oxidized alginate; oxCS: Oxidized chitosan; oxDEX: Oxidized dextran; oxPLN: Oxidized pullulan; PBCS: Phenylboronic-modified chitosan; PEG: Polyethylene glycol; PEGDMA: Polyethilenen glycol dimethacrylate; PEI: Poly(ethylene imine); PF: Pluronic F; PLGA: Poly(lactide-co-glycolide); PPLG: 4-Arm PEG-polyglutammic acid; PPLL: 4-arm poly(ethylene glycol)-poly(L-lysine); SCS: Succinate chitosan; STC: Starch; TA-ZnPc: Tetra-aldehyde functionalized zinc phthalocyanine; TCS: Thiolated chitosan; TMPyP: Meso-tetrakis(1-methylpyridinium-4-yl) porphyrin; PP: Tetrakis(4-aminophenyl)porphyrin; TPT: Topotecan; TRIPOD: 2,4,6-tris(p-formylphenoxy)-1,3,5-triazine; VEGF: Vascular endothelial growth factor

In more detais, carboxymethyl chitosan (CMCS) was copolymerized with NIPAAm [213] to obtain pH- and thermo-responsive depots for the on-off release of 5-FU to cervix and breast cancers. hydroxypropyl chitosan (HPCS) was condensed with PPLL dendrimers by Schiff's bases and subjected to an ionic gelation process in the presence of PEG dendrimers and oxDEX nanoparticles bearing DOX, IL-2, and IFN-γ for a synergistic anticancer therapy.

In another approach, CS [218], PBCS [219], CS-DA [220], or CS alkyl derivatives [214,216,217] were condensed with oxidized polysaccharides, including DEX [218,219], ALG [214,216], HA [217], and PLN [220].

SCS was combined with oxCS [221] or oxALG [222] to obtain pH-responsive injectable hydrogels for DOX sustained release. Other examples of CS derivatives include GTMACS [223] and CS-CAT [224], used for DTX or DOX/DTX combination therapy, respectively. TCS was employed to produce an enzyme-responsive CUR delivery vehicle [226], and CS-TPP was proposed for photothermal therapy in breast and liver cancers [225].

Hyaluronic acid (HA, Figure 12), a non-sulfated glycosaminoglycan, is one of the major components of connective tissues and synovial fluid.

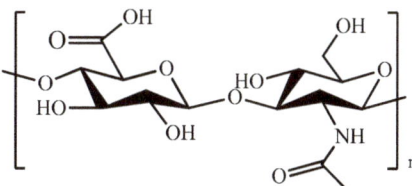

**Figure 12.** Schematic representation of hyaluronic acid (HA).

It is able to interact with cell surface receptors (e.g., CD44), thus promoting cell migration, and, in virtue of its high biocompatibility, has been extensively exploited as a starting material for the fabrication of different injectable hydrogel systems (Table 8) [227–230]. The thermo-gelation of HA in the presence of PF127 carried out to injectable hydrogels suitable for DOX release to breast [231] and colon cancers [68], or for the DOX-DTX synergistic treatment of CT26 cancer cells [67]. Oxidized HA was chemically cross-linked to obtain an injectable biomaterial mimicking embryonic microenvironments, thus exerting and controlling the phenotype of aggressive cancer cells [232]. Injectable HA hydrogels obtained with the same approach were either physically loaded with, or chemically conjugated to, CisPt-loaded HA nanogels for gastric cancer treatment [233]. Different cross-linking strategies involved the preliminary derivatization of HA with Tyr residues [234–236], or the insertion of thiol groups [237]. In the first case, horseradish peroxidase (HRP) catalyzed the coupling reaction between HA–Tyr chains with the formation of injectable hydrogels for the delivery of IFN-α to Kidney cancer (Figure 13) [236], while the incorporation of hyaluronidase allowed the selective vectorization of conjugated IFN-α [234] and loaded TZB [235] to liver and breast cancer, respectively.

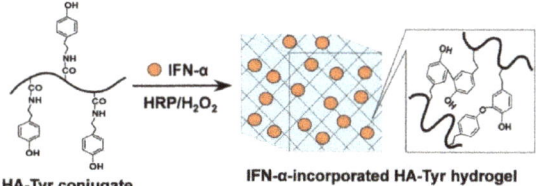

**Figure 13.** Schematic illustration of in situ formation of IFN-α-incorporated HA–Tyr hydrogels through enzymatic cross-linking reaction. HRP: horseradish peroxidase. With permission from [236]. Elsevier, [2016].

Table 8. Composition and anticancer performance of injectable hydrogels based on hyaluronic acid.

| Ref | Composition | | Carrier Properties | | Delivery Properties | | | Cancer Model | | | Health Model | |
|---|---|---|---|---|---|---|---|---|---|---|---|---|
| | Hydrogel (Gelation Process) | Composite Component | Degradation Time (Days) | Smart Responsivity | Bioactive Agent (DL% w/w) | Release Time (Days) | Type | In Vitro | In Vivo | | In Vitro | In Vivo |
| | **Naturals** | | | | | | | | | | | |
| [232] | oxHA (Chemical – C) | --- | 10 | --- | Anti-2B11 | 24# | Breast | MCF7 | --- | | --- | --- |
| | | | | | | | | MDA-MB-231 | MDA-MB-231 | | | |
| | | | | | | | Murine Breast | --- | BT-474 | | | |
| | | | | | | | Murine Melanoma | --- | B16 | | | |
| [233] | oxHA (Chemical – C) | HA-IDA HA-MA | --- | --- | CisPt* (200) | 7.5 | Stomach | --- | MKN45P | | --- | --- |
| [236] | HA–Tyr (Chemical – HRP) | --- | --- | --- | IFN-α | 1 | Kidney | ACHN | ACHN | | --- | --- |
| [234] | HA–Tyr (Chemical – HRP) | --- | --- | Enzyme (HAse) | IFN-α | 1 | Liver | HAK-1B | HAK-1B | | --- | --- |
| [235] | HA–Tyr (Chemical – HRP) | --- | 28 | Enzyme (HAse) | TZB (0.3) | 28 | Breast | BT474 | BT474 | | --- | --- |
| [237] | HA-SH (Chemical – Red) | --- | --- | Redox | DOX (1.0) | 21 | Breast | MCF7 | --- | | --- | --- |
| | | | | | DOX/SRB (1.0/1.0) | | | MDA-MB-231 | | | | |
| | | | | | DOX/SRB/MTF (1.0/1.0/1.0) | | Murine Breast | 4T1 | 4T1 | | | |
| [238] | HA (Physical – pH) | MSNs | --- | Enzyme (HAse) | DOX* (27) | 6 | Breast | SKBR3 | --- | | 293T | --- |
| [239] | HA–αCD (Physical) | AuBNs-MSNs | 7 | Enzyme (HAse) | DOX* (11.1–32.0) | 7 | Squamous Carcinoma | SCC | --- | | HaCaT | --- |
| | **N/S Hybrids** | | | | | | | | | | | |
| [231] | HA/PF127 (Physical – T) | --- | 31 | --- | DOX (0.5) | 31 | Breast | MCF7 | --- | | --- | mice |

Table 8. Cont.

| Ref | Composition | Carrier Properties | | Delivery Properties | | Cancer Model | | | Health Model | |
|---|---|---|---|---|---|---|---|---|---|---|
| | Hydrogel (Gelation Process) | Composite Component | Degradation Time (Days) | Smart Responsivity | Bioactive Agent (DL% w/w) | Release Time (Days) | Type | In Vitro | In Vivo | In Vitro | In Vivo |
| [68] | PF127/HA (Physical – T) | - - - | - - - | - - - | DOX (1.0) | 1 | Murine colon | C26 | C26 | - - - | - - - |
| | | | | | | | Colon | HT29 | - - - | | |
| | | | | | | | Breast | MCF7 | - - - | | |
| [67] | PF127/HA (Physical – T) | PF127_PL121 | - - - | - - - | DOX (1.0) DTX* (1.6) | 3 | Colorectal | - - - | CT26 | - - - | - - - |
| [240] | HA–Gln/PEG-8-SH-Lys (Chemical – E) | - - - | - - - | - - - | - - - | - - - | Breast | MCF7 | - - - | C2C12 | - - - |

* Loaded in composite component; # from in vivo experiments; C: Condensation; E: Enzymatic; Red: Redox; T: Temperature; CD: Cyclodextrin; AuBNs: Gold nanobipyramids; CisPt: Cisplatin; DOX: Doxorubicin; DTX: Docetaxel; Gln: Glutamine substrate peptide; HA: Hyaluronic acid; HAase: Hyaluronidase; HA-IDA: HA-iminodiacetic Acid; HA-MA: HA-malonic acid; HRP: Horseradish peroxidase; IFN: Interferon; Lys: Lysine; MSNs: Mesoporous silica nanoparticles; MTF: Metformin; oxHA: Oxidized HA; PEG-8-SH: 8-arm PEG; PF: Pluronic F; PL: Pluronic L; SRB: Sorafenib; Tyr: Tyramine; TZB: Trastuzumab.

On the other hand, the oxidation of thiol groups was exploited to generate disulfide bonds acting as cross-links of the hydrogel. The resulting redox-responsive material was employed as a delivery vehicle of DOX and the combinations of DOX–SRB and DOX–SRB–MTF [237]. HA was also employed as a functional element for the enzymatic synthesis of PEGylated dendrimers able to modulate the cellular phenotype of human mammary cancer epithelial cells and mouse myoblasts [240].

Finally, the incorporation of MSNs [238] and α-CD–AuBNs–MSNs [239] within HA hydrogels allowed the fabrication of hybrid systems suitable for photothermal DOX combination therapy of mammary and squamous carcinoma, respectively.

Cellulose (CL, Figure 14) is a polysaccharide consisting of repeating β-$D$-glucopyranose units obtained from different sources, including wood pulp, cotton, tunicates, fungi, bacteria, and algae [241].

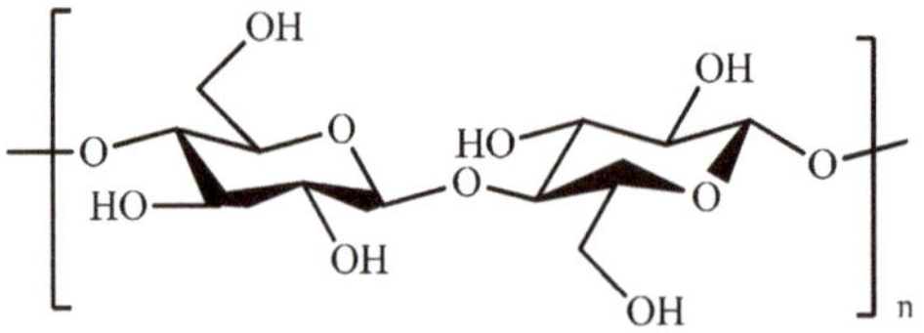

**Figure 14.** Schematic representation of cellulose (CL).

The superior biological features, together with the large availability and low cost, make CL-based materials suitable for a wide range of applications, including biomedicine (Table 9) [242].

Hydrophilic CL derivatives, such as quaternized cellulose [243] and hydroxypropyl methyl cellulose [244], were investigated for the DOX-based and PTX/TMZ therapy of hepatocellular carcinoma [243] and glioma [244], respectively. Pristine CL was also tested as a base material for the fabrication of hybrid hydrogels for the photothermal treatment of melanoma and hepatic cancer, both in vitro and in vivo [245], with black phosphorus nanosheets acting as active agent.

Alginate (ALG, Figure 15), an anionic biopolymer consisting of units of mannuronic acid and guluronic acid in irregular blocks [246], is widely used in biomedical field due to its several favorable properties, including biocompatibility, hydrophobicity, and availability of hydroxyl and carboxyl groups for tailored chemical modifications (Table 9) [247].

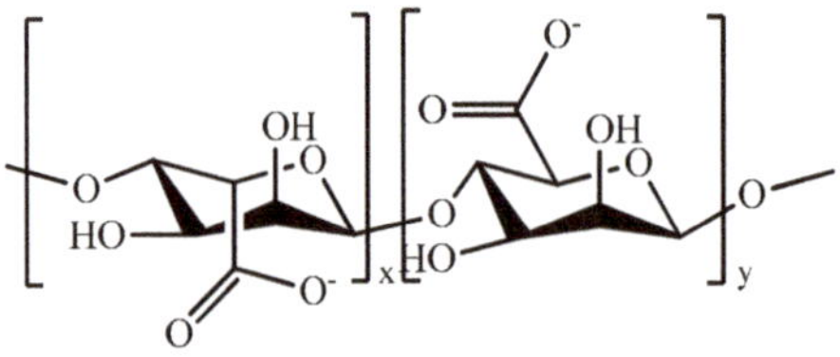

**Figure 15.** Schematic representation of alginate (ALG).

Injectable hydrogels prepared by ionic gelation were proposed for the delivery of CisPt dendrimers to breast and lung cancer cells with high efficiency [248], as well as for the incorporation of magnetic nanoparticles for the thermal ablation of different types of cancers, including breast, ovary, glioblastoma, and colon [201]. The insertion of NIPAAm moieties carried out the formation of thermo-responsive vehicles of gene [249] and DOX@micelles [250] to prostate cancer and osteosarcoma. Further modifications of ALG chains involved the oxidation to aldehyde derivatives, suitable for coupling with PEI polymers. The obtained in situ gelling systems were proposed as delivery systems for core-shell nanoparticles loaded with CisPt and PTX, and found to be effective in the treatment of breast, skin, and liver neoplasia [251,252].

Table 9. Composition and anticancer performance of injectable hydrogels based on other polysaccharides.

| Ref | Composition | Carrier Properties | | | Delivery Properties | | | Cancer Model | | | Health Model | |
|---|---|---|---|---|---|---|---|---|---|---|---|---|
| | Hydrogel (Gelation Process) | Composite Component | Degradation Time (Days) | Smart Responsivity | Bioactive Agent (DL% w/w) | Release Time (Days) | Type | In Vitro | In Vivo | In Vitro | In Vivo |
| | | | | | **Naturals** | | | | | | | |
| [243] | QCL-CCNCs (Physical – I) | --- | 18 | --- | DOX (0.5) | 21 | Murine Liver | --- | H22 | COS-7 | --- |
| [245] | CL (Chemical – C) | BPNs | --- | --- | --- | --- | Murine Melanoma Liver | B16 SMMC-7721 | --- SMMC-7721 | J774A.1 | --- |
| [248] | ALG (Physical – I) | PAMAM | --- | --- | CisPt* (37.0) | 30# | Breast Lung | MFC7 MDA-MB-231 PC9 | --- PC9 | 3T3 | --- |
| [253] | oxDEX-SRC (Chemical – C) | --- | 70 | --- | HRP (0.39–1.36) DOX (0.39–1.36) | 50 30 | Melanoma | --- | B16F10 | C2C12 HL7702 | mice |
| [254] | oxDEX (Chemical – C) | PAMAM | 14 | --- | Pt* | 9# | Breast | MDA-MB-231 | MDA-MB-231 | --- | mice |
| [255] | MADEX-SH/MAHA (Chemical – Red) | Bi NPs | --- | --- | DOX* (3.1) | 7.5 | Murine Breast | 4T1 | 4T1 | --- | --- |
| [256] | GG (Physical) | Liposome | --- | --- | PTX (33) | 2 | Bladder Murine Bladder | T24 NBT-II | --- | --- | mice |
| [257] | GG (Physical) | CuS NPs | --- | NIR | DOX* (0.1) | 0.2 | Murine Breast | 4T1 | 4T1 | --- | --- |
| [258] | AGR (Physical) | --- | 7 | pH NIR | DOX (4.5) | 2 | Murine Breast Cervix | 4T1 HeLa | 4T1 --- | L929 HUVEC | --- |
| [259] | AGR (Physical) | DEX-SH | --- | --- | DOX* (20–50) | 80 | Breast | MDA-MB-231 | --- | 3T3 | --- |
| | | | | | **N/S Hybrids** | | | | | | | |
| [244] | HPMCL/PF127/ALG (Physical) | MPEG-DPPE | --- | --- | PTX* (5.1) TMZ* (5.3) | 3 | Murine Glioma | C6 | C6 | --- | --- |
| [249] | ALG–NIPAAm (Physical – T) | --- | 365 | Thermo | DNA | 29 | Prostate | PC3 | --- | --- | --- |
| [250] | ALG–NIPAAm (Physical – T) | --- | --- | Thermo | DOX (1.2) | 21 | Prostate | AT3B-1N AT3B-1 | --- | --- | --- |

173

Table 9. Cont.

| Ref | Composition | | Carrier Properties | | | Delivery Properties | | | Cancer Model | | | Health Model | |
|---|---|---|---|---|---|---|---|---|---|---|---|---|---|
| | Hydrogel (Gelation Process) | Composite Component | Degradation Time (Days) | Smart Responsivity | Bioactive Agent (DL% w/w) | Release Time (Days) | Type | In Vitro | In Vivo | In Vitro | In Vivo |
| [251] | oxALG–PEI (Physical) | PLGA-PLA | --- | --- | CisPt* (0.01–2.48) PTX* (1.0–1.7) | 45 | Breast | MDA-MB-231 | --- | --- | --- |
| [252] | oxALG–PEI (Physical) | PLGA-PLA | --- | --- | CisPt* (0.01–2.48) PTX* (1.49–1.70) | 45 | Liver | HepG2 | --- | MRC-5 | --- |
| [260] | DEX–HEMA/PEI-MA (Chemical – RP) | --- | 9–17 | --- | sRNA | 9–17 | --- | --- | --- | HEK 293 | --- |

* Loaded in composite component; # from in vivo experiments; C: Condensation; I: Ionic; T: Temperature; Red: Redox; RP: Radical polymerization; AGR: Agarose; ALG: Alginate; BPNs: Black phosphorus nanosheets; CCNCs: Cationic cellulose nanocrystals; CisPt: Cisplatin; CL: Cellulose; DEX: Dextran; DOX: Doxorubicin; DPPE: Dipalmitoylphosphatidyle-thanoiamine; GG: Gellan gum; HPMCL: Hydroxypropyl methyl cellulose; HRP: Horseradish peroxidase; MADEX: Methacrylated DEX; MAHA: Methacrylated HA; MPEG: Monomethoxy poly(ethylene glycol); NIPAAm: N-isopropyl acrylamide; NIR: Near-infrared; NPs: Nanoparticles; oxALG: Oxidized alginate; oxDEX: Oxidized dextran; PAMAM: Polyamidoamine dendrimer; PEI: Poly(ethylene imine); PF: Pluronic F; PLA: Polylactide; PLGA: Poly(lactide-co-glycolide); PTX: Paclitaxel; QCL: Quaternized cellulose; SRC: Sericin; TMZ: Temozolomide.

Dextran (DEX, Figure 16) consists of glucose monomers linked via α-1,6 glycosidic bonds, with branches originating from α-1,3 linkages. It finds a wide range of applications in the biomedical field, due to its high availability, low cost, and easy chemical modification.

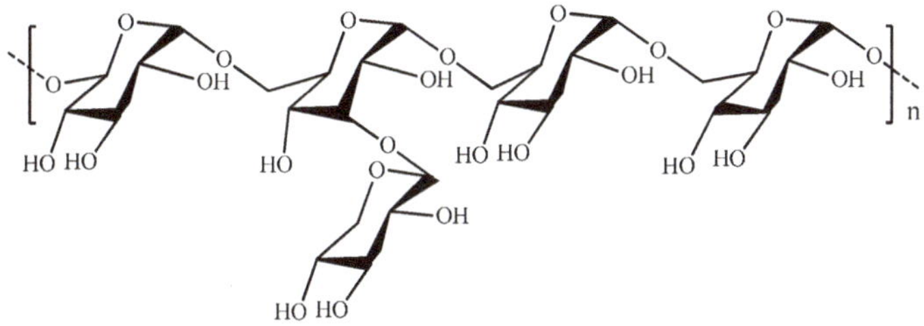

**Figure 16.** Schematic representation of dextran (DEX).

Moreover, its high stability, hydrophilicity, absence of toxicity, and biodegradability make this polysaccharide an ideal drug delivery carrier (Table 9) [261]. It is able to promote the penetration of chemotherapeutic agents in tumor masses [262], thus allowing the fabrication of effective delivery vehicles for cancer treatment [263]. Preliminary derivatization of dextran, including oxidation [253,254] and conjugation to acrylic [260] or thiol groups [255], was carried out to obtain effective carriers for the delivery of cytotoxic drugs [253,254], gene [260], or DOX in combination with Bismuth Nanoparticles in a combined X-ray radio- and chemo-therapy [255].

Gellan gum (GG, Figure 17) is a linear anionic polysaccharide approved by the FDA as an additive in food and pharmaceutical formulations (Table 9) [264].

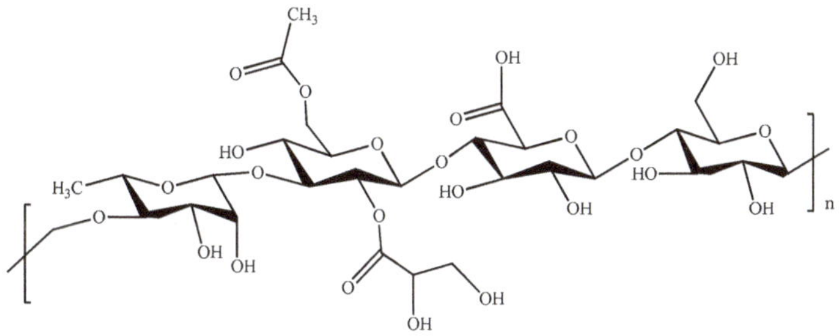

**Figure 17.** Schematic representation of gellan gum (GG).

Its biodegradability, mucoadhesivity, and thermo-reversible gelling properties make it the ideal candidate for the preparation of injectable matrices to be employed in tissue engineering and wound healing. Injectable nanocomposites, consisting of GG hydrogels incorporating drug-loaded nanoparticles, were proposed for the treatment of different cancer diseases. More closely, PTX-loaded liposomes were loaded on a GG hydrogel matrix and the overall system directly instilled in the urinary bladder [256]; whereas, in another work, DOX-loaded CuS nanoparticles were embedded in GG injectable hydrogels for NIR-triggered chemo-photothermal therapy of breast cancer [257].

Agarose (AGR, Figure 18) is an FDA-approved linear polysaccharide derived from marine algae. A robust injectable thermo-responsive AGR hydrogel incorporating sodium humate and DOX was proposed as a valuable tool for chemo-photothermal treatment of breast cancer [258]. Furthermore, DOX@nanoparticles were encapsulated in AGR injectable hydrogels for sustained local drug delivery (Table 9) [259].

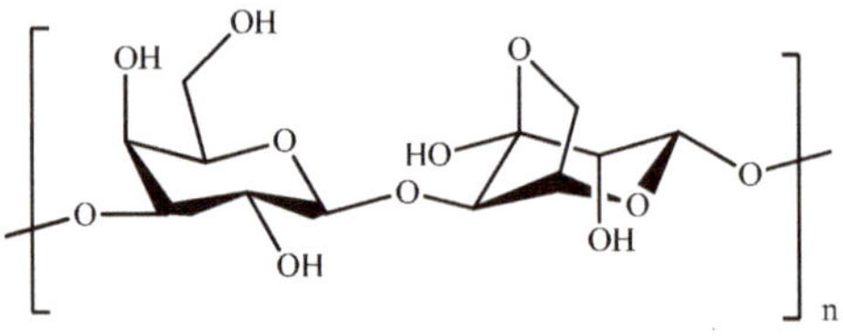

**Figure 18.** Schematic representation of agarose (AGR).

*3.2. Proteins*

The integration of the structural and functional properties of proteins in injectable hydrogels was also tested, thanks to the high biocompatibility, biodegradability, non-toxicity, and non-immunogenicity of such materials, as well as by virtue of their similarity to naturally occurring components of organs, tissues, and cells (Table 10) [25,265–268].

Serum albumins, from both bovine and human serum, are the most abundant protein in blood plasma (40–50 mg/mL) and the primary transport proteins of various endogenous and exogenous substances in plasma, including cations, bilirubin, fatty acids, and drugs [269,270].

Albumin from bovine serum (BSA) was proposed as polymeric support in the synthesis of injectable hydrogels for cancer therapy. BSA was added to the cross-linking agent epichlorohydrin to prepare a gel with suitable mechanical strength, viscoelastic behavior, shear thinning, injectability, and self-healing properties useful as DOX delivery vehicles to cervix and breast cancer [269]. Alternatively, an injectable hydrogel consisting of PEG-modified BSA- and PTX-encapsulated red blood cell membrane nanoparticles was proposed to improve the intraperitoneal retention of PTX in the treatment of human gastric cancer [271]. Finally, human serum albumin (HSA) chemically conjugated to PEG dendrimers was suggested as a functional biomaterial for the induction of apoptosis in pancreatic cancer [270].

Gelatin (GEL) represents another interesting protein material able to spontaneously undergo the gel–sol transition process at body temperature. Despite its good biological properties, gelatin hydrogel cannot be used in biomedical applications without chemical modifications, due to its instability under physiological conditions and, also, poor mechanical properties [272]. Different approaches were proposed to improve its performance in the biomedical field [273]: GEL–dendrimer [274], GEL–pectin [275], and GEL–CS [276] composites, cross-linked by means of HRP chemistry [274,275] or ionic gelation [276], were successfully employed in lung and skin carcinomas studies, and for the controlled release of DOX@Liposome. In addition, GEL injectable hydrogels were proposed as DOX carriers in the treatment of prostate cancer [277] in a multifunctional system, also acting as regenerative matrix with pronounced adhesion to abdominal tissue that, by in situ polymerization, allow to overcome the inconvenience usually related to radical prostatectomy. Moreover, due to its surfactant properties [278], GEL was also employed for the fabrication of thermo-responsive hybrid hydrogels for the controlled release of DOX to gastric cancer [279], with improved efficiency due to the incorporation of rod-like-shaped nanoparticles, such as carbon nanotubes [280,281]. Finally, an injectable and colloidal hydrogel composed of amphoteric GEL nanoparticles and polydopamine (PDA) nanoparticles was developed to realize multi-stimuli (pH, enzymes, and near-infrared light)-responsive drug delivery properties and combined chemo-photothermal cancer treatment [282]. Due to the sensitivity of GEL nanoparticles to the tumor microenvironment and PDA nanoparticles to the NIR laser, DOX-loaded hydrogel could show multiple responsivity to acidic pH and NIR laser irradiation, resulting in controlled and sustained anticancer release profiles.

Silk fibroin (SF) was proved to be a biodegradable and biocompatible native natural material derived from Bombyx mori silkworm with safe record in vivo [283,284]. SF hydrogels developed by the protein conformation transition from amorphous to β-sheet induced by physical cross-linking, including the ultrasound assisted processes, possess injectability as well as biocompatibility and safety features [285]. SAL–PTX-loaded silk fibroin hydrogel was fabricated by ultrasound-assisted cross-linkage, without toxic organic solvents and surfactants, for loco-regional tumor treatment and cancer stem cell inhibition in vivo [286]. Additionally, self-assembling pH-responsive silk nanofiber hydrogels with thixotropic properties were proposed to support the injectable delivery DOX for the treatment of breast cancers in mouse models [287]. The possibility to obtain benefits from a photothermal treatment was exploited in the synthesis of SF nanofiber hydrogel systems complexed with lanthanide-doped rare-earth up-conversion nanoparticles and nano-graphene oxide for breast cancer treatment [288]. In this case, a synergistic effect of combined up-conversion luminescence imaging diagnosis and photothermal therapy was confirmed to decrease dosage-limiting toxicity and tissue damage by over-heating and improve the therapeutic efficiency. An innovative approach that drastically reduces gelation times involved an enzyme-mediated cross-linking strategy to produce fast-gelled SF-based injectable hydrogels at physiological conditions [289].

Finally, silk–elastin-like protein (SELP), genetically engineered materials composed of tandem repeats of a six amino acid sequence commonly found in silkworm silk fibroin and a five amino acid sequence commonly found in mammalian elastin, was proposed in the synthesis of injectable hydrogels. This combination of silk and elastin molecular properties results in a polymer which is responsive to temperature increases and irreversibly forms hydrogels at physiological temperature. Gelation occurs without the need of chemically-induced cross-linking, because this phase transition spontaneously occurs when elastin-like units collapse thermodynamically aligning the silk-like units that form hydrogen-bonded beta sheets, and results in a physically cross-linked matrix. SELP-based carriers were applied as a platform for drug delivery with negligible toxicity for the radiation treatment of prostate and pancreas cancers [288,290], for localized delivery in transarterial chemoembolization to treat intermediate stage hepatocellular carcinoma [290,291], or as gene-directed enzyme prodrug therapy [292]. In particular, injectable brachytherapy polymers [290,291] composed of SELP labeled with the radionuclide $^{131}$I exhibit a gelling transition as a result of two independent mechanisms, firstly involving SELP moieties that, at the body temperature, are rapidly converted into an insoluble material. Afterwards, the high energy β-emissions of $^{131}$I further stabilize the depot by introducing cross-links within the SELP depot over 24 h. Additionally, SELP-based hydrogel was proposed to overcome the limitations usually associated with the commercial embolic liquids that discourage their employment in transarterial chemoembolization. To this regard, DOX and SRB, two chemotherapeutics used in the treatment of hepatic carcinoma, were incorporated into the in situ gelling liquid embolic composed of SELP polymer [290,291]. Due to its pore size and in vivo gelation properties, SELP restricts the distribution and controls the release of therapeutic viruses, such as herpes simplex virus, for up to one month, representing a valuable approach which may also have significant potential for increasing the safety of adenoviral gene delivery, while not sacrificing efficacy is spatial and temporal delivery of viruses following injection into a localized area [292]. In this way, gene expression levels at the site of interest were localized, prolonged, and significantly increased.

## 4. Conclusions and Future Perspectives

Hydrogel systems represent a relevant class of healthcare products with applications ranging from tissue engineering, bio-sensing, and bio-imaging, to drug delivery [293]. The huge interest in hydrogels is underlined by the worldwide market, estimated at around US$10 billion in 2017 and expected to grow up to US$15 billion by 2020 [294]. Injectable hydrogels have been proved to be a valuable tool for the delivery of anticancer drugs, providing temporal and spatial control over the releasing rate, thus improving the therapeutic index of commonly used chemotherapeutics [29]. To date, a few products are currently available on the market, including CS/Organophosphate (BST-Gel ®), PLGA–PEG–PLGA (ReGel ®), Poloxamer 407 (LeGOO ®), Poly(vinyl methyl ether co maleic anhydride) (Gantrez ®) hydrogels, available as cartilage repair [295] hydrogel market, tumors [296], vascular injury [297], and vaccine adjuvants [298].

Table 10. Composition and anticancer performance of injectable hydrogels based on proteins.

| Ref | Composition | Carrier Properties | | Delivery Properties | | | Cancer Model | | | Health Model | |
|---|---|---|---|---|---|---|---|---|---|---|---|
| | Hydrogel (Gelation Process) | Composite Component | Degradation Time (Days) | Smart Responsivity | Bioactive Agent (DL% w/w) | Release Time (Days) | Type | In Vitro | In Vivo | In Vitro | In Vivo |
| | | | | | Naturals | | | | | | |
| [269] | BSA (Chemical – CR) | --- | 3 | --- | DOX (0.11–0.14) | 5 | Cervix | HeLa | --- | --- | --- |
| | | | | | | | Breast | MCF-7 | | | |
| | | | | | | | | MDA-MB 231 | | | |
| [277] | GEL (Chemical – E) | --- | --- | --- | AraC | --- | Prostate | DU-145 | --- | L929 | mice |
| [274] | GEL–HPA (Chemical – HRP) | --- | --- | Enzyme (COLase) | DCs Ad | 19# | Murine Lung | --- | LL | --- | --- |
| [275] | GEL–SBP (Chemical – HRP) | --- | --- | --- | DOX (0.9) | 6 | Murine Melanoma | --- | B16F10 | --- | --- |
| [276] | GEL–CS (Chemical – C) | Liposome | --- | --- | CAL (0.39–0.47) CAL* | 7 21 | --- | --- | --- | --- | --- |
| [282] | GEL (Physical – T) | PDA | 7 | pH Enzyme NIR | DOX* | 7 | Murine Breast | --- | 4T1 | --- | --- |
| [286] | SF (Physical – US) | SF NPs | --- | --- | SAL* (12) PTX* (12) | 5 28 | Murine Liver | --- | H22 | --- | mice |
| [287] | SF (Physical – CDP) | --- | --- | pH | DOX (8–24) | 56 | Breast | MDA-MB-231 | MDA-MB-231 | --- | --- |
| [288] | SF (Physical – T) | GO | --- | --- | NaLuF$_4$:Er$^{3+}$, Yb$^{3+}$ | --- | Murine Breast | 4T1 | 4T1 | --- | --- |
| [290] | SELP (Physical – T) | --- | --- | --- | $^{131}$I | --- | Prostate | --- | PC3 | --- | --- |
| | | | | | | | Pancreas | | BxPc3 | | |
| [291] | SELP (Physical – T) | --- | --- | --- | DOX (21–28) SRB (21–28) | 15-30 | --- | --- | --- | --- | --- |
| [292] | SELP (Physical – T) | --- | --- | --- | HSVtk/GCV | --- | --- | --- | --- | --- | mice |
| | | | | | N/S Hybrids | | | | | | |
| [270] | HSA-SH/PEG-4-SH (Physical – T) | --- | 21 | --- | TRAIL (5.8) | 7 | Pancreas | Mia PAca-2 | Mia PAca-2 | --- | --- |
| [271] | PEG–BSA (Physical – T) | PRNP | 50 | --- | PTX* (22.1) | 6 | Stomach | MKN45 | MKN-45 | --- | --- |
| [279] | GEL–SWCNT–pNIPAAm (Chemical – RP) | --- | --- | Temperature | DOX (1.11) | 28 | Stomach | BGC-823 | BGC-823 | --- | --- |

* Loaded in composite component; # from in vivo experiments; C: Condensation; CDP: Concentration dilution process; CR: Cross-linking; E: Enzyme; US: Ultrasound; RP: Radical polymerization; T: Temperature; Ad: Oncolytic adenovirus; AraC: Cytosine arabinoside; BSA: Bovine serum albumin; CAL: Calcein; COLase: Collagenase; CS: Chitosan; DCs: Dendritic cells; DOX: Doxorubicin; GCV: Ganciclovir; GEL: Gelatin; HPA: Hydroxyphenyl propionic acid; GO: Graphene oxide; HPA: Hydroxyphenyl propionic acid; HRP: Horseradish peroxidase; HSA: Human serum albumin; HSVtk: Herpes simplex virus thymidine kinase; PDA: Polydopamine; PEG-4-SH: 4-arm PEG; pNIPAAm: Poly(N-isopropyl acrylamide); PRNP: Red blood cell membrane nanoparticles; PTX: Paclitaxel; SAL: Salinomycin; SBP: Sugar beet pectin; SELP: Silk-elastin-like protein; SF: Silk fibroin; SRB: Sorafenib; SWCNT: Single-walled carbon nanotubes; TRAIL: Tumor necrosis factor-related apoptosis-inducing ligand.

The main limiting issues, concerning sterilization, scale-up, shelf-life, and user compliance (professional and/or patient), must be addressed before the benefits afforded by injectable hydrogels can be translated into clinical practice. Some formulations are currently in clinical trials, mainly consisting in radiopaque PEG hydrogels (TraceIT ® and SpaceOAR ®) useful to improve the target definition of radiotherapy, thus reducing the radiation doses [299,300].

The scientific community recognizes great potential to the use of injectable systems for anticancer delivery, but to definitely replace the conventional therapies with the injectable systems, continuous innovation in the development of new architectures and design strategies is required. For a more effective translation of injectable hydrogels from research into clinical reality, future attempts should be done to explore the possibility of combining chemotherapy, hyperthermia therapy, immunotherapy, and radiotherapy, by selecting appropriate materials and evaluating the biological effects on metabolic and cellular mechanisms, both in the normal and diseased states.

**Funding:** This research received no external funding.

**Conflicts of Interest:** The authors declare no conflict of interest.

## References

1. Li, Y.; Rodrigues, J.; Tomás, H. Injectable and biodegradable hydrogels: Gelation, biodegradation and biomedical applications. *Chem. Soc. Rev.* **2012**, *41*, 2193–2221. [CrossRef] [PubMed]
2. Norouzi, M.; Nazari, B.; Miller, D.W. Injectable hydrogel-based drug delivery systems for local cancer therapy. *Drug Discov. Today* **2016**, *21*, 1835–1849. [CrossRef] [PubMed]
3. Mathew, A.P.; Uthaman, S.; Cho, K.H.; Cho, C.S.; Park, I.K. Injectable hydrogels for delivering biotherapeutic molecules. *Int. J. Biol. Macromol.* **2018**, *110*, 17–29. [CrossRef] [PubMed]
4. Yu, S.; He, C.; Chen, X. Injectable Hydrogels as Unique Platforms for Local Chemotherapeutics-Based Combination Antitumor Therapy. *Macromol. Biosci.* **2018**, *18*. [CrossRef]
5. Ko, D.Y.; Shinde, U.P.; Yeon, B.; Jeong, B. Recent progress of in situ formed gels for biomedical applications. *Prog. Polym. Sci.* **2013**, *38*, 672–701. [CrossRef]
6. Qi, C.; Liu, J.; Jin, Y.; Xu, L.; Wang, G.; Wang, Z.; Wang, L. Photo-crosslinkable, injectable sericin hydrogel as 3D biomimetic extracellular matrix for minimally invasive repairing cartilage. *Biomaterials* **2018**, *163*, 89–104. [CrossRef] [PubMed]
7. Wu, X.; He, C.; Wu, Y.; Chen, X.; Cheng, J. Nanogel-Incorporated Physical and Chemical Hybrid Gels for Highly Effective Chemo-Protein Combination Therapy. *Adv. Funct. Mater.* **2015**, *25*, 6744–6755. [CrossRef]
8. Zhang, Z.; He, C.L.; Xu, Q.H.; Zhuang, X.L.; Chen, X.S. Preparation of Poly(L-glutamic acid)-based Hydrogels via Diels-Alder Reaction and Study on Their Biomolecule-responsive Properties. *Acta Polym. Sin.* **2018**, *1*, 99–108. [CrossRef]
9. Xu, Q.; Guo, L.; Sigen, A.; Gao, Y.; Zhou, D.; Greiser, U.; Creagh-Flynn, J.; Zhang, H.; Dong, Y.; Cutlar, L.; et al. Injectable hyperbranched poly(β-amino ester) hydrogels with on-demand degradation profiles to match wound healing processes. *Chem. Sci.* **2018**, *9*, 2179–2187. [CrossRef]
10. Castro, V.; Rodríguez, H.; Albericio, F. CuAAC: An Efficient Click Chemistry Reaction on Solid Phase. *ACS Comb. Sci.* **2016**, *18*, 1–14. [CrossRef]
11. Dey, P.; Hemmati-Sadeghi, S.; Haag, R. Hydrolytically degradable, dendritic polyglycerol sulfate based injectable hydrogels using strain promoted azide-alkyne cycloaddition reaction. *Polym. Chem.* **2016**, *7*, 375–383. [CrossRef]
12. Yang, Z.; Gao, D.; Cao, Z.; Zhang, C.; Cheng, D.; Liu, J.; Shuai, X. Drug and gene co-delivery systems for cancer treatment. *Biomater. Sci.* **2015**, *3*, 1035–1049. [CrossRef] [PubMed]
13. Nezhad-Mokhtari, P.; Ghorbani, M.; Roshangar, L.; Soleimani Rad, J. Chemical gelling of hydrogels-based biological macromolecules for tissue engineering: Photo- and enzymatic-crosslinking methods. *Int. J. Biol. Macromol.* **2019**, *139*, 760–772. [CrossRef] [PubMed]
14. Xu, Q.; He, C.; Zhang, Z.; Ren, K.; Chen, X. Injectable, Biomolecule-Responsive Polypeptide Hydrogels for Cell Encapsulation and Facile Cell Recovery through Triggered Degradation. *ACS Appl. Mater. Interfaces* **2016**, *8*, 30692–30702. [CrossRef] [PubMed]

15. Yu, S.; Wang, C.; Yu, J.; Wang, J.; Lu, Y.; Zhang, Y.; Zhang, X.; Hu, Q.; Sun, W.; He, C.; et al. Injectable Bioresponsive Gel Depot for Enhanced Immune Checkpoint Blockade. *Adv. Mater.* **2018**, *30*. [CrossRef]
16. Kim, S.H.; Tan, J.P.K.; Nederberg, F.; Fukushima, K.; Colson, J.; Yang, C.; Nelson, A.; Yang, Y.Y.; Hedrick, J.L. Hydrogen bonding-enhanced micelle assemblies for drug delivery. *Biomaterials* **2010**, *31*, 8063–8071. [CrossRef]
17. Huebsch, N.; Kearney, C.J.; Zhao, X.; Kim, J.; Cezar, C.A.; Suo, Z.; Mooney, D.J. Ultrasound-triggered disruption and self-healing of reversibly cross-linked hydrogels for drug delivery and enhanced chemotherapy. *Proc. Natl. Acad. Sci. USA* **2014**, *111*, 9762–9767. [CrossRef]
18. Appel, E.A.; Del Barrio, J.; Loh, X.J.; Scherman, O.A. Supramolecular polymeric hydrogels. *Chem. Soc. Rev.* **2012**, *41*, 6195–6214. [CrossRef]
19. Bai, Y.; Li, S.; Li, X.; Han, X.; Li, Y.; Zhao, J.; Zhang, J.; Hou, X.; Yuan, X. An injectable robust denatured albumin hydrogel formed via double equilibrium reactions. *J. Biomater. Sci. Polym. Ed.* **2019**, *30*, 662–678. [CrossRef]
20. Gačanin, J.; Kovtun, A.; Fischer, S.; Schwager, V.; Quambusch, J.; Kuan, S.L.; Liu, W.; Boldt, F.; Li, C.; Yang, Z.; et al. Spatiotemporally Controlled Release of Rho-Inhibiting C3 Toxin from a Protein–DNA Hybrid Hydrogel for Targeted Inhibition of Osteoclast Formation and Activity. *Adv. Healthc. Mater.* **2017**, *6*. [CrossRef]
21. Yan, C.; Pochan, D.J. Rheological properties of peptide-based hydrogels for biomedical and other applications. *Chem. Soc. Rev.* **2010**, *39*, 3528–3540. [CrossRef] [PubMed]
22. Slaughter, B.V.; Khurshid, S.S.; Fisher, O.Z.; Khademhosseini, A.; Peppas, N.A. Hydrogels in regenerative medicine. *Adv. Mater.* **2009**, *21*, 3307–3329. [CrossRef] [PubMed]
23. Bakaic, E.; Smeets, N.M.B.; Hoare, T. Injectable hydrogels based on poly(ethylene glycol) and derivatives as functional biomaterials. *RSC Adv.* **2015**, *5*, 35469–35486. [CrossRef]
24. Singh, N.K.; Lee, D.S. In situ gelling pH- and temperature-sensitive biodegradable block copolymer hydrogels for drug delivery. *J. Control. Release* **2014**, *193*, 214–227. [CrossRef] [PubMed]
25. Nguyen, M.K.; Lee, D.S. Injectable biodegradable hydrogels. *Macromol. Biosci.* **2010**, *10*, 563–579. [CrossRef] [PubMed]
26. Tran, R.T.; Gyawali, D.; Nair, P.; Yang, J. Biodegradable injectable systems for bone tissue engineering. In *A Handbook of Applied Biopolymer Technology: Synthesis, Degradation and Applications*; Sharma, S., Mudhoo, A., Eds.; RSC: London, UK, 2011; pp. 419–451.
27. Srinivasan, C.; Weight, A.K.; Bussemer, T.; Klibanov, A.M. Non-aqueous suspensions of antibodies are much less viscous than equally concentrated aqueous solutions. *Pharm. Res.* **2013**, *30*, 1749–1757. [CrossRef]
28. Sun, S.; Cao, H.; Su, H.; Tan, T. Preparation and characterization of a novel injectable in situ cross-linked hydrogel. *Polym. Bull.* **2009**, *62*, 699–711. [CrossRef]
29. Tu, Y.; Chen, N.; Li, C.; Liu, H.; Zhu, R.; Chen, S.; Xiao, Q.; Liu, J.; Ramakrishna, S.; He, L. Advances in injectable self-healing biomedical hydrogels. *Acta. Biomater.* **2019**, *90*, 1–20. [CrossRef]
30. Kretlow, J.D.; Klouda, L.; Mikos, A.G. Injectable matrices and scaffolds for drug delivery in tissue engineering. *Adv. Drug Deliv. Rev.* **2007**, *59*, 263–273. [CrossRef]
31. Říhová, B. Immunocompatibility and biocompatibility of cell delivery systems. *Adv. Drug Deliv. Rev.* **2000**, *42*, 65–80. [CrossRef]
32. Fu, C.X.; Lin, X.X.; Wang, J.; Zheng, X.Q.; Li, X.Y.; Lin, Z.F.; Lin, G.Y. Injectable micellar supramolecular hydrogel for delivery of hydrophobic anticancer drugs. *J. Mater. Sci.* **2016**, *27*, 1–7. [CrossRef] [PubMed]
33. Brigger, I.; Dubernet, C.; Couvreur, P. Nanoparticles in cancer therapy and diagnosis. *Adv. Drug Deliv. Rev.* **2012**, *64*, 24–36. [CrossRef]
34. Wu, X.; He, C.; Wu, Y.; Chen, X. Synergistic therapeutic effects of Schiff's base cross-linked injectable hydrogels for local co-delivery of metformin and 5-fluorouracil in a mouse colon carcinoma model. *Biomaterials* **2016**, *75*, 148–162. [CrossRef] [PubMed]
35. Cirillo, G.; Nicoletta, F.P.; Curcio, M.; Spizzirri, U.G.; Picci, N.; Iemma, F. Enzyme immobilization on smart polymers: Catalysis on demand. *React. Funct. Polym.* **2014**, *83*, 62–69. [CrossRef]
36. Allcock, H.R.; Morozowich, N.L. Bioerodible polyphosphazenes and their medical potential. *Polym. Chem.* **2012**, *3*, 578–590. [CrossRef]
37. Baillargeon, A.L.; Mequanint, K. Biodegradable polyphosphazene biomaterials for tissue engineering and delivery of therapeutics. *BioMed Res. Int.* **2014**, *2014*. [CrossRef] [PubMed]

38. Hindenlang, M.D.; Soudakov, A.A.; Imler, G.H.; Laurencin, C.T.; Nair, L.S.; Allcock, H.R. Iodine-containing radio-opaque polyphosphazenes. *Polym. Chem.* **2010**, *1*, 1467–1474. [CrossRef]
39. Singh, A.; Krogman, N.R.; Sethuraman, S.; Nair, L.S.; Sturgeon, J.L.; Brown, P.W.; Laurencin, C.T.; Allcock, H.R. Effect of side group chemistry on the properties of biodegradable l-alanine cosubstituted polyphosphazenes. *Biomacromolecules* **2006**, *7*, 914–918. [CrossRef]
40. Cho, J.K.; Hong, K.Y.; Park, J.W.; Yang, H.K.; Song, S.C. Injectable delivery system of 2-methoxyestradiol for breast cancer therapy using biodegradable thermosensitive poly(organophosphazene) hydrogel. *J. Drug Target.* **2011**, *19*, 270–280. [CrossRef]
41. Allcock, H.R.; Pucher, S.R.; Scopelianos, A.G. Poly[(amino acid ester)phosphazenes] as substrates for the controlled release of small molecules. *Biomaterials* **1994**, *15*, 563–569. [CrossRef]
42. Teasdale, I.; Brüggemann, O. Polyphosphazenes: Multifunctional, biodegradable vehicles for drug and gene delivery. *Polymers* **2013**, *5*, 161–187. [CrossRef] [PubMed]
43. Ogueri, K.S.; Allcock, H.R.; Laurencin, C.T. Polyphosphazene Polymer. *Encycl. Polym. Sci. Technol.* **2019**. [CrossRef]
44. Kwak, M.K.; Hur, K.; Yu, J.E.; Han, T.S.; Yanagihara, K.; Kim, W.H.; Lee, S.M.; Song, S.C.; Yang, H.K. Suppression of in vivo tumor growth by using a biodegradable thermosensitive hydrogel polymer containing chemotherapeutic agent. *Investig. New Drugs* **2010**, *28*, 284–290. [CrossRef] [PubMed]
45. Al-Abd, A.M.; Hong, K.Y.; Song, S.C.; Kuh, H.J. Pharmacokinetics of doxorubicin after intratumoral injection using a thermosensitive hydrogel in tumor-bearing mice. *J. Control. Release* **2010**, *142*, 101–107. [CrossRef] [PubMed]
46. Wang, J.; Wang, D.; Yan, H.; Tao, L.; Wei, Y.; Li, Y.; Wang, X.; Zhao, W.; Zhang, Y.; Zhao, L.; et al. An injectable ionic hydrogel inducing high temperature hyperthermia for microwave tumor ablation. *J. Mater. Chem. B* **2017**, *5*, 4110–4120. [CrossRef]
47. Cho, J.K.; Hong, J.M.; Han, T.; Yang, H.K.; Song, S.C. Injectable and biodegradable poly(organophosphazene) hydrogel as a delivery system of docetaxel for cancer treatment. *J. Drug Target.* **2013**, *21*, 564–573. [CrossRef] [PubMed]
48. Kim, J.H.; Lee, J.H.; Kim, K.S.; Na, K.; Song, S.C.; Lee, J.; Kuh, H.J. Intratumoral delivery of paclitaxel using a thermosensitive hydrogel in human tumor xenografts. *Arch. Pharmacal Res.* **2013**, *36*, 94–101. [CrossRef] [PubMed]
49. Cho, J.K.; Kuh, H.J.; Song, S.C. Injectable poly(organophosphazene) hydrogel system for effective paclitaxel and doxorubicin combination therapy. *J. Drug Target.* **2014**, *22*, 761–767. [CrossRef] [PubMed]
50. Kim, Y.M.; Park, M.R.; Song, S.C. An injectable cell penetrable nano-polyplex hydrogel for localized siRNA delivery. *Biomaterials* **2013**, *34*, 4493–4500. [CrossRef] [PubMed]
51. Cho, J.K.; Chun, C.; Kuh, H.J.; Song, S.C. Injectable poly(organophosphazene)-camptothecin conjugate hydrogels: Synthesis, characterization, and antitumor activities. *Eur. J. Pharm. Biopharm.* **2012**, *81*, 582–590. [CrossRef] [PubMed]
52. Kim, J.I.; Kim, B.; Chun, C.; Lee, S.H.; Song, S.C. MRI-monitored long-term therapeutic hydrogel system for brain tumors without surgical resection. *Biomaterials* **2012**, *33*, 4836–4842. [CrossRef] [PubMed]
53. Zhang, Z.Q.; Song, S.C. Thermosensitive/superparamagnetic iron oxide nanoparticle-loaded nanocapsule hydrogels for multiple cancer hyperthermia. *Biomaterials* **2016**, *106*, 13–23. [CrossRef] [PubMed]
54. Akash, M.S.H.; Rehman, K. Recent progress in biomedical applications of pluronic (PF127): Pharmaceutical perspectives. *J. Control. Release* **2015**, *209*, 120–138. [CrossRef] [PubMed]
55. Moebus, K.; Siepmann, J.; Bodmeier, R. Alginate-poloxamer microparticles for controlled drug delivery to mucosal tissue. *Eur. J. Pharm. Biopharm.* **2009**, *72*, 42–53. [CrossRef] [PubMed]
56. Klouda, L. Thermoresponsive hydrogels in biomedical applications A seven-year update. *Eur. J. Pharm. Biopharm.* **2015**, *97*, 338–349. [CrossRef]
57. Cabana, A.; Aït-Kadi, A.; Juhász, J. Study of the gelation process of polyethylene oxide(a)-polypropylene oxide(b)-polyethylene oxide, copolymer (poloxamer 407) aqueous solutions. *J. Colloid Interface Sci.* **1997**, *190*, 307–312. [CrossRef]
58. Thimmaraju, M.K.; Bheemanapally, K.; Dharavath, R.; Kakarla, L.; Botlagunta, M. Improved anticancer activity of meloxicam hydrogels in K562 and HL60 cell lines. *J. Young Pharm.* **2017**, *9*, 209–213. [CrossRef]

59. Hu, H.; Lin, Z.; He, B.; Dai, W.; Wang, X.; Wang, J.; Zhang, X.; Zhang, H.; Zhang, Q. A novel localized co-delivery system with lapatinib microparticles and paclitaxel nanoparticles in a peritumorally injectable in situ hydrogel. *J. Control. Release* **2015**, *220*, 189–200. [CrossRef]
60. Xu, G.; Li, B.; Wang, T.; Wan, J.; Zhang, Y.; Huang, J.; Shen, Y. Enhancing the anti-ovarian cancer activity of quercetin using a self-assembling micelle and thermosensitive hydrogel drug delivery system. *RSC Adv.* **2018**, *8*, 21229–21242. [CrossRef]
61. Kim, D.Y.; Kwon, D.Y.; Kwon, J.S.; Park, J.H.; Park, S.H.; Oh, H.J.; Kim, J.H.; Min, B.H.; Park, K.; Kim, M.S. Synergistic anti-tumor activity through combinational intratumoral injection of an in-situ injectable drug depot. *Biomaterials* **2016**, *85*, 232–245. [CrossRef]
62. Zhang, N.; Xu, X.; Zhang, X.; Qu, D.; Xue, L.; Mo, R.; Zhang, C. Nanocomposite hydrogel incorporating gold nanorods and paclitaxel-loaded chitosan micelles for combination photothermal-chemotherapy. *Int. J. Pharm.* **2016**, *497*, 210–221. [CrossRef] [PubMed]
63. Fu, J.J.; Chen, M.Y.; Li, J.X.; Zhou, J.H.; Xie, S.N.; Yuan, P.; Tang, B.; Liu, C.C. Injectable hydrogel encapsulating $Cu_2MnS_2$ nanoplates for photothermal therapy against breast cancer. *J. Nanobiotechnology* **2018**, *16*. [CrossRef] [PubMed]
64. Bruschi, M.L.; Borghi-Pangoni, F.B.; Junqueira, M.V.; de Souza Ferreira, S.B.; Ficai, D.; Grumezescu, A.M. Chapter 12—Nanostructured therapeutic systems with bioadhesive and thermoresponsive properties. In *Nanostructures for Novel Therapy*; Ficai, D., Grumezescu, A., Eds.; Elsevier: Amsterdam, The Netherlands, 2017; pp. 313–342.
65. Lin, H.R.; Tseng, C.C.; Lin, Y.J.; Ling, M.H. A novel in-situ-gelling liquid suppository for site-targeting delivery of anti-colorectal cancer drugs. *J. Biomater. Sci. Polym. Ed.* **2012**, *23*, 807–822. [CrossRef] [PubMed]
66. Gao, L.; Wang, X.; Ma, J.; Hao, D.; Wei, P.; Zhou, L.; Liu, G. Evaluation of TPGS-modified thermo-sensitive Pluronic PF127 hydrogel as a potential carrier to reverse the resistance of P-gp-overexpressing SMMC-7721 cell lines. *Colloids Surf. B Biointerfaces* **2016**, *140*, 307–316. [CrossRef] [PubMed]
67. Sheu, M.T.; Jhan, H.J.; Su, C.Y.; Chen, L.C.; Chang, C.E.; Liu, D.Z.; Ho, H.O. Codelivery of doxorubicin-containing thermosensitive hydrogels incorporated with docetaxel-loaded mixed micelles enhances local cancer therapy. *Colloids Surf. B Biointerfaces* **2016**, *143*, 260–270. [CrossRef]
68. Jhan, H.J.; Liu, J.J.; Chen, Y.C.; Liu, D.Z.; Sheu, M.T.; Ho, H.O. Novel injectable thermosensitive hydrogels for delivering hyaluronic acid-doxorubicin nanocomplexes to locally treat tumors. *Nanomedicine* **2015**, *10*, 1263–1274. [CrossRef] [PubMed]
69. Khan, S.; Minhas, M.U.; Ahmad, M.; Sohail, M. Self-assembled supramolecular thermoreversible β-cyclodextrin/ethylene glycol injectable hydrogels with difunctional Pluronic®127 as controlled delivery depot of curcumin. Development, characterization and in vitro evaluation. *J. Biomater. Sci. Polym. Ed.* **2018**, *29*, 1–34. [CrossRef]
70. Hu, X.; Li, D.; Tan, H.; Pan, C.; Chen, X. Injectable graphene oxide/graphene composite supramolecular hydrogel for delivery of anti-cancer drugs. *J. Macromol. Sci. A Pure Appl. Chem.* **2014**, *51*, 378–384. [CrossRef]
71. Moon, H.J.; Ko, D.Y.; Park, M.H.; Joo, M.K.; Jeong, B. Temperature-responsive compounds as in situ gelling biomedical materials. *Chem. Soc. Rev.* **2012**, *41*, 4860–4883. [CrossRef]
72. Kang, Y.M.; Kim, G.H.; Kim, J.I.; Kim, D.Y.; Lee, B.N.; Yoon, S.M.; Kim, J.H.; Kim, M.S. In vivo efficacy of an intratumorally injected in situ-forming doxorubicin/poly(ethylene glycol)-b-polycaprolactone diblock copolymer. *Biomaterials* **2011**, *32*, 4556–4564. [CrossRef]
73. Lei, N.; Gong, C.; Qian, Z.; Luo, F.; Wang, C.; Wang, H.; Wei, Y. Therapeutic application of injectable thermosensitive hydrogel in preventing local breast cancer recurrence and improving incision wound healing in a mouse model. *Nanoscale* **2012**, *4*, 5686–5693. [CrossRef] [PubMed]
74. Kondiah, P.J.; Choonara, Y.E.; Kondiah, P.P.D.; Marimuthu, T.; Kumar, P.; Du Toit, L.C.; Pillay, V. A review of injectable polymeric hydrogel systems for application in bone tissue engineering. *Molecules* **2016**, *21*. [CrossRef] [PubMed]
75. Choi, B.; Lee, M. Injectable Hydrogels for Articular Cartilage Regeneration. In *Injectable Hydrogels for Regenerative Engineering*; Nair, L.S., Ed.; Imperial College Press: London, UK, 2016; pp. 355–376.
76. Shi, K.; Wang, Y.L.; Qu, Y.; Liao, J.F.; Chu, B.Y.; Zhang, H.P.; Luo, F.; Qian, Z.Y. Synthesis, characterization, and application of reversible PDLLA-PEG-PDLLA copolymer thermogels in vitro and in vivo. *Sci. Rep.* **2016**, *6*. [CrossRef] [PubMed]

77. Shi, K.; Xue, B.; Jia, Y.; Yuan, L.; Han, R.; Yang, F.; Peng, J.; Qian, Z. Sustained co-delivery of gemcitabine and cis-platinum via biodegradable thermo-sensitive hydrogel for synergistic combination therapy of pancreatic cancer. *Nano Res.* **2019**. [CrossRef]
78. Fan, R.; Tong, A.; Li, X.; Gao, X.; Mei, L.; Zhou, L.; Zhang, X.; You, C.; Guo, G. Enhanced antitumor effects by docetaxel/LL 37-loaded thermosensitive hydrogel nanoparticles in peritoneal carcinomatosis of colorectal cancer. *Int. J. Nanomed.* **2015**, *10*, 7291–7305. [CrossRef]
79. Li, X.; Fan, R.; Wang, Y.; Wu, M.; Tong, A.; Shi, J.; Xiang, M.; Zhou, L.; Guo, G. In situ gel-forming dual drug delivery system for synergistic combination therapy of colorectal peritoneal carcinomatosis. *RSC Adv.* **2015**, *5*, 101494–101506. [CrossRef]
80. Liang, Y.; Dong, C.; Zhang, J.; Deng, L.; Dong, A. A reconstituted thermosensitive hydrogel system based on paclitaxel-loaded amphiphilic copolymer nanoparticles and antitumor efficacy. *Drug Dev. Ind. Pharm.* **2017**, *43*, 972–979. [CrossRef]
81. Park, M.H.; Joo, M.K.; Choi, B.G.; Jeong, B. Biodegradable thermogels. *Acc. Chem. Res.* **2012**, *45*, 424–433. [CrossRef]
82. Qiu, B.; Stefanos, S.; Ma, J.; Lalloo, A.; Perry, B.A.; Leibowitz, M.J.; Sinko, P.J.; Stein, S. A hydrogel prepared by in situ cross-linking of a thiol-containing poly(ethylene glycol)-based copolymer: A new biomaterial for protein drug delivery. *Biomaterials* **2003**, *24*, 11–18. [CrossRef]
83. Ma, H.; He, C.; Cheng, Y.; Yang, Z.; Zang, J.; Liu, J.; Chen, X. Localized Co-delivery of Doxorubicin, Cisplatin, and Methotrexate by Thermosensitive Hydrogels for Enhanced Osteosarcoma Treatment. *ACS Appl. Mater. Interfaces* **2015**, *7*, 27040–27048. [CrossRef]
84. He, C.; Kim, S.W.; Lee, D.S. In situ gelling stimuli-sensitive block copolymer hydrogels for drug delivery. *J. Control. Release* **2008**, *127*, 189–207. [CrossRef] [PubMed]
85. Chang, G.; Ci, T.; Yu, L.; Ding, J. Enhancement of the fraction of the active form of an antitumor drug topotecan via an injectable hydrogel. *J. Control. Release* **2011**, *156*, 21–27. [CrossRef] [PubMed]
86. Yang, Z.; Yu, S.; Li, D.; Gong, Y.; Zang, J.; Liu, J.; Chen, X. The effect of PLGA-based hydrogel scaffold for improving the drug maximum-tolerated dose for in situ osteosarcoma treatment. *Colloids Surf. B Biointerfaces* **2018**, *172*, 387–394. [CrossRef] [PubMed]
87. Gong, C.; Wang, C.; Wang, Y.; Wu, Q.; Zhang, D.; Luo, F.; Qian, Z. Efficient inhibition of colorectal peritoneal carcinomatosis by drug loaded micelles in thermosensitive hydrogel composites. *Nanoscale* **2012**, *4*, 3095–3104. [CrossRef] [PubMed]
88. Liu, J.; Jiang, Y.; Cui, Y.; Xu, C.; Ji, X.; Luan, Y. Cytarabine-AOT catanionic vesicle-loaded biodegradable thermosensitive hydrogel as an efficient cytarabine delivery system. *Int. J. Pharm.* **2014**, *473*, 560–571. [CrossRef] [PubMed]
89. Xing, Y.; Chen, H.; Li, S.; Guo, X. In vitro and in vivo investigation of a novel two-phase delivery system of 2-methoxyestradiol liposomes hydrogel. *J. Liposome Res.* **2014**, *24*, 10–16. [CrossRef] [PubMed]
90. Jiang, L.; Ding, Y.; Xue, X.; Zhou, S.; Li, C.; Zhang, X.; Jiang, X. Entrapping multifunctional dendritic nanoparticles into a hydrogel for local therapeutic delivery and synergetic immunochemotherapy. *Nano Res.* **2018**, *11*, 6062–6073. [CrossRef]
91. Guo, X.; Cui, F.; Xing, Y.; Mei, Q.; Zhang, Z. Investigation of a new injectable thermosensitive hydrogel loading solid lipid nanoparticles. *Pharmazie* **2011**, *66*, 948–952. [CrossRef]
92. Ma, H.; He, C.; Cheng, Y.; Li, D.; Gong, Y.; Liu, J.; Tian, H.; Chen, X. PLK1shRNA and doxorubicin co-loaded thermosensitive PLGA-PEG-PLGA hydrogels for osteosarcoma treatment. *Biomaterials* **2014**, *35*, 8723–8734. [CrossRef]
93. Shen, W.; Chen, X.; Luan, J.; Wang, D.; Yu, L.; Ding, J. Sustained Codelivery of Cisplatin and Paclitaxel via an Injectable Prodrug Hydrogel for Ovarian Cancer Treatment. *ACS Appl. Mater. Interfaces* **2017**, *9*, 40031–40046. [CrossRef]
94. Liu, Y.; Xiao, L.; Joo, K.I.; Hu, B.; Fang, J.; Wang, P. In situ modulation of dendritic cells by injectable thermosensitive hydrogels for cancer vaccines in mice. *Biomacromolecules* **2014**, *15*, 3836–3845. [CrossRef] [PubMed]
95. Wang, Y.; Gong, C.; Yang, L.; Wu, Q.; Shi, S.; Shi, H.; Qian, Z.; Wei, Y. 5-FU-hydrogel inhibits colorectal peritoneal carcinomatosis and tumor growth in mice. *BMC Cancer* **2010**, *10*. [CrossRef] [PubMed]
96. Lin, X.; Deng, L.; Xu, Y.; Dong, A. Thermosensitive in situ hydrogel of paclitaxel conjugated poly(ε-caprolactone)-poly(ethylene glycol)-poly(ε-caprolactone). *Soft Matter* **2012**, *8*, 3470–3477. [CrossRef]

97. Liu, L.; Wu, Q.; Ma, X.; Xiong, D.; Gong, C.; Qian, Z.; Zhao, X.; Wei, Y. Camptothecine encapsulated composite drug delivery system for colorectal peritoneal carcinomatosis therapy: Biodegradable microsphere in thermosensitive hydrogel. *Colloids Surf. B Biointerfaces* **2013**, *106*, 93–101. [CrossRef] [PubMed]
98. Liu, M.; Huang, P.; Wang, W.; Feng, Z.; Zhang, J.; Deng, L.; Dong, A. An injectable nanocomposite hydrogel co-constructed with gold nanorods and paclitaxel-loaded nanoparticles for local chemo-photothermal synergetic cancer therapy. *J. Mater. Chem. B* **2019**, *7*, 2667–2677. [CrossRef]
99. Peng, M.; Xu, S.; Zhang, Y.; Zhang, L.; Huang, B.; Fu, S.; Xue, Z.; Da, Y.; Dai, Y.; Qiao, L.; et al. Thermosensitive injectable hydrogel enhances the antitumor effect of embelin in mouse hepatocellular carcinoma. *J. Pharm. Sci.* **2014**, *103*, 965–973. [CrossRef] [PubMed]
100. Wang, W.; Deng, L.; Xu, S.; Zhao, X.; Lv, N.; Zhang, G.; Gu, N.; Hu, R.; Zhang, J.; Liu, J.; et al. A reconstituted "two into one" thermosensitive hydrogel system assembled by drug-loaded amphiphilic copolymer nanoparticles for the local delivery of paclitaxel. *J. Mater. Chem. B* **2013**, *1*, 552–563. [CrossRef]
101. Huang, P.; Song, H.; Zhang, Y.; Liu, J.; Zhang, J.; Wang, W.; Li, C.; Kong, D. Bridging the Gap between Macroscale Drug Delivery Systems and Nanomedicines: A Nanoparticle-Assembled Thermosensitive Hydrogel for Peritumoral Chemotherapy. *ACS Appl. Mater. Interfaces* **2016**, *8*, 29323–29333. [CrossRef]
102. Huang, P.; Zhang, Y.; Wang, W.; Zhou, J.; Sun, Y.; Liu, J.; Kong, D.; Dong, A. Co-delivery of doxorubicin and 131 I by thermosensitive micellar-hydrogel for enhanced in situ synergetic chemoradiotherapy. *J. Control. Release* **2015**, *220*, 456–464. [CrossRef]
103. Zhu, W.; Li, Y.; Liu, L.; Chen, Y.; Xi, F. Supramolecular hydrogels as a universal scaffold for stepwise delivering Dox and Dox/cisplatin loaded block copolymer micelles. *International, J. Pharm.* **2012**, *437*, 11–19. [CrossRef]
104. Ren, L.; He, L.; Sun, T.; Dong, X.; Chen, Y.; Huang, J.; Wang, C. Dual-responsive supramolecular hydrogels from water-soluble PEG-grafted copolymers and cyclodextrin. *Macromol. Biosci.* **2009**, *9*, 902–910. [CrossRef] [PubMed]
105. Xu, S.; Wang, W.; Li, X.; Liu, J.; Dong, A.; Deng, L. Sustained release of PTX-incorporated nanoparticles synergized by burst release of DOX·HCl from thermosensitive modified PEG/PCL hydrogel to improve anti-tumor efficiency. *European, J. Pharm. Sci.* **2014**, *62*, 267–273. [CrossRef] [PubMed]
106. Zhu, W.; Li, Y.; Liu, L.; Chen, Y.; Wang, C.; Xi, F. Supramolecular hydrogels from cisplatin-loaded block copolymer nanoparticles and α-cyclodextrins with a stepwise delivery property. *Biomacromolecules* **2010**, *11*, 3086–3092. [CrossRef] [PubMed]
107. Kuang, H.; He, H.; Zhang, Z.; Qi, Y.; Xie, Z.; Jing, X.; Huang, Y. Injectable and biodegradable supramolecular hydrogels formed by nucleobase-terminated poly(ethylene oxide)s and α-cyclodextrin. *J. Mater. Chem. B* **2014**, *2*, 659–667. [CrossRef]
108. Liu, X.; Li, Z.; Loh, X.J.; Chen, K.; Wu, Y.L. Targeted and Sustained Corelease of Chemotherapeutics and Gene by Injectable Supramolecular Hydrogel for Drug-Resistant Cancer Therapy. *Macromol. Rapid Commun.* **2019**, *40*. [CrossRef] [PubMed]
109. Shahin, M.; Lavasanifar, A. Novel self-associating poly(ethylene oxide)-b-poly(ε-caprolactone) based drug conjugates and nano-containers for paclitaxel delivery. *Int. J. Pharm.* **2010**, *389*, 213–222. [CrossRef]
110. Ma, G.; Miao, B.; Song, C. Thermosensitive PCL-PEG-PCL hydrogels: Synthesis, characterization, and delivery of proteins. *J. Appl. Polym. Sci.* **2010**, *116*, 1985–1993. [CrossRef]
111. Wang, W.; Deng, L.; Liu, S.; Li, X.; Zhao, X.; Hu, R.; Zhang, J.; Han, H.; Dong, A. Adjustable degradation and drug release of a thermosensitive hydrogel based on a pendant cyclic ether modified poly(e-caprolactone) and poly(ethylene glycol)co-polymer. *Acta. Biomater.* **2012**, *8*, 3963–3973. [CrossRef]
112. Wang, W.; Song, H.; Zhang, J.; Li, P.; Li, C.; Wang, C.; Kong, D.; Zhao, Q. An injectable, thermosensitive and multicompartment hydrogel for simultaneous encapsulation and independent release of a drug cocktail as an effective combination therapy platform. *J. Control. Release* **2015**, *203*, 57–66. [CrossRef]
113. Yin, L.; Xu, S.; Feng, Z.; Deng, H.; Zhang, J.; Gao, H.; Deng, L.; Tang, H.; Dong, A. Supramolecular hydrogel based on high-solid-content mPECT nanoparticles and cyclodextrins for local and sustained drug delivery. *Biomater. Sci.* **2017**, *5*, 698–706. [CrossRef]
114. Kunz-Schughart, L.A.; Dubrovska, A.; Peitzsch, C.; Ewe, A.; Aigner, A.; Schellenburg, S.; Muders, M.H.; Hampel, S.; Cirillo, G.; Iemma, F.; et al. Nanoparticles for radiooncology: Mission, vision, challenges. *Biomaterials* **2017**, *120*, 155–184. [CrossRef] [PubMed]

115. Peng, C.L.; Shih, Y.H.; Liang, K.S.; Chiang, P.F.; Yeh, C.H.; Tang, I.C.; Yao, C.J.; Lee, S.Y.; Luo, T.Y.; Shieh, M.J. Development of in situ forming thermosensitive hydrogel for radiotherapy combined with chemotherapy in a mouse model of hepatocellular carcinoma. *Mol. Pharm.* **2013**, *10*, 1854–1864. [CrossRef] [PubMed]
116. Biolato, M.; Marrone, G.; Racco, S.; Di Stasi, C.; Miele, L.; Gasbarrini, G.; Landolfi, R.; Grieco, A. Transarterial chemoembolization (TACE) for unresectable HCC: A new life begins? *Eur. Rev. Med. Pharmacol. Sci.* **2010**, *14*, 356–362.
117. Lym, J.S.; Nguyen, Q.V.; Ahn, D.W.; Huynh, C.T.; Jae, H.J.; Kim, Y.I.; Lee, D.S. Sulfamethazine-based pH-sensitive hydrogels with potential application for transcatheter arterial chemoembolization therapy. *Acta. Biomater.* **2016**, *41*, 253–263. [CrossRef] [PubMed]
118. Huynh, C.T.; Nguyen, Q.V.; Lym, J.S.; Kim, B.S.; Huynh, D.P.; Jae, H.J.; Kim, Y.I.; Lee, D.S. Intraarterial gelation of injectable cationic pH/temperature-sensitive radiopaque embolic hydrogels in a rabbit hepatic tumor model and their potential application for liver cancer treatment. *RSC Adv.* **2016**, *6*, 47687–47697. [CrossRef]
119. Gil, M.S.; Thambi, T.; Phan, V.H.G.; Kim, S.H.; Lee, D.S. Injectable hydrogel-incorporated cancer cell-specific cisplatin releasing nanogels for targeted drug delivery. *J. Mater. Chem. B* **2017**, *5*, 7140–7152. [CrossRef]
120. Varghese, O.P.; Liu, J.; Sundaram, K.; Hilborn, J.; Oommen, O.P. Chondroitin sulfate derived theranostic nanoparticles for targeted drug delivery. *Biomater. Sci.* **2016**, *4*, 1310–1313. [CrossRef] [PubMed]
121. Andrgie, A.T.; Mekuria, S.L.; Addisu, K.D.; Hailemeskel, B.Z.; Hsu, W.H.; Tsai, H.C.; Lai, J.Y. Non-Anticoagulant Heparin Prodrug Loaded Biodegradable and Injectable Thermoresponsive Hydrogels for Enhanced Anti-Metastasis Therapy. *Macromol. Biosci.* **2019**, *19*. [CrossRef]
122. Phan, V.H.G.; Lee, E.; Maeng, J.H.; Thambi, T.; Kim, B.S.; Lee, D.; Lee, D.S. Pancreatic cancer therapy using an injectable nanobiohybrid hydrogel. *RSC Adv.* **2016**, *6*, 41644–41655. [CrossRef]
123. Nguyen, Q.V.; Lym, J.S.; Huynh, C.T.; Kim, B.S.; Jae, H.J.; Kim, Y.I.; Lee, D.S. A novel sulfamethazine-based pH-sensitive copolymer for injectable radiopaque embolic hydrogels with potential application in hepatocellular carcinoma therapy. *Polym. Chem.* **2016**, *7*, 5805–5818. [CrossRef]
124. Bobbala, S.; Tamboli, V.; McDowell, A.; Mitra, A.K.; Hook, S. Novel Injectable Pentablock Copolymer Based Thermoresponsive Hydrogels for Sustained Release Vaccines. *AAPS J.* **2016**, *18*, 261–269. [CrossRef]
125. Kim, S.H.; Tan, J.P.K.; Fukushima, K.; Nederberg, F.; Yang, Y.Y.; Waymouth, R.M.; Hedrick, J.L. Thermoresponsive nanostructured polycarbonate block copolymers as biodegradable therapeutic delivery carriers. *Biomaterials* **2011**, *32*, 5505–5514. [CrossRef] [PubMed]
126. Zawaneh, P.N.; Singh, S.P.; Padera, R.F.; Henderson, P.W.; Spector, J.A.; Putnam, D. Design of an injectable synthetic and biodegradable surgical biomaterial. *Proc. Natl. Acad. Sci. USA* **2010**, *107*, 11014–11019. [CrossRef] [PubMed]
127. Lee, A.L.Z.; Ng, V.W.L.; Gao, S.; Hedrick, J.L.; Yang, Y.Y. Injectable hydrogels from triblock copolymers of vitamin E-functionalized polycarbonate and poly(ethylene glycol) for subcutaneous delivery of antibodies for cancer therapy. *Adv. Funct. Mater.* **2014**, *24*, 1538–1550. [CrossRef]
128. Yang, C.; Lee, A.; Gao, S.; Liu, S.; Hedrick, J.L.; Yang, Y.Y. Hydrogels with prolonged release of therapeutic antibody: Block junction chemistry modification of 'ABA' copolymers provides superior anticancer efficacy. *J. Control. Release* **2019**, *293*, 193–200. [CrossRef] [PubMed]
129. Leprince, J.G.; Palin, W.M.; Hadis, M.A.; Devaux, J.; Leloup, G. Progress in dimethacrylate-based dental composite technology and curing efficiency. *Dent. Mater.* **2013**, *29*, 139–156. [CrossRef]
130. Cirillo, G.; Spataro, T.; Curcio, M.; Spizzirri, U.G.; Nicoletta, F.P.; Picci, N.; Iemma, F. Tunable thermo-responsive hydrogels: Synthesis, structural analysis and drug release studies. *Mater. Sci. Eng. C* **2015**, *48*, 499–510. [CrossRef]
131. Pal, A.; Vernon, B.L.; Nikkhah, M. Therapeutic neovascularization promoted by injectable hydrogels. *Bioact. Mater.* **2018**, *3*, 389–400. [CrossRef]
132. Fourniols, T.; Randolph, L.D.; Staub, A.; Vanvarenberg, K.; Leprince, J.G.; Préat, V.; Des Rieux, A.; Danhier, F. Temozolomide-loaded photopolymerizable PEG-DMA-based hydrogel for the treatment of glioblastoma. *J. Control. Release* **2015**, *210*, 95–104. [CrossRef]
133. Zhao, M.; Danhier, F.; Bastiancich, C.; Joudiou, N.; Ganipineni, L.P.; Tsakiris, N.; Gallez, B.; Rieux, A.D.; Jankovski, A.; Bianco, J.; et al. Post-resection treatment of glioblastoma with an injectable nanomedicine-loaded photopolymerizable hydrogel induces long-term survival. *Int. J. Pharm.* **2018**, *548*, 522–529. [CrossRef]

134. Zhang, H.; Zhu, X.; Ji, Y.; Jiao, X.; Chen, Q.; Hou, L.; Zhang, Z. Near-infrared-triggered in situ hybrid hydrogel system for synergistic cancer therapy. *J. Mater. Chem. B* **2015**, *3*, 6310–6326. [CrossRef]
135. Wu, H.; Song, L.; Chen, L.; Huang, Y.; Wu, Y.; Zang, F.; An, Y.; Lyu, H.; Ma, M.; Chen, J.; et al. Injectable thermosensitive magnetic nanoemulsion hydrogel for multimodal-imaging-guided accurate thermoablative cancer therapy. *Nanoscale* **2017**, *9*, 16175–16182. [CrossRef] [PubMed]
136. Khang, M.K.; Zhou, J.; Huang, Y.; Hakamivala, A.; Tang, L. Preparation of a novel injectable in situ-gelling nanoparticle with applications in controlled protein release and cancer cell entrapment. *RSC Adv.* **2018**, *8*, 34625–34633. [CrossRef]
137. Ishii, S.; Kaneko, J.; Nagasaki, Y. Development of a long-acting, protein-loaded, redox-active, injectable gel formed by a polyion complex for local protein therapeutics. *Biomaterials* **2016**, *84*, 210–218. [CrossRef]
138. Xu, X.; Huang, Z.; Zhang, X.; He, S.; Sun, X.; Shen, Y.; Yan, M.; Zhao, C. Injectable, NIR/pH-Responsive Nanocomposite Hydrogel as Long-Acting Implant for Chemophotothermal Synergistic Cancer Therapy. *ACS Appl. Mater. Interfaces* **2017**, *9*, 20361–20375. [CrossRef]
139. Wu, Y.; Wang, H.; Gao, F.; Xu, Z.; Dai, F.; Liu, W. An Injectable Supramolecular Polymer Nanocomposite Hydrogel for Prevention of Breast Cancer Recurrence with Theranostic and Mammoplastic Functions. *Adv. Funct. Mater.* **2018**, *28*. [CrossRef]
140. Song, Z.; Han, Z.; Lv, S.; Chen, C.; Chen, L.; Yin, L.; Cheng, J. Synthetic polypeptides: From polymer design to supramolecular assembly and biomedical application. *Chem. Soc. Rev.* **2017**, *46*, 6570–6599. [CrossRef]
141. Shen, Y.; Fu, X.; Fu, W.; Li, Z. Biodegradable stimuli-responsive polypeptide materials prepared by ring opening polymerization. *Chem. Soc. Rev.* **2015**, *44*, 612–622. [CrossRef]
142. Deming, T.J. Synthesis of Side-Chain Modified Polypeptides. *Chem. Rev.* **2016**, *116*, 786–808. [CrossRef]
143. Maude, S.; Ingham, E.; Aggeli, A. Biomimetic self-assembling peptides as scaffolds for soft tissue engineering. *Nanomedicine* **2013**, *8*, 823–847. [CrossRef]
144. Szkolar, L.; Guilbaud, J.B.; Miller, A.F.; Gough, J.E.; Saiani, A. Enzymatically triggered peptide hydrogels for 3D cell encapsulation and culture. *J. Pept. Sci.* **2014**, *20*, 578–584. [CrossRef] [PubMed]
145. Rodriguez, A.L.; Wang, T.Y.; Bruggeman, K.F.; Li, R.; Williams, R.J.; Parish, C.L.; Nisbet, D.R. Tailoring minimalist self-assembling peptides for localized viral vector gene delivery. *Nano Res.* **2016**, *9*, 674–684. [CrossRef]
146. Abbas, M.; Xing, R.; Zhang, N.; Zou, Q.; Yan, X. Antitumor Photodynamic Therapy Based on Dipeptide Fibrous Hydrogels with Incorporation of Photosensitive Drugs. *ACS Biomater. Sci. Eng.* **2018**, *4*, 2046–2052. [CrossRef]
147. Xing, R.; Li, S.; Zhang, N.; Shen, G.; Möhwald, H.; Yan, X. Self-Assembled Injectable Peptide Hydrogels Capable of Triggering Antitumor Immune Response. *Biomacromolecules* **2017**, *18*, 3514–3523. [CrossRef] [PubMed]
148. Weiden, J.; Voerman, D.; Dölen, Y.; Das, R.K.; Van Duffelen, A.; Hammink, R.; Eggermont, L.J.; Rowan, A.E.; Tel, J.; Figdor, C.G. Injectable biomimetic hydrogels as tools for efficient T Cell expansion and delivery. *Front. Immunol.* **2018**, *9*. [CrossRef] [PubMed]
149. Yamada, Y.; Chowdhury, A.; Schneider, J.P.; Stetler-Stevenson, W.G. Macromolecule-Network Electrostatics Controlling Delivery of the Biotherapeutic Cell Modulator TIMP-2. *Biomacromolecules* **2018**, *19*, 1285–1293. [CrossRef]
150. Qi, Y.; Min, H.; Mujeeb, A.; Zhang, Y.; Han, X.; Zhao, X.; Anderson, G.J.; Zhao, Y.; Nie, G. Injectable Hexapeptide Hydrogel for Localized Chemotherapy Prevents Breast Cancer Recurrence. *ACS Appl. Mater. Interfaces* **2018**, *10*, 6972–6981. [CrossRef]
151. Mei, L.; Xu, K.; Zhai, Z.; He, S.; Zhu, T.; Zhong, W. Doxorubicin-reinforced supramolecular hydrogels of RGD-derived peptide conjugates for pH-responsive drug delivery. *Org. Biomol. Chem.* **2019**, *17*, 3853–3860. [CrossRef]
152. Leach, D.G.; Dharmaraj, N.; Piotrowski, S.L.; Lopez-Silva, T.L.; Lei, Y.L.; Sikora, A.G.; Young, S.; Hartgerink, J.D. STINGel: Controlled release of a cyclic dinucleotide for enhanced cancer immunotherapy. *Biomaterials* **2018**, *163*, 67–75. [CrossRef]
153. Jin, H.; Wan, C.; Zou, Z.; Zhao, G.; Zhang, L.; Geng, Y.; Chen, T.; Huang, A.; Jiang, F.; Feng, J.P.; et al. Tumor Ablation and Therapeutic Immunity Induction by an Injectable Peptide Hydrogel. *ACS Nano* **2018**, *12*, 3295–3310. [CrossRef]

154. Hu, C.; Liu, X.; Ran, W.; Meng, J.; Zhai, Y.; Zhang, P.; Yin, Q.; Yu, H.; Zhang, Z.; Li, Y. Regulating cancer associated fibroblasts with losartan-loaded injectable peptide hydrogel to potentiate chemotherapy in inhibiting growth and lung metastasis of triple negative breast cancer. *Biomaterials* **2017**, *144*, 60–72. [CrossRef] [PubMed]
155. Li, L.; Gu, J.; Zhang, J.; Xie, Z.; Lu, Y.; Shen, L.; Dong, Q.; Wang, Y. Injectable and biodegradable pH-responsive hydrogels for localized and sustained treatment of human fibrosarcoma. *ACS Appl. Mater. Interfaces* **2015**, *7*, 8033–8040. [CrossRef] [PubMed]
156. Yu, S.; Zhang, D.; He, C.; Sun, W.; Cao, R.; Cui, S.; Deng, M.; Gu, Z.; Chen, X. Injectable Thermosensitive Polypeptide-Based CDDP-Complexed Hydrogel for Improving Localized Antitumor Efficacy. *Biomacromolecules* **2017**, *18*, 4341–4348. [CrossRef] [PubMed]
157. Song, H.; Huang, P.; Niu, J.; Shi, G.; Zhang, C.; Kong, D.; Wang, W. Injectable polypeptide hydrogel for dual-delivery of antigen and TLR3 agonist to modulate dendritic cells in vivo and enhance potent cytotoxic T-lymphocyte response against melanoma. *Biomaterials* **2018**, *159*, 119–129. [CrossRef]
158. Wei, L.; Chen, J.; Zhao, S.; Ding, J.; Chen, X. Thermo-sensitive polypeptide hydrogel for locally sequential delivery of two-pronged antitumor drugs. *Acta. Biomater.* **2017**, *58*, 44–53. [CrossRef]
159. Cheng, Y.; He, C.; Ding, J.; Xiao, C.; Zhuang, X.; Chen, X. Thermosensitive hydrogels based on polypeptides for localized and sustained delivery of anticancer drugs. *Biomaterials* **2013**, *34*, 10338–10347. [CrossRef]
160. Wu, C.; Li, R.; Yin, Y.; Wang, J.; Zhang, L.; Zhong, W. Redox-responsive supramolecular hydrogel based on 10-hydroxy camptothecin-peptide covalent conjugates with high loading capacity for drug delivery. *Mater. Sci. Eng. C* **2017**, *76*, 196–202. [CrossRef]
161. Wang, H.; Lv, L.; Xu, G.; Yang, C.; Sun, J.; Yang, Z. Molecular hydrogelators consist of Taxol and short peptides/amino acids. *J. Mater. Chem.* **2012**, *22*, 16933–16938. [CrossRef]
162. Singh, M.; Kundu, S.; Reddy, M.A.; Sreekanth, V.; Motiani, R.K.; Sengupta, S.; Srivastava, A.; Bajaj, A. Injectable small molecule hydrogel as a potential nanocarrier for localized and sustained in vivo delivery of doxorubicin. *Nanoscale* **2014**, *6*, 12849–12855. [CrossRef]
163. Lin, Q.; Yang, Y.; Hu, Q.; Guo, Z.; Liu, T.; Xu, J.; Wu, J.; Kirk, T.B.; Ma, D.; Xue, W. Injectable supramolecular hydrogel formed from α-cyclodextrin and PEGylated arginine-functionalized poly(L-lysine) dendron for sustained MMP-9 shRNA plasmid delivery. *Acta. Biomater.* **2017**, *49*, 456–471. [CrossRef]
164. Ma, Y.; Fu, X.; Shen, Y.; Fu, W.; Li, Z. Irreversible low critical solution temperature behaviors of thermal-responsive OEGylated poly(L-cysteine) containing disulfide bonds. *Macromolecules* **2014**, *47*, 4684–4689. [CrossRef]
165. Zhang, S.; Fu, W.; Li, Z. Supramolecular hydrogels assembled from nonionic poly(ethylene glycol)-b-polypeptide diblocks containing OEGylated poly-l-glutamate. *Polym. Chem.* **2014**, *5*, 3346–3351. [CrossRef]
166. Singh, S.K.; Singh, S.; Wlillard, J.; Singh, R. Drug delivery approaches for breast cancer. *Int. J. Nanomed.* **2017**, *12*, 6205–6218. [CrossRef] [PubMed]
167. Kesharwani, P.; Gothwal, A.; Iyer, A.K.; Jain, K.; Chourasia, M.K.; Gupta, U. Dendrimer nanohybrid carrier systems: An expanding horizon for targeted drug and gene delivery. *Drug Discov. Today* **2018**, *23*, 300–314. [CrossRef] [PubMed]
168. Kitchens, K.M.; El-Sayed, M.E.H.; Ghandehari, H. Transepithelial and endothelial transport of poly(amidoamine) dendrimers. *Adv. Drug Deliv. Rev.* **2005**, *57*, 2163–2176. [CrossRef] [PubMed]
169. Northfelt, D.W.; Dezube, B.J.; Thommes, J.A.; Miller, B.J.; Fischl, M.A.; Friedman-Kien, A.; Kaplan, L.D.; Du Mond, C.; Mamelok, R.D.; Henry, D.H. Pegylated-liposomal doxorubicin versus doxorubicin, bleomycin, and vincristine in the treatment of AIDS-related Kaposi's sarcoma: Results of a randomized phase III clinical trial. *J. Clin. Oncol.* **1998**, *16*, 2445–2451. [CrossRef] [PubMed]
170. Lo, Y.W.; Sheu, M.T.; Chiang, W.H.; Chiu, Y.L.; Tu, C.M.; Wang, W.Y.; Wu, M.H.; Wang, Y.C.; Lu, M.; Ho, H.O. In situ chemically crosslinked injectable hydrogels for the subcutaneous delivery of trastuzumab to treat breast cancer. *Acta. Biomater.* **2019**, *86*, 280–290. [CrossRef]
171. Zhang, H.; Zhao, C.; Cao, H.; Wang, G.; Song, L.; Niu, G.; Yang, H.; Ma, J.; Zhu, S. Hyperbranched poly(amine-ester) based hydrogels for controlled multi-drug release in combination chemotherapy. *Biomaterials* **2010**, *31*, 5445–5454. [CrossRef]
172. Xu, L.; Cooper, R.C.; Wang, J.; Yeudall, W.A.; Yang, H. Synthesis and Application of Injectable Bioorthogonal Dendrimer Hydrogels for Local Drug Delivery. *ACS Biomater. Sci. Eng.* **2017**, *3*, 1641–1653. [CrossRef]

173. Patil, S.S.; Shinde, V.S.; Misra, R.D.K. pH and reduction dual-stimuli-responsive PEGDA/PAMAM injectable network hydrogels via aza-michael addition for anticancer drug delivery. *J. Polym. Sci. A Polym. Chem.* **2018**, *56*, 2080–2095. [CrossRef]
174. Yang, W.J.; Zhou, P.; Liang, L.; Cao, Y.; Qiao, J.; Li, X.; Teng, Z.; Wang, L. Nanogel-Incorporated Injectable Hydrogel for Synergistic Therapy Based on Sequential Local Delivery of Combretastatin-A4 Phosphate (CA4P) and Doxorubicin (DOX). *ACS Appl. Mater. Interfaces* **2018**, *10*, 18560–18573. [CrossRef] [PubMed]
175. Kharkar, P.M.; Kloxin, A.M.; Kiick, K.L. Dually degradable click hydrogels for controlled degradation and protein release. *J. Mater. Chem. B* **2014**, *2*, 5511–5521. [CrossRef] [PubMed]
176. Huang, Z.; Delparastan, P.; Burch, P.; Cheng, J.; Cao, Y.; Messersmith, P.B. Injectable dynamic covalent hydrogels of boronic acid polymers cross-linked by bioactive plant-derived polyphenols. *Biomater. Sci.* **2018**, *6*, 2487–2495. [CrossRef] [PubMed]
177. Gao, W.; Liang, Y.; Peng, X.; Hu, Y.; Zhang, L.; Wu, H.; He, B. In situ injection of phenylboronic acid based low molecular weight gels for efficient chemotherapy. *Biomaterials* **2016**, *105*, 1–11. [CrossRef] [PubMed]
178. Seib, F.P.; Tsurkan, M.; Freudenberg, U.; Kaplan, D.L.; Werner, C. Heparin-Modified Polyethylene Glycol Microparticle Aggregates for Focal Cancer Chemotherapy. *ACS Biomater. Sci. Eng.* **2016**, *2*, 2287–2293. [CrossRef]
179. Fang, Y.; Xue, J.; Gao, S.; Lu, A.; Yang, D.; Jiang, H.; He, Y.; Shi, K. Cleavable PEGylation: A strategy for overcoming the "PEG dilemma" in efficient drug delivery. *Drug Deliv.* **2017**, *24*, 22–32. [CrossRef] [PubMed]
180. Bastiancich, C.; Bianco, J.; Vanvarenberg, K.; Ucakar, B.; Joudiou, N.; Gallez, B.; Bastiat, G.; Lagarce, F.; Préat, V.; Danhier, F. Injectable nanomedicine hydrogel for local chemotherapy of glioblastoma after surgical resection. *J. Control. Release* **2017**, *264*, 45–54. [CrossRef] [PubMed]
181. Ye, Y.; Hu, X. A pH-sensitive injectable nanoparticle composite hydrogel for anticancer drug delivery. *J. Nanomater.* **2016**, *2016*. [CrossRef]
182. Makharza, S.; Cirillo, G.; Bachmatiuk, A.; Ibrahim, I.; Ioannides, N.; Trzebicka, B.; Hampel, S.; Ruemmeli, M.H. Graphene oxide-based drug delivery vehicles: Functionalization, characterization, and cytotoxicity evaluation. *J. Nanoparticle Res.* **2013**, *15*. [CrossRef]
183. Vittorio, O.; Le Grand, M.; Makharza, S.A.; Curcio, M.; Tucci, P.; Iemma, F.; Nicoletta, F.P.; Hampel, S.; Cirillo, G. Doxorubicin synergism and resistance reversal in human neuroblastoma BE(2)C cell lines: An in vitro study with dextran-catechin nanohybrids. *Eur. J. Pharm. Biopharm.* **2018**, *122*, 176–185. [CrossRef]
184. Lerra, L.; Farfalla, A.; Sanz, B.; Cirillo, G.; Vittorio, O.; Voli, F.; Grand, M.L.; Curcio, M.; Nicoletta, F.P.; Dubrovska, A.; et al. Graphene oxide functional nanohybrids with magnetic nanoparticles for improved vectorization of doxorubicin to neuroblastoma cells. *Pharmaceutics* **2019**, *11*. [CrossRef] [PubMed]
185. Thambi, T.; Phan, V.H.G.; Lee, D.S. Stimuli-Sensitive Injectable Hydrogels Based on Polysaccharides and Their Biomedical Applications. *Macromol. Rapid Commun.* **2016**, *37*, 1881–1896. [CrossRef] [PubMed]
186. Spizzirri, U.G.; Altimari, I.; Puoci, F.; Parisi, O.I.; Iemma, F.; Picci, N. Innovative antioxidant thermo-responsive hydrogels by radical grafting of catechin on inulin chain. *Carbohydr. Polym.* **2011**, *84*, 517–523. [CrossRef]
187. Ahsan, S.M.; Thomas, M.; Reddy, K.K.; Sooraparaju, S.G.; Asthana, A.; Bhatnagar, I. Chitosan as biomaterial in drug delivery and tissue engineering. *Int. J. Biol. Macromol.* **2018**, *110*, 97–109. [CrossRef] [PubMed]
188. Kozen, B.G.; Kircher, S.J.; Henao, J.; Godinez, F.S.; Johnson, A.S. An alternative hemostatic dressing: Comparison of CELOX, HemCon, and QuikClot. *Acad. Emerg. Med.* **2008**, *15*, 74–81. [CrossRef] [PubMed]
189. Ueno, H.; Mori, T.; Fujinaga, T. Topical formulations and wound healing applications of chitosan. *Adv. Drug Deliv. Rev.* **2001**, *52*, 105–115. [CrossRef]
190. Karimi, A.R.; Khodadadi, A.; Hadizadeh, M. A nanoporous photosensitizing hydrogel based on chitosan cross-linked by zinc phthalocyanine: An injectable and pH-stimuli responsive system for effective cancer therapy. *RSC Adv.* **2016**, *6*, 91445–91452. [CrossRef]
191. Abdel-Bar, H.M.; Abdel-Reheem, A.Y.; Osman, R.; Awad, G.A.S.; Mortada, N. Defining cisplatin incorporation properties in thermosensitive injectable biodegradable hydrogel for sustained delivery and enhanced cytotoxicity. *Int. J. Pharm.* **2014**, *477*, 623–630. [CrossRef]
192. Fathi, M.; Alami-Milani, M.; Geranmayeh, M.H.; Barar, J.; Erfan-Niya, H.; Omidi, Y. Dual thermo-and pH-sensitive injectable hydrogels of chitosan/(poly(N-isopropylacrylamide-co-itaconic acid)) for doxorubicin delivery in breast cancer. *Int. J. Biol. Macromol.* **2019**, *128*, 957–964. [CrossRef]
193. Zhang, W.; Jin, X.; Li, H.; Zhang, R.R.; Wu, C.W. Injectable and body temperature sensitive hydrogels based on chitosan and hyaluronic acid for pH sensitive drug release. *Carbohydr. Polym.* **2018**, *186*, 82–90. [CrossRef]

194. Saeednia, L.; Yao, L.; Cluff, K.; Asmatulu, R. Sustained Releasing of Methotrexate from Injectable and Thermosensitive Chitosan-Carbon Nanotube Hybrid Hydrogels Effectively Controls Tumor Cell Growth. *ACS Omega* **2019**, *4*, 4040–4048. [CrossRef] [PubMed]
195. Huang, F.Y.J.; Hung, C.C.; Chang, C.W.; Chao, J.H.; Hsieh, B.T. Evaluation of injectable chitosan-based co-crosslinking hydrogel for local delivery of 188Re-LIPO-DOX to breast-tumor-bearing mouse model. *Anticancer. Res.* **2018**, *38*, 4651–4659. [CrossRef] [PubMed]
196. Alexander, A.; Ajazuddin, A.; Khan, J.; Saraf, S. Formulation and evaluation of chitosan-based long-acting injectable hydrogel for PEGylated melphalan conjugate. *J. Pharm. Pharmacol.* **2014**, *66*, 1240–1250. [CrossRef] [PubMed]
197. López-Noriega, A.; Hastings, C.L.; Ozbakir, B.; O'Donnell, K.E.; O'Brien, F.J.; Storm, G.; Hennink, W.E.; Duffy, G.P.; Ruiz-Hernández, E. Hyperthermia-Induced Drug Delivery from Thermosensitive Liposomes Encapsulated in an Injectable Hydrogel for Local Chemotherapy. *Adv. Healthc. Mater.* **2014**, *3*, 854–859. [CrossRef] [PubMed]
198. Xing, J.; Qi, X.; Jiang, Y.; Zhu, X.; Zhang, Z.; Qin, X.; Wu, Z. Topotecan hydrochloride liposomes incorporated into thermosensitive hydrogel for sustained and efficient in situ therapy of H22 tumor in Kunming mice. *Pharm. Dev. Technol.* **2015**, *20*, 812–819. [CrossRef] [PubMed]
199. Huang, F.Y.J.; Gan, G.Y.; Lin, W.Y.; Huang, L.K.; Luo, T.Y.; Hong, J.J.; Hsieh, B.T. Investigation of the local delivery of an intelligent chitosan-based 188Re thermosensitive in situ-forming hydrogel in an orthotopic hepatoma-bearing rat model. *J. Radioanal. Nucl. Chem.* **2014**, *299*, 31–40. [CrossRef]
200. Zhang, D.; Sun, P.; Li, P.; Xue, A.; Zhang, X.; Zhang, H.; Jin, X. A magnetic chitosan hydrogel for sustained and prolonged delivery of Bacillus Calmette-Guérin in the treatment of bladder cancer. *Biomaterials* **2013**, *34*, 10258–10266. [CrossRef] [PubMed]
201. Le Renard, P.E.; Jordan, O.; Faes, A.; Petri-Fink, A.; Hofmann, H.; Rüfenacht, D.; Bosman, F.; Buchegger, F.; Doelker, E. The in vivo performance of magnetic particle-loaded injectable, in situ gelling, carriers for the delivery of local hyperthermia. *Biomaterials* **2010**, *31*, 691–705. [CrossRef]
202. Zhu, X.; Zhang, H.; Huang, H.; Zhang, Y.; Hou, L.; Zhang, Z. Functionalized graphene oxide-based thermosensitive hydrogel for magnetic hyperthermia therapy on tumors. *Nanotechnology* **2015**, *26*. [CrossRef] [PubMed]
203. Saeednia, L.; Yao, L.; Berndt, M.; Cluff, K.; Asmatulu, R. Structural and biological properties of thermosensitive chitosan–graphene hybrid hydrogels for sustained drug delivery applications. *J. Biomed. Mater. Res. A* **2017**, *105*, 2381–2390. [CrossRef]
204. Fletcher, N.A.; Krebs, M.D. Sustained delivery of anti-VEGF from injectable hydrogel systems provides a prolonged decrease of endothelial cell proliferation and angiogenesis: In vitro. *RSC Adv.* **2018**, *8*, 8999–9005. [CrossRef]
205. Chen, C.H.; Kuo, C.Y.; Chen, S.H.; Mao, S.H.; Chang, C.Y.; Shalumon, K.T.; Chen, J.P. Thermosensitive injectable hydrogel for simultaneous intraperitoneal delivery of doxorubicin and prevention of peritoneal adhesion. *Int. J. Mol. Sci.* **2018**, *19*. [CrossRef] [PubMed]
206. Fong, Y.T.; Chen, C.H.; Chen, J.P. Intratumoral delivery of doxorubicin on folate-conjugated graphene oxide by in-situ forming thermo-sensitive hydrogel for breast cancer therapy. *Nanomaterials* **2017**, *7*. [CrossRef] [PubMed]
207. Hyun, H.; Park, M.H.; Lim, W.; Kim, S.Y.; Jo, D.; Jung, J.S.; Jo, G.; Um, S.; Lee, D.W.; Yang, D.H. Injectable visible light-cured glycol chitosan hydrogels with controlled release of anticancer drugs for local cancer therapy in vivo: A feasible study. *Artif. Cells Nanomed. Biotechnol.* **2018**, *46*, 874–882. [CrossRef] [PubMed]
208. Zhou, X.; Li, Y.; Chen, S.; Fu, Y.N.; Wang, S.; Li, G.; Tao, L.; Wei, Y.; Wang, X.; Liang, J.F. Dynamic agent of an injectable and self-healing drug-loaded hydrogel for embolization therapy. *Colloids Surf. B Biointerfaces* **2018**, *172*, 601–607. [CrossRef] [PubMed]
209. Xia, L.Y.; Zhang, X.; Cao, M.; Chen, Z.; Wu, F.G. Enhanced Fluorescence Emission and Singlet Oxygen Generation of Photosensitizers Embedded in Injectable Hydrogels for Imaging-Guided Photodynamic Cancer Therapy. *Biomacromolecules* **2017**, *18*, 3073–3081. [CrossRef]
210. Xie, W.; Gao, Q.; Guo, Z.; Wang, D.; Gao, F.; Wang, X.; Wei, Y.; Zhao, L. Injectable and self-healing thermosensitive magnetic hydrogel for asynchronous control release of doxorubicin and docetaxel to treat triple-negative breast cancer. *ACS Appl. Mater. Interfaces* **2017**, *9*, 33660–33673. [CrossRef] [PubMed]

211. Liu, Z.; Xu, G.; Wang, C.; Li, C.; Yao, P. Shear-responsive injectable supramolecular hydrogel releasing doxorubicin loaded micelles with pH-sensitivity for local tumor chemotherapy. *Int. J. Pharm.* **2017**, *530*, 53–62. [CrossRef]
212. Wang, Q.Q.; Kong, M.; An, Y.; Liu, Y.; Li, J.J.; Zhou, X.; Feng, C.; Li, J.; Jiang, S.Y.; Cheng, X.J.; et al. Hydroxybutyl chitosan thermo-sensitive hydrogel: A potential drug delivery system. *J. Mater. Sci.* **2013**, *48*, 5614–5623. [CrossRef]
213. Khan, S.; Akhtar, N.; Minhas, M.U.; Badshah, S.F. pH/Thermo-Dual Responsive Tunable In Situ Cross-Linkable Depot Injectable Hydrogels Based on Poly(N-Isopropylacrylamide)/Carboxymethyl Chitosan with Potential of Controlled Localized and Systemic Drug Delivery. *AAPS Pharm. Sci. Tech.* **2019**, *20*. [CrossRef]
214. Wang, H.; Song, F.; Chen, Q.; Hu, R.; Jiang, Z.; Yang, Y.; Han, B. Antitumor and antimetastasis effects of macerating solutions from an injectable chitosan-based hydrogel on hepatocarcinoma. *J. Biomed. Mater. Res.-A* **2015**, *103*, 3879–3885. [CrossRef] [PubMed]
215. Qu, J.; Zhao, X.; Ma, P.X.; Guo, B. pH-responsive self-healing injectable hydrogel based on N-carboxyethyl chitosan for hepatocellular carcinoma therapy. *Acta. Biomater.* **2017**, *58*, 168–180. [CrossRef] [PubMed]
216. Chen, X.; Fan, M.; Tan, H.; Ren, B.; Yuan, G.; Jia, Y.; Li, J.; Xiong, D.; Xing, X.; Niu, X.; et al. Magnetic and self-healing chitosan-alginate hydrogel encapsulated gelatin microspheres via covalent cross-linking for drug delivery. *Mater. Sci. Eng. C* **2019**, *101*, 619–629. [CrossRef] [PubMed]
217. Qian, C.; Zhang, T.; Gravesande, J.; Baysah, C.; Song, X.; Xing, J. Injectable and self-healing polysaccharide-based hydrogel for pH-responsive drug release. *Int. J. Biol. Macromol.* **2019**, *123*, 140–148. [CrossRef] [PubMed]
218. Gao, N.; Lü, S.; Gao, C.; Wang, X.; Xu, X.; Bai, X.; Feng, C.; Liu, M. Injectable shell-crosslinked F127 micelle/hydrogel composites with pH and redox sensitivity for combined release of anticancer drugs. *Chem. Eng. J.* **2016**, *287*, 20–29. [CrossRef]
219. Li, J.; Hu, W.; Zhang, Y.; Tan, H.; Yan, X.; Zhao, L.; Liang, H. PH and glucose dually responsive injectable hydrogel prepared by in situ crosslinking of phenylboronic modified chitosan and oxidized dextran. *J. Polym. Sci. A Polym. Chem.* **2015**, *53*, 1235–1244. [CrossRef]
220. Liang, Y.; Zhao, X.; Ma, P.X.; Guo, B.; Du, Y.; Han, X. pH-responsive injectable hydrogels with mucosal adhesiveness based on chitosan-grafted-dihydrocaffeic acid and oxidized pullulan for localized drug delivery. *J. Colloid Interface Sci.* **2019**, *536*, 224–234. [CrossRef] [PubMed]
221. Jalalvandi, E.; Shavandi, A. In situ-forming and pH-responsive hydrogel based on chitosan for vaginal delivery of therapeutic agents. *J. Mater. Sci. Mater. Med.* **2018**, *29*. [CrossRef] [PubMed]
222. Shi, J.; Guobao, W.; Chen, H.; Zhong, W.; Qiu, X.; Xing, M.M.Q. Schiff based injectable hydrogel for in situ pH-triggered delivery of doxorubicin for breast tumor treatment. *Polym. Chem.* **2014**, *5*, 6180–6189. [CrossRef]
223. Zahedi, P.; De Souza, R.; Piquette-Miller, M.; Allen, C. Docetaxel distribution following intraperitoneal administration in mice. Journal of pharmacy & pharmaceutical sciences: A publication of the Canadian Society for Pharmaceutical Sciences. *J. Pharm. Pharm. Sci.* **2011**, *14*, 90–99.
224. Yavvari, P.S.; Pal, S.; Kumar, S.; Kar, A.; Awasthi, A.K.; Naaz, A.; Srivastava, A.; Bajaj, A. Injectable, Self-Healing Chimeric Catechol-Fe(III) Hydrogel for Localized Combination Cancer Therapy. *ACS Biomater. Sci. Eng.* **2017**, *3*, 3404–3413. [CrossRef]
225. Belali, S.; Karimi, A.R.; Hadizadeh, M. Cell-specific and pH-sensitive nanostructure hydrogel based on chitosan as a photosensitizer carrier for selective photodynamic therapy. *Int. J. Biol. Macromol.* **2018**, *110*, 437–448. [CrossRef] [PubMed]
226. Ning, P.; Lü, S.; Bai, X.; Wu, X.; Gao, C.; Wen, N.; Liu, M. High encapsulation and localized delivery of curcumin from an injectable hydrogel. *Mater. Sci. Eng. C* **2018**, *83*, 121–129. [CrossRef] [PubMed]
227. Burdick, J.A. Injectable gels for tissue/organ repair. *Biomed. Mater.* **2012**, *7*. [CrossRef] [PubMed]
228. Seliktar, D. Designing cell-compatible hydrogels for biomedical applications. *Science* **2012**, *336*, 1124–1128. [CrossRef] [PubMed]
229. Mitragotri, S.; Burke, P.A.; Langer, R. Overcoming the challenges in administering biopharmaceuticals: Formulation and delivery strategies. *Nat. Rev. Drug Discov.* **2014**, *13*, 655–672. [CrossRef] [PubMed]
230. Burdick, J.A.; Prestwich, G.D. Hyaluronic acid hydrogels for biomedical applications. *Adv. Mater.* **2011**, *23*, H41–H56. [CrossRef]

231. Chen, Y.Y.; Wu, H.C.; Sun, J.S.; Dong, G.C.; Wang, T.W. Injectable and thermoresponsive self-assembled nanocomposite hydrogel for long-term anticancer drug delivery. *Langmuir* **2013**, *29*, 3721–3729. [CrossRef]
232. Zhao, Y.; Yan, H.; Qiao, S.; Zhang, L.; Wang, T.; Meng, Q.; Chen, X.; Lin, F.H.; Guo, K.; Li, C.; et al. Hydrogels bearing bioengineered mimetic embryonic microenvironments for tumor reversion. *J. Mater. Chem. B* **2016**, *4*, 6183–6191. [CrossRef]
233. Ohta, S.; Hiramoto, S.; Amano, Y.; Emoto, S.; Yamaguchi, H.; Ishigami, H.; Kitayama, J.; Ito, T. Intraperitoneal Delivery of Cisplatin via a Hyaluronan-Based Nanogel/in Situ Cross-Linkable Hydrogel Hybrid System for Peritoneal Dissemination of Gastric Cancer. *Mol. Pharm.* **2017**, *14*, 3105–3113. [CrossRef]
234. Xu, K.; Lee, F.; Gao, S.J.; Chung, J.E.; Yano, H.; Kurisawa, M. Injectable hyaluronic acid-tyramine hydrogels incorporating interferon-α2a for liver cancer therapy. *J. Control. Release* **2013**, *166*, 203–210. [CrossRef] [PubMed]
235. Xu, K.; Lee, F.; Gao, S.; Tan, M.H.; Kurisawa, M. Hyaluronidase-incorporated hyaluronic acid-tyramine hydrogels for the sustained release of trastuzumab. *J. Control. Release* **2015**, *216*, 47–55. [CrossRef] [PubMed]
236. Ueda, K.; Akiba, J.; Ogasawara, S.; Todoroki, K.; Nakayama, M.; Sumi, A.; Kusano, H.; Sanada, S.; Suekane, S.; Xu, K.; et al. Growth inhibitory effect of an injectable hyaluronic acid-tyramine hydrogels incorporating human natural interferon-α and sorafenib on renal cell carcinoma cells. *Acta. Biomater.* **2016**, *29*, 103–111. [CrossRef] [PubMed]
237. He, M.; Sui, J.; Chen, Y.; Bian, S.; Cui, Y.; Zhou, C.; Sun, Y.; Liang, J.; Fan, Y.; Zhang, X. Localized multidrug co-delivery by injectable self-crosslinking hydrogel for synergistic combinational chemotherapy. *J. Mater. Chem. B* **2017**, *5*, 4852–4862. [CrossRef]
238. Chen, X.; Liu, Z. A pH-Responsive Hydrogel Based on a Tumor-Targeting Mesoporous Silica Nanocomposite for Sustained Cancer Labeling and Therapy. *Macromol. Rapid Commun.* **2016**, *37*, 1533–1539. [CrossRef] [PubMed]
239. Chen, X.; Liu, Z.; Parker, S.G.; Zhang, X.; Gooding, J.J.; Ru, Y.; Liu, Y.; Zhou, Y. Light-Induced Hydrogel Based on Tumor-Targeting Mesoporous Silica Nanoparticles as a Theranostic Platform for Sustained Cancer Treatment. *ACS Appl. Mater. Interfaces* **2016**, *8*, 15857–15863. [CrossRef] [PubMed]
240. Ranga, A.; Lutolf, M.P.; Hilborn, J.; Ossipov, D.A. Hyaluronic Acid Hydrogels Formed in Situ by Transglutaminase-Catalyzed Reaction. *Biomacromolecules* **2016**, *17*, 1553–1560. [CrossRef]
241. Moon, R.J.; Martini, A.; Nairn, J.; Simonsen, J.; Youngblood, J. Cellulose nanomaterials review: Structure, properties and nanocomposites. *Chem. Soc. Rev.* **2011**, *40*, 3941–3994. [CrossRef]
242. Ngwabebhoh, F.A.; Yildiz, U. Nature-derived fibrous nanomaterial toward biomedicine and environmental remediation: Today's state and future prospects. *J. Appl. Polym. Sci.* **2019**, *136*. [CrossRef]
243. You, J.; Cao, J.; Zhao, Y.; Zhang, L.; Zhou, J.; Chen, Y. Improved Mechanical Properties and Sustained Release Behavior of Cationic Cellulose Nanocrystals Reinforeced Cationic Cellulose Injectable Hydrogels. *Biomacromolecules* **2016**, *17*, 2839–2848. [CrossRef]
244. Ding, L.; Wang, Q.; Shen, M.; Sun, Y.; Zhang, X.; Huang, C.; Chen, J.; Li, R.; Duan, Y. Thermoresponsive nanocomposite gel for local drug delivery to suppress the growth of glioma by inducing autophagy. *Autophagy* **2017**, *13*, 1176–1190. [CrossRef] [PubMed]
245. Xing, C.; Chen, S.; Qiu, M.; Liang, X.; Liu, Q.; Zou, Q.; Li, Z.; Xie, Z.; Wang, D.; Dong, B.; et al. Conceptually Novel Black Phosphorus/Cellulose Hydrogels as Promising Photothermal Agents for Effective Cancer Therapy. *Adv. Healthc. Mater.* **2018**, *7*. [CrossRef] [PubMed]
246. Yang, J.S.; Xie, Y.J.; He, W. Research progress on chemical modification of alginate: A review. *Carbohydr. Polym.* **2011**, *84*, 33–39. [CrossRef]
247. Wróblewska-Krepsztul, J.; Rydzkowski, T.; Michalska-Pożoga, I.; Thakur, V.K. Biopolymers for biomedical and pharmaceutical applications: Recent advances and overview of alginate electrospinning. *Nanomaterials* **2019**, *9*. [CrossRef] [PubMed]
248. Wang, C.; Wang, X.; Dong, K.; Luo, J.; Zhang, Q.; Cheng, Y. Injectable and responsively degradable hydrogel for personalized photothermal therapy. *Biomaterials* **2016**, *104*, 129–137. [CrossRef] [PubMed]
249. Chalanqui, M.J.; Pentlavalli, S.; McCrudden, C.; Chambers, P.; Ziminska, M.; Dunne, N.; McCarthy, H.O. Influence of alginate backbone on efficacy of thermo-responsive alginate-g-P(NIPAAm) hydrogel as a vehicle for sustained and controlled gene delivery. *Mater. Sci. Eng. C* **2019**, *95*, 409–421. [CrossRef] [PubMed]

250. Liu, M.; Song, X.; Wen, Y.; Zhu, J.L.; Li, J. Injectable Thermoresponsive Hydrogel Formed by Alginate-g-Poly(N-isopropylacrylamide) That Releases Doxorubicin-Encapsulated Micelles as a Smart Drug Delivery System. *ACS Appl. Mater. Interfaces* **2017**, *9*, 35673–35682. [CrossRef] [PubMed]

251. Davoodi, P.; Ng, W.C.; Srinivasan, M.P.; Wang, C.H. Codelivery of anti-cancer agents via double-walled polymeric microparticles/injectable hydrogel: A promising approach for treatment of triple negative breast cancer. *Biotechnol. Bioeng.* **2017**, *114*, 2931–2946. [CrossRef]

252. Davoodi, P.; Ng, W.C.; Yan, W.C.; Srinivasan, M.P.; Wang, C.H. Double-walled microparticles-embedded self-cross-linked, injectable, and antibacterial hydrogel for controlled and sustained release of chemotherapeutic agents. *ACS Appl. Mater. Interfaces* **2016**, *8*, 22785–22800. [CrossRef]

253. Liu, J.; Qi, C.; Tao, K.; Zhang, J.; Xu, L.; Jiang, X.; Zhang, Y.; Huang, L.; Li, Q.; Xie, H.; et al. Sericin/Dextran Injectable Hydrogel as an Optically Trackable Drug Delivery System for Malignant Melanoma Treatment. *ACS Appl. Mater. Interfaces* **2016**, *8*, 6411–6422. [CrossRef]

254. Li, L.; Wang, C.; Huang, Q.; Xiao, J.; Zhang, Q.; Cheng, Y. A degradable hydrogel formed by dendrimer-encapsulated platinum nanoparticles and oxidized dextran for repeated photothermal cancer therapy. *J. Mater. Chem. B* **2018**, *6*, 2474–2480. [CrossRef]

255. Deng, J.; Xun, X.; Zheng, W.; Su, Y.; Zheng, L.; Wang, C.; Su, M. Sequential delivery of bismuth nanoparticles and doxorubicin by injectable macroporous hydrogels for combined anticancer kilovoltage X-ray radio- and chemo-therapy. *J. Mater. Chem. B* **2018**, *6*, 7966–7973. [CrossRef]

256. GuhaSarkar, S.; More, P.; Banerjee, R. Urothelium-adherent, ion-triggered liposome-in-gel system as a platform for intravesical drug delivery. *J. Control. Release* **2017**, *245*, 147–156. [CrossRef] [PubMed]

257. Zheng, Y.; Liang, Y.; Zhang, D.; Zhou, Z.; Li, J.; Sun, X.; Liu, Y.N. Fabrication of injectable CuS nanocomposite hydrogels based on UCST-type polysaccharides for NIR-triggered chemo-photothermal therapy. *Chem. Commun.* **2018**, *54*, 13805–13808. [CrossRef] [PubMed]

258. Hou, M.; Yang, R.; Zhang, L.; Liu, G.; Xu, Z.; Kang, Y.; Xue, P. Injectable and Natural Humic Acid/Agarose Hybrid Hydrogel for Localized Light-Driven Photothermal Ablation and Chemotherapy of Cancer. *ACS Biomater. Sci. Eng.* **2018**, *4*, 4266–4277. [CrossRef]

259. Niu, X.; Zhang, Z.; Zhong, Y. Hydrogel loaded with self-assembled dextran sulfate-doxorubicin complexes as a delivery system for chemotherapy. *Mater. Sci. Eng. C* **2017**, *77*, 888–894. [CrossRef]

260. Nguyen, K.; Dang, P.N.; Alsberg, E. Functionalized, biodegradable hydrogels for control over sustained and localized siRNA delivery to incorporated and surrounding cells. *Acta Biomater.* **2013**, *9*, 4487–4495. [CrossRef] [PubMed]

261. Vittorio, O.; Cirillo, G.; Iemma, F.; Di Turi, G.; Jacchetti, E.; Curcio, M.; Barbuti, S.; Funel, N.; Parisi, O.I.; Puoci, F.; et al. Dextran-catechin conjugate: A potential treatment against the pancreatic ductal adenocarcinoma. *Pharm. Res.* **2012**, *29*, 2601–2614. [CrossRef]

262. Vittorio, O.; Brandl, M.; Cirillo, G.; Kimpton, K.; Hinde, E.; Gaus, K.; Yee, E.; Kumar, N.; Duong, H.; Fleming, C.; et al. Dextran-Catechin: An anticancer chemically-modified natural compound targeting copper that attenuates neuroblastoma growth. *Oncotarget* **2016**, *7*, 47479–47493. [CrossRef]

263. Agarwal, A.; Gupta, U.; Asthana, A.; Jain, N.K. Dextran conjugated dendritic nanoconstructs as potential vectors for anti-cancer agent. *Biomaterials* **2009**, *30*, 3588–3596. [CrossRef]

264. Abdo Qasem, A.A.; Alamri, M.S.; Mohamed, A.A.; Hussain, S.; Mahmood, K.; Ibraheem, M.A. High Soluble-Fiber Pudding: Formulation, Processing, Texture and Sensory Properties. *J. Food Process. Preserv.* **2017**, *41*. [CrossRef]

265. Carlini, A.S.; Gaetani, R.; Braden, R.L.; Luo, C.; Christman, K.L.; Gianneschi, N.C. Enzyme-responsive progelator cyclic peptides for minimally invasive delivery to the heart post-myocardial infarction. *Nat. Commun.* **2019**, *10*. [CrossRef] [PubMed]

266. Haines-Butterick, L.; Rajagopal, K.; Branco, M.; Salick, D.; Rughani, R.; Pilarz, M.; Lamm, M.S.; Pochan, D.J.; Schneider, J.P. Controlling hydrogelation kinetics by peptide design for three-dimensional encapsulation and injectable delivery of cells. *Proc. Natl. Acad. Sci. USA* **2007**, *104*, 7791–7796. [CrossRef] [PubMed]

267. Mano, J.F. Stimuli-responsive polymeric systems for biomedical applications. *Adv. Eng. Mater.* **2008**, *10*, 515–527. [CrossRef]

268. Xing, R.; Liu, K.; Jiao, T.; Zhang, N.; Ma, K.; Zhang, R.; Zou, Q.; Ma, G.; Yan, X. An Injectable Self-Assembling Collagen-Gold Hybrid Hydrogel for Combinatorial Antitumor Photothermal/Photodynamic Therapy. *Adv. Mater.* **2016**, *28*, 3669–3676. [CrossRef] [PubMed]

269. Upadhyay, A.; Kandi, R.; Rao, C.P. Injectable, Self-Healing, and Stress Sustainable Hydrogel of BSA as a Functional Biocompatible Material for Controlled Drug Delivery in Cancer Cells. *ACS Sustain. Chem. Eng.* **2018**, *6*, 3321–3330. [CrossRef]
270. Kim, I.; Choi, J.S.; Lee, S.; Byeon, H.J.; Lee, E.S.; Shin, B.S.; Choi, H.G.; Lee, K.C.; Youn, Y.S. In situ facile-forming PEG cross-linked albumin hydrogels loaded with an apoptotic TRAIL protein. *J. Control. Release* **2015**, *214*, 30–39. [CrossRef]
271. Qian, H.Q.; Qian, K.Y.; Cai, J.; Yang, Y.; Zhu, L.J.; Liu, B.R. Therapy for Gastric Cancer with Peritoneal Metastasis Using Injectable Albumin Hydrogel Hybridized with Paclitaxel-Loaded Red Blood Cell Membrane Nanoparticles. *ACS Biomater. Sci. Eng.* **2019**, *5*, 1100–1112. [CrossRef]
272. Curcio, M.; Altimari, I.; Spizzirri, U.G.; Cirillo, G.; Vittorio, O.; Puoci, F.; Picci, N.; Iemma, F. Biodegradable gelatin-based nanospheres as pH-responsive drug delivery systems. *J. Nanoparticle Res.* **2013**, *15*. [CrossRef]
273. Curcio, M.; Spizzirri, U.G.; Iemma, F.; Puoci, F.; Cirillo, G.; Parisi, O.I.; Picci, N. Grafted thermo-responsive gelatin microspheres as delivery systems in triggered drug release. *Eur. J. Pharm. Biopharm.* **2010**, *76*, 48–55. [CrossRef]
274. Oh, E.; Oh, J.E.; Hong, J.; Chung, Y.; Lee, Y.; Park, K.D.; Kim, S.; Yun, C.O. Optimized biodegradable polymeric reservoir-mediated local and sustained co-delivery of dendritic cells and oncolytic adenovirus co-expressing IL-12 and GM-CSF for cancer immunotherapy. *J. Control. Release* **2017**, *259*, 115–127. [CrossRef] [PubMed]
275. Takei, T.; Sugihara, K.; Yoshida, M.; Kawakami, K. Injectable and biodegradable sugar beet pectin/gelatin hydrogels for biomedical applications. *J. Biomater. Sci. Polym. Ed.* **2013**, *24*, 1333–1342. [CrossRef] [PubMed]
276. Ciobanu, B.C.; Cadinoiu, A.N.; Popa, M.; Desbrières, J.; Peptu, C.A. Modulated release from liposomes entrapped in chitosan/gelatin hydrogels. *Mater. Sci. Eng. C* **2014**, *43*, 383–391. [CrossRef] [PubMed]
277. Franke, K.; Baur, M.; Daum, L.; Vaegler, M.; Sievert, K.D.; Schlosshauer, B. Prostate carcinoma cell growth-inhibiting hydrogel supports axonal regeneration in vitro. *Neurosci. Lett.* **2013**, *541*, 248–252. [CrossRef] [PubMed]
278. Cirillo, G.; Vittorio, O.; Hampel, S.; Spizzirri, U.G.; Picci, N.; Iemma, F. Incorporation of carbon nanotubes into a gelatin-catechin conjugate: Innovative approach for the preparation of anticancer materials. *Int. J. Pharm.* **2013**, *446*, 176–182. [CrossRef] [PubMed]
279. Zhou, M.; Liu, S.; Jiang, Y.; Ma, H.; Shi, M.; Wang, Q.; Zhong, W.; Liao, W.; Xing, M.M.Q. Doxorubicin-Loaded Single Wall Nanotube Thermo-Sensitive Hydrogel for Gastric Cancer Chemo-Photothermal Therapy. *Adv. Funct. Mater.* **2015**, *25*, 4730–4739. [CrossRef]
280. Cirillo, G.; Hampel, S.; Spizzirri, U.G.; Parisi, O.I.; Picci, N.; Iemma, F. Carbon Nanotubes Hybrid Hydrogels in Drug Delivery: A Perspective Review. *Biomed. Res. Int.* **2014**. [CrossRef]
281. Cirillo, G.; Caruso, T.; Hampel, S.; Haase, D.; Puoci, F.; Ritschel, M.; Leonhardt, A.; Curcio, M.; Iemma, F.; Khavrus, V.; et al. Novel carbon nanotube composites by grafting reaction with water-compatible redox initiator system. *Colloid Polym. Sci.* **2013**, *291*, 699–708. [CrossRef]
282. He, G.; Chen, S.; Xu, Y.J.; Miao, Z.H.; Ma, Y.; Qian, H.S.; Lu, Y.; Zha, Z.B. Charge reversal induced colloidal hydrogel acts as a multi-stimuli responsive drug delivery platform for synergistic cancer therapy. *Mater. Horiz.* **2019**, *6*, 711–716. [CrossRef]
283. Maitz, M.F.; Sperling, C.; Wongpinyochit, T.; Herklotz, M.; Werner, C.; Seib, F.P. Biocompatibility assessment of silk nanoparticles: Hemocompatibility and internalization by human blood cells. *Nanomedicine* **2017**, *13*, 2633–2642. [CrossRef]
284. Omenetto, F.G.; Kaplan, D.L. New opportunities for an ancient material. *Science* **2010**, *329*, 528–531. [CrossRef] [PubMed]
285. Seib, F.P.; Pritchard, E.M.; Kaplan, D.L. Self-assembling doxorubicin silk hydrogels for the focal treatment of primary breast cancer. *Adv. Funct. Mater.* **2013**, *23*, 58–65. [CrossRef] [PubMed]
286. Wu, P.; Liu, Q.; Wang, Q.; Qian, H.; Yu, L.; Liu, B.; Li, R. Novel silk fibroin nanoparticles incorporated silk fibroin hydrogel for inhibition of cancer stem cells and tumor growth. *Int. J. Nanomed.* **2018**, *13*, 5405–5418. [CrossRef] [PubMed]
287. Wu, H.; Liu, S.; Xiao, L.; Dong, X.; Lu, Q.; Kaplan, D.L. Injectable and pH-Responsive Silk Nanofiber Hydrogels for Sustained Anticancer Drug Delivery. *ACS Appl. Mater. Interfaces* **2016**, *8*, 17118–17126. [CrossRef] [PubMed]

288. He, W.; Li, P.; Zhu, Y.; Liu, M.; Huang, X.; Qi, H. An injectable silk fibroin nanofiber hydrogel hybrid system for tumor upconversion luminescence imaging and photothermal therapy. *New J. Chem.* **2019**, *43*, 2213–2219. [CrossRef]
289. Ribeiro, V.P.; Silva-Correia, J.; Goncalves, C.; Pina, S.; Radhouani, H.; Montonen, T.; Hyttinen, J.; Roy, A.; Oliveira, A.L.; Reis, R.L.; et al. Rapidly responsive silk fibroin hydrogels as an artificial matrix for the programmed tumor cells death. *PLoS ONE* **2018**, *13*. [CrossRef] [PubMed]
290. Schaal, J.L.; Li, X.; Mastria, E.; Bhattacharyya, J.; Zalutsky, M.R.; Chilkoti, A.; Liu, W. Injectable polypeptide micelles that form radiation crosslinked hydrogels in situ for intratumoral radiotherapy. *J. Control. Release* **2016**, *228*, 58–66. [CrossRef]
291. Poursaid, A.; Jensen, M.M.; Nourbakhsh, I.; Weisenberger, M.; Hellgeth, J.W.; Sampath, S.; Cappello, J.; Ghandehari, H. Silk-Elastinlike Protein Polymer Liquid Chemoembolic for Localized Release of Doxorubicin and Sorafenib. *Mol. Pharm.* **2016**, *13*, 2736–2748. [CrossRef]
292. Gustafson, J.A.; Price, R.A.; Greish, K.; Cappello, J.; Ghandehari, H. Silk-elastin-like hydrogel improves the safety of adenovirus-mediated gene-directed enzyme-'prodrug therapy. *Mol. Pharm.* **2010**, *7*, 1050–1056. [CrossRef]
293. Hoffman, A.S. Hydrogels for biomedical applications. *Adv. Drug Deliv. Rev.* **2012**, *64*, 18–23. [CrossRef]
294. Naahidi, S.; Jafari, M.; Logan, M.; Wang, Y.; Yuan, Y.; Bae, H.; Dixon, B.; Chen, P. Biocompatibility of hydrogel-based scaffolds for tissue engineering applications. *Biotechnol. Adv.* **2017**, *35*, 530–544. [CrossRef] [PubMed]
295. Steinwachs, M.; Cavalcanti, N.; Mauuva Venkatesh Reddy, S.; Werner, C.; Tschopp, D.; Choudur, H.N. Arthroscopic and open treatment of cartilage lesions with BST-CARGEL scaffold and microfracture: A cohort study consecutive patients. *Knee* **2019**, *26*, 174–184. [CrossRef] [PubMed]
296. Elstad, N.L.; Fowers, K.D. OncoGel (ReGel/paclitaxel) - Clinical applications for a novel paclitaxel delivery system. *Adv. Drug Deliv. Rev.* **2009**, *61*, 785–794. [CrossRef] [PubMed]
297. Shalhoub, J.; Hinchliffe, R.J.; Powell, J.T. The world of legoo assessed: A short systematic and critical review. *Eur. J. Vasc. Endovasc. Surg.* **2013**, *45*, 44–45. [CrossRef] [PubMed]
298. Moreno, E.; Schwartz, J.; Larrañeta, E.; Nguewa, P.A.; Sanmartín, C.; Agüeros, M.; Irache, J.M.; Espuelas, S. Thermosensitive hydrogels of poly(methyl vinyl ether-co-maleic anhydride) - Pluronic® F127 copolymers for controlled protein release. *Int. J. Pharm.* **2014**, *459*, 1–9. [CrossRef]
299. Hwang, M.E.; Black, P.J.; Elliston, C.D.; Wolthuis, B.A.; Smith, D.R.; Wu, C.C.; Wenske, S.; Deutsch, I. A novel model to correlate hydrogel spacer placement, perirectal space creation, and rectum dosimetry in prostate stereotactic body radiotherapy. *Radiat. Oncol.* **2018**, *13*. [CrossRef] [PubMed]
300. Rao, A.D.; Feng, Z.; Shin, E.J.; He, J.; Waters, K.M.; Coquia, S.; DeJong, R.; Rosati, L.M.; Su, L.; Li, D.; et al. A Novel Absorbable Radiopaque Hydrogel Spacer to Separate the Head of the Pancreas and Duodenum in Radiation Therapy for Pancreatic Cancer. *Int. J. Radiat. Oncol. Biol. Phys.* **2017**, *99*, 1111–1120. [CrossRef]

 © 2019 by the authors. Licensee MDPI, Basel, Switzerland. This article is an open access article distributed under the terms and conditions of the Creative Commons Attribution (CC BY) license (http://creativecommons.org/licenses/by/4.0/).

MDPI  
St. Alban-Anlage 66  
4052 Basel  
Switzerland  
Tel. +41 61 683 77 34  
Fax +41 61 302 89 18  
www.mdpi.com  

*Pharmaceutics* Editorial Office  
E-mail: pharmaceutics@mdpi.com  
www.mdpi.com/journal/pharmaceutics